Laser Photobiology and Photomedicine

ETTORE MAJORANA INTERNATIONAL SCIENCE SERIES

Series Editor:
Antonino Zichichi
European Physical Society
Geneva, Switzerland

(PHYSICAL SCIENCES)

Recent volumes in the series:

A Continuation Order Plan is available for this series. A continuation order will bring delivery of each new volume immediately upon publication. Volumes are billed only upon actual shipment. For further information please contact the publisher.

Laser Photobiology and Photomedicine

Edited by

S. Martellucci

Second University of Rome
Rome, Italy

and

A. N. Chester

Hughes Aircraft Company
El Segundo, California

Plenum Press • New York and London

Library of Congress Cataloging in Publication Data

Main entry under title:

Laser photobiology and photomedicine.

(Ettore Majorana international science series. Physical sciences; v. 22)
"Proceedings of a two-week course on laser applications to biology and medicine,
held September 4–16, 1983, in Erice, Italy"—T.p. verso.
Bibliography: p.
Includes index.
1. Lasers in biology—Congresses. 2. Lasers in medicine—Congresses. I. Martel-
lucci, S. II. Chester, A. N. III. Series.
QH324.9.L37L37 1985   574'.028   84-26627
ISBN-13: 978-1-4612-9494-8   e-ISBN-13: 978-1-4613-2461-4
DOI: 10.1007/978-1-4613-2461-4

Proceedings of a two-week course on Laser Applications to Biology and Medicine,
held September 4–16, 1983, in Erice, Italy

©1985 Plenum Press, New York
Softcover reprint of the hardcover 1st edition 1985
A Division of Plenum Publishing Corporation
233 Spring Street, New York, N.Y. 10013

PREFACE

This volume contains the Proceedings of a two-week course on "Laser Applications to Biology and Medicine" held from September 4 to 16, 1983 in Erice, Italy. This is the 10th annual course of the International School of Quantum Electronics organized under the auspices of the "E. Majorana" Center for Scientific Culture.

Among the possible approaches to a course on Laser Applications to Biology and Medicine, the one which emphasizes the scientific and technological aspects of the advanced laser techniques when applied to laboratory and clinical tests has been chosen. In fact, it reflects the new policy of the School to stress the advanced scientific and technological achievements in the field of Quantum Electronics. Accordingly, the Course has given the broadest information on the ultimate performances already achieved and the perspectives of their applications.

Because of the great variety of applications of laser in biology, medicine, chemistry, engineering and related branches of science, this school addressed a subject of interdisciplinary interest. The formal sessions have been balanced between tutorial presentations and lectures focusing on unsolved problems and future directions. In addition, wide time has been provided for the participants to meet together informally for additional discussions on the forefront of current work. Therefore the character of the Course was a blend of current research and tutorial reviews.

We have brought together some of the world's acknowledged masters in the field to summarize both the present state of their researches and the background behind them. Most of the lecturers attended all the lectures and devoted their spare hours to stimulating discussions and the organization of the panel discussions, conducted by Professor G. Salvatore, on a questions-answers basis. We would like to thank them all for their admirable contributions. The Course also has taken advantage from the very active audience; most of the students were experts in the field and contributed with panel discussions and seminars. Some students' seminars are also included in these Proceedings.

"Laser Photobiology and Photomedicine" could hardly be a more appropriate and timely title to be given. The papers in these proceedings give a fairly complete accounting of the Course lectures with the exception of the informal panel discussions. In editing this material we did not modify the original manuscripts except to assist in uniformity of style. The papers are not ordered exactly according to the chronology of the Course but in the following topical arrangement:

A   "Physical and Biological Basis"
B   "Biological Effects and Applications"
C   "Photochemotherapy"
D   "Photobiology and Dermatology"
E   "Surgical and Ophthalmological Applications"
F   "Laser Safety"
G   "Diagnostics and Technological Aspects"

Several reasons have prevented the inclusion of the contributions by some busy authors. The lectures given by F. Hillenkamp and M. Wolbarsht were based on papers already published elsewhere and have not been included in these proceedings. The tutorials on laser sources given by O. Svelto have also been omitted because we have assumed that the main properties of the "laser" are well known.

This volume would, of course, not be possible without the considerable efforts put forth by the authors represented here; all of them are cordially acknowledged. We also wish to mention with sincere thanks the very specialized assistance of Miss Madeleine Carter and Dr. R.H. Andrews of Plenum Press in the preparation of this volume.

Before concluding, we acknowledge the organizations who sponsored the School, especially the E. Majorana Center for Scientific Culture, whose continuous support made this Course possible.

The Directors of the I.S.Q.E.   S. Martellucci
A. N. Chester   The Second University
Hughes A. C.   Rome (Italy)
El Segundo, CA (USA)

April 5, 1984

CONTENTS

LASER SAFETY

DIAGNOSTICS AND TECHNOLOGICAL ASPECTS

PHYSICAL AND BIOLOGICAL BASIS

PHOTOBIOLOGY AND PHOTOMEDICINE

J. A. Parrish

Department of Dermatology, Harvard Medical School
Massachusetts General Hospital
Boston, MA 02114, USA

INTRODUCTION

The scope of modern photomedicine is broad. It includes the study of molecular and cellular mechanisms of pharmacology, chemistry, physiology, and pathology as they apply to photobiology. It is the application of the principles of photobiology to the diagnosis treatment, and understanding of health and disease. Photiobiology is the study of the interactions of nonionizing electromagnetic radiation with biomolecules, and the ensuing biologic responses. It is concerned predominantly with electromagnetic radiation wavelengths in the range of 200-800 nm. Photobiology is concerned mainly with photochemical alterations in biomolecules that eventually affect the visability or function of living cells. Photomedicine includes the study, diagnosis, treatment, and prevention of photodermatoses, skin cancer, and chronic actinic changes of skin.

Living tissue is affected in a variety of ways by electromagnetic radiation. The energy of photons in the ultraviolet (UV) and visible wavelengths sufficient to cause electronic excitation of specific chromophore molecules leading to specific chemical reactions. Use of UV and visible radiation therefore offers the possibility of causing specific photochemical reactions by selecting among a wide variety of specific target molecules.

Longer and shorter wavelengths are less specific in their actions compared with the relatively specific effects of the UV and visible wavebands of electromagnetic radiation. The low quantum energy of infrared photons and microwaves excites specific vibrational or rotational modes and simultaneously affects many target molecules. The most significant biologic effects of these wavebands

3

is the heating caused by such kinetic excitation. A high intensity
is usually required to be effective and the resultant effects on
biologic systems is often not specific. Very high-energy short-
wavelength photons such as X-rays and γ rays affect the highly organized and complex human tissues and any other matter by relatively
indiscriminate ionization of molecules. The ionized molecules are
highly reactive but the absorption is relatively nonspecific. Therefore, the ability to selectively affect "target" molecules is
limited.

Maximizing the beneficial and/or minimizing the adverse effects
of nonionizing electromagnetic radiation by manipulation of exposure
dose and by the alteration of the host with exogenous agents necessi-
tates identification and quantification of the effects of that radiation on abnormal and normal human tissue. One specific aspect of
photomedicine is the use of nonionizing electromagnetic radiation,
with and without the addition of photoactive drugs, to treat disease.

ULTRAVIOLET RADIATION

Medical scientists, dermatologists and photobiologists often
divide the ultraviolet spectrum into three portions, UVA, UVB, and
UVC, in order of decreasing wavelength (Figure 1). Radiation of
wavelengths shorter than 200 nm is mostly absorbed by air. The
wavelength range from 200 to 290 nm is called UVC. Solar radiation
of wavelengths below 290 nm does not reach the earth's surface because of absorption by ozone formed in the stratosphere. The band
from 290 to 320 nm is called UVB. UVA is the band from 320 to 400
nm. The divisions between UVA, UVB, and UVC are not phenomenologi-
cally exact nor agreed upon. One should always define UV radiation
in more rigorous spectroidiometric terms for precise biological
studies.

UVA is sometimes referred to as long-wave UV and is also called
near-UV because of its proximity to the visible spectrum. It is also
called the blacklight region because its principal use for many years
was to excite fluorescent and phosphorescent substances that re-
radiate the absorbed energy as light in the visible spectrum. UVA
may potentiate or add to the biologic effects of UVB. There is
experimental and epidemiologic evidence to suggest that solar UVA is
one of the possible etiologic agents for certain kinds of cataracts
in humans. UVA-induced photopolymerization and photochemical reactions are used in industry to alter rubber, plastic, glass, metal,
paper, and photographs. Many chemical photosensitizers have action
spectra in the UVA in conjunction with photosensitizing drugs has
opened up new therapeutic possibilities in chronic skin disorders.
UVA is both melanogenic and erythemogenic, but the amount of energy
required to produce an effect is orders of magnitude higher than for
the UVB region.

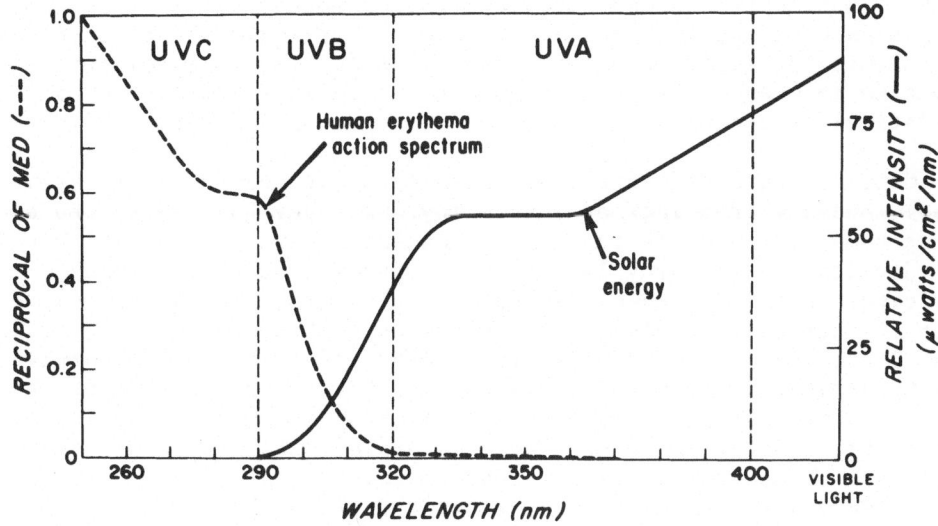

Fig. 1.   The A, B, and C portions of the ultraviolet spectrum as used
in photobiology and photomedicine.

UVB is also called "mid-UV" because of its relative spectral
position. Compared to UVA, solar UV radiation of wavelengths between
290 and 320 nm reaches the earth in relatively small quantities but
is very efficient in causing sunburning of human skin. Therefore,
UVB is often referred to as the sunburn spectrum or the erythema
band. The UVB portion of the spectrum has been shown to induce skin
cancer in laboratory animals and mutations in bacteria. Epidemiologic evidence strongly suggests that solar UVB causes skin cancer in
humans. Long-term UVB exposure is thought to be at least partly
responsible for producing the changes of exposed human skin commonly
termed "premature" aging or actinic degeneration.

UVC is also called germicidal radiation because of its effectiveness in killing one-celled organisms. It is often called short-
wave UV because the wavelengths in this region are the shortest
transmitted through air, and because this region is the furthest from
the visible spectrum it is also called far-UV. Radiation in the UVC
band causes erythema of normal skin very efficiently and can cause
photokeratitis.

THE SKIN

The skin is the organ most often injured by UV radiation and is
the organ of access for most forms of phototherapy. The skin
accounts for about 15% of the total body weight and can be considered
the largest organ of the body. In an adult, it is a living tissue

system almost 6ft long and 3ft wide (approximately $2m^2$). The skin absorbs much of the mechanical stresses of our world and also shields from chemicals, sunlight, and bacteria. Acting as an insulator and selective membrane, the skin keeps the environment within the body at a relatively constant temperature and salt water content.

Compared to most other land mammals, human beings are relatively naked. Because of a lack of insulating fur, they have developed a unique combination of features: a thick outer layer of skin with a well-developed dead horny layer, a widespread system of thermal-sensitive sweat glands and vascular overperfusion, and an extensive layer of thermally insulating fatty tissue at the undersurface of the skin (Figure 2). This complex arrangement allows humans to survive in a wide range of temperatures and humidities. The purpose of the skin is to protect the host from a noxious environment and maintain a homeostatic internal milieu. The skin has a vast network of nerve endings which mediate the sensations of touch, heat and cold to provide an environmental testing facility. When the skin performs these life supporting functions normally, we notice it only for its esthetic qualities. But if any of the protective mechanisms mal-function or become overwhelmed, we suffer embarrassment, discomfort, disfigurement, and possibly death.

From the point of view of metabolic need, the skin is vastly overperfused with blood. The mean blood flow is many times greater than the minimum flow necessary for skin cell nutrition because cutaneous blood flow serves as a heat regulator of the entire organ-ism and is not governed solely by metabolic requirements of the organ. Depending on body and ambient temperature, as much as 10% of the total blood volume is in the skin and this available for exposure to UV radiation. Prolonged exposure of the skin may make it possible to irradiate a larger portion of blood and blood cells as they course through superficial skin vessels. Effects of in vitro UV radiation on blood cells and blood-borne metabolites have been demonstrated in animals and humans.

The thin, outermost epidermis is composed of tightly packed sheets of cells called keratinocytes beneath a very thin but tough outer layer called the stratum corneum (Figure 3). The stratum corneum provides protection against water loss and surface abrasion, and attenuates UV radiation before it reaches living cells. Kera-tinocytes stem from a single layer of germinative cells (basal cells). After these cells divide, the daughter cells are pushed toward the surface. They no longer divide, but differentiate to form the precursors of keratin. As differentiation and outward migration continue, the keratinocytes lose their nuclei, dehydrate, extrude lipids, and flatten out into dead, polygonal cells with a surface area about 25 times that of the basal cells. This closely packed, cemented, flat, dead cell layer laden with keratin and lipids forms tough, protective stratum corneum.

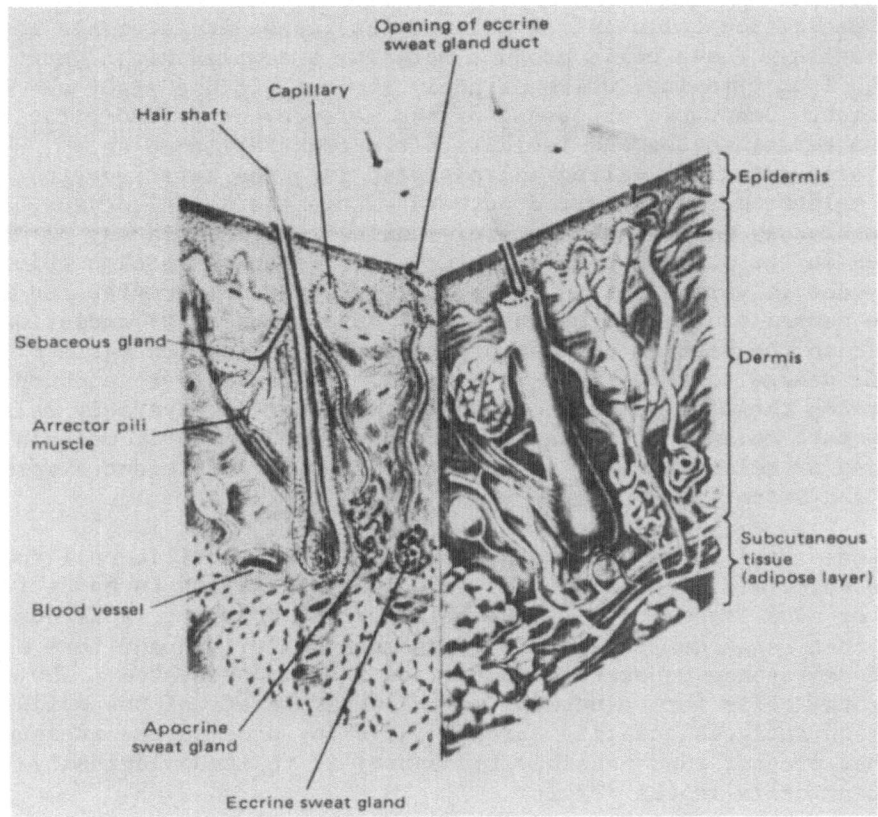

Fig. 2. Cross-section of the skin of human beings.

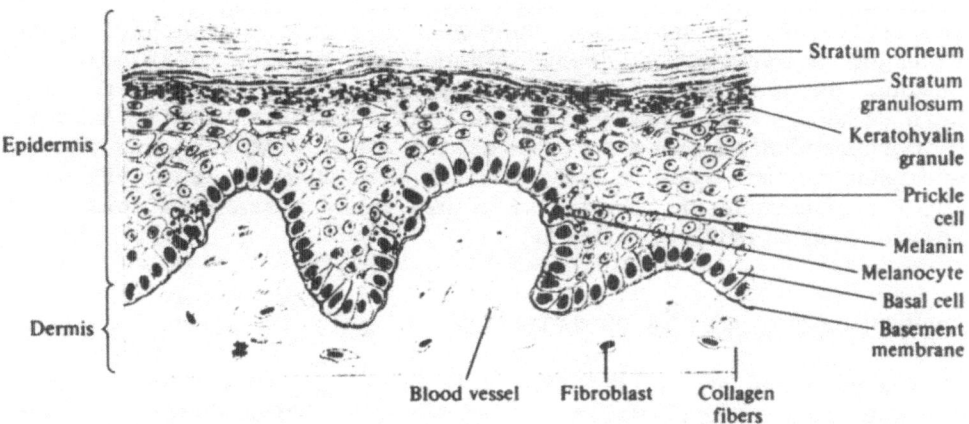

Fig. 3. Structural composition of epidermis and dermis.

Specialized cells called melanocytes reside at intervals between basal cells. These cells produce melanin, a complex biopolymer derived from tyrosine, which strongly absorbs visible light and UV radiation. Dendritic processes of the melanocytes interdigitate between keratinocytes and facilitate the transfer of melanin-containing granules, called melanosomes, into the keratinocytes. These melanosomes are carried outward within the keratinocytes, and ultimately, in more pigmented individuals, some melanin may be deposited in the stratum corneum. Racial differences in skin color are mostly due to variations in the productivity of melanocytes and not in the number or size of melanocytes. Absorption of UV radiation by melanin in the stratum corneum provides some protection against actinic damage to the skin. In general, the tendency to sunburn or to develop the most common forms of skin cancer is inversely related to how much melanin is present. Increased production of melanin (tanning or melanogenesis) is induced following sufficient exposure to UV radiation.

Langerhans cells are another specialized, dendritic cell found in the epidermis. In fact, they were long considered to be derived from, or even precursors of, the melanocyte. However, it is now known that Langerhans cells are of mesenchymal origin and form part of the macrophage-monocyte component of the immune system. The Langerhans cells form a network in the upper layers of the epidermis to screen antigens entering through the skin, process the antigen, and then present (and possibly transport) it to immunologically competent cells in the dermis.

The dermis is much thicker than the epidermis (up to 4 mm), has fewer cells, and is mostly connective tissue or fibers. Blood vessels, lymphatics, and nerves course through the dermis. The dermis is mostly a semisolid mixture of fibers, water, and a viscous gel containing mucopolysaccharides and three types of fibers: collagen, reticulum, and elastin. Collagen constitutes about 70% of the dry weight of dermis. Scattered fibroblasts produce the fibers, proteins, and viscous materials of the dermis. The complex gel and fibers of the dermis create a tissue with very high tensile strength and impressive resistance to compressive force. At the same time, the tissue remains pliable and movable. The deepest layer of the skin, the subcutaneous tissue, is mainly fatty tissue and acts as an insulator and a shock absorber.

RESPONSES OF THE SKIN TO UV RADIATION

The response of the skin to UV exposure is, in general, a reparative and protective reaction. Over hours to days, changes in blood flow, cell kinetics, and pigment production cause grossly observable changes in the whole organ. Many of the immediate and most of the subsequent intracellular events remain unknown. Sunburn is an ex-

ample of inflammation, a generalized, primitive, protective, patho-
physiologic response designed to remove injurious agents. Many
features of the response are similar to those caused by other
irritating or toxic agents. However, UV exposure is one cause of
inflammation in which no matter actually enters the tissue, often
producing little heating and no physical trauma. The injurious
agents produced by UV exposure must be certain photoproducts formed
within the tissue.

Repeated episodes of UV-induced skin injury over many years
eventually cause skin cancer and appearance of prematurely aged skin.
The evidence that UV radiation causes skin cancer is very convincing.
Repeated experiments with animals and epidemiologic studies of humans
have documented the causal relationship. Human skin cancer occurs
mostly on sites of the body that are habitually exposed to sun and is
more common in outdoor than indoor workers. Skin cancer in white
persons is greatest in tropical climates and least in northern lat-
itudes with little sunlight. Skin cancer is the most common malignant tumor affecting humans and comprises more than one-third of all
cancers. If properly diagnosed, most skin cancer can be removed or
cured by a variety of techniques. However, these tumors still result
in substantial discomfort and permanent disfigurement, and are responsible for millions of dollars in direct and indirect treatment
costs.

Repeated sun exposure over many years gradually alters the cells
and fibers in the skin and leads to changes in the appearance of the
skin: wrinkling and furrowing, irregular pigmentation, dilated tor-
tuous blood vessels, and irregularly thin outer layer. It has not
been shown that the microscopic, chemical, and molecular alterations
of chronic sun exposure are the same as those occurring in the normal
aging process. In fact, most evidence suggests that these are different processes. But because the cumulative sun-induced changes have
come to be associated with aged appearance, the sun has been said to
cause "premature aging" of skin. Any UV exposure sufficient to
injure cells adds a small but finite contribution of this process.
These changes occur most rapidly in persons with maximum sun exposure
and the least protective pigmentation.

Multiple photochemical events occur upon absorption of ultraviolet photons within the skin. Alterations of bases in DNA have
received much attention because of the importance of DNA in cell
regulation and replication, the presence of nonredundant genetic
sequences, and the relative stability of some of the DNA photoproducts. Many other photochemical changes occur in cells, affecting
RNA, structural and enzymatic proteins, and membranes. Some of these
alterations may have little effect on cells, while others may change
cellular metabolism or survival, or lead to the release of chemicals
that later affect adjacent cells or tissues. In the first hours
after radiation of skin cells, synthesis of DNA, RNA, and proteins is

decreased, metabolism is altered, and histochemical evidence of cell injury is present. It is early in this period of decreased macro-molecular synthesis that DNA repair is initiated. Subsequently, the cell either recovers, mutates, or dies. Expression of mutation or death may be immediate or delayed for several cell cycles.

Chemical photosensitivity describes enhanced photosensitivity resulting from excitation of an identified chemical by electro-magnetic radiation. A wide variety of photosensitizing chemicals of therapeutic, industrial, agricultural, or other origin may reach the skin directly or via the bloodstream, each having its own pattern of absorption, metabolism, and binding to skin components. A variety of photochemical and molecular mechanisms is involved. In general, the compounds possess highly resonant structures with molecular weights less than 500, and absorb radiation in the UV and visible range. Many are planar molecules. Abnormal responses of the skin to the UV and visible radiation are many and include a special challenge to the photobiologist. Exaggerated or qualitatively abnormal responses occur if persons have too little protective pigment, have abnormal DNA repair, or produce or retain excess or abnormal endogenous photo-sensitizing chemicals such as porphyrins. Some persons also exhibit abnormal inflammation or immunologic reactions to UV or visible radiation.

Melanin pigmentation is a major defense of the skin against the acute and chronic effects of sun exposure. Constitutive pigmentation describes the individual's baseline color, and faculative pigmen-tation is the ability to tan in response to UV exposure. These characteristics are both generically determined. Thickening of the epidermis is a generalized protective response often associated with inflammation. Increased proliferation of keratinocytes following UV exposure of skin leads to thickening of the epidermis and subsequent-ly may be associated with a noticeable increase in desquamation. The thickened epidermis and stratum corneum provides increased protein barrier against photons.

OPTICAL PROPERTIES OF SKIN AND BLOOD

When UV and visible radiation impinge on the surface of the epidermis, the radiation may be reflected from the surface, scattered, absorbed within any layer of skin, or transmitted to deeper layers. About 7-8% of UV and visible radiation is reflected off the surface. Most of the visible radiation returning back from the skin has reached the dermis, where it is back-scattered. Absorp-tion of radiation by the stratum corneum and by the whole epidermis is high for wavelengths shorter than 290 nm and is often maximum around 275 nm. This is because proteins, DNA, RNA, amino acids, and numerous other biomolecules absorb in the shortwave UV region. Chromophores absorbing within the visible range include hemoglobin,

oxyhemoglobin and bilirubin in the dermis, and melanin in the epidermis (Figure 4). Melanin has increasing absorbance at shorter
wavelengths.

The dominant optical interaction in the dermis is scattering.
Scattering increases the average pathlength the photons must take,
and therefore increases their chances of being absorbed before pen-
etrating deeper into the tissue. Because of differential absorption,
scattering, and the site and layering of some skin pigments which
absorb within the UV and visible range, the longer the wavelength the
deeper the transmission into skin.

PHOTOIMMUNOLOGY

Cutaneous responses to UVR are complex and involve not only skin
and eyes but also circulating and noncirculating components of the
immune system. Photoimmunology is the study of the effects of non-
ionizing electromagnetic radiation on normal and abnormal immune
function, and involves photobiology, dermatology, and immunology.
The possibility that UVR might induce or alter immune responses is
the basis of many of the hypotheses about the etiology of certain
abnormal reactions to sun. Immune responses are involved in certain
forms of photosensitivity in which both chemicals and photons part-
icipated in the development of specific altered reactivity. UVR can
inactivate certain antibodies and alter specific biomolecules to make
them more or less antigenic. Certain skin diseases with associated
immunologic abnormalities (i.e., atopic eczema) and with abnormal

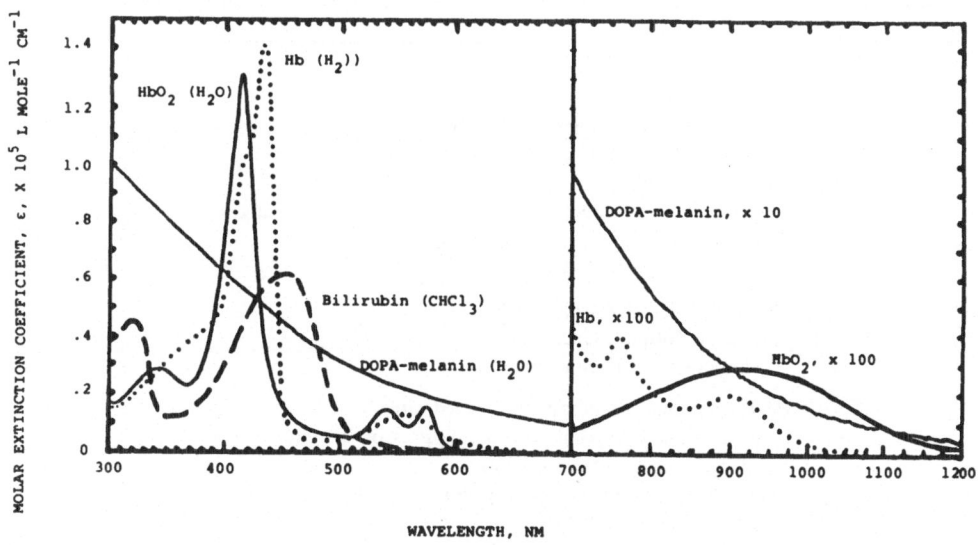

Fig. 4. Absorption curves of chromophores in the visible range.

lymphocytic infiltrates (i.e., lichen planus) have been successfully
treated with UVR.

The most impressive new aspect of photoimmunology is the sheer
volume of information resulting from recent, well-controlled studies
in experimental photoimmunology. A partial list of results of in
vivo photoimmunology studies is found in Table 1 and each entry can
be supported by several to many recent references. Several observations direct new interest into the study of photoimmunology. Collective evidence show skin to be an immune organ with features unique
to its role as protective barrier. Certain membrane characteristics
and functions of the Langerhans cell may be selectively sensitive to
UVR. The introduction of oral psoralen photochemotherapy (PUVA) led
to a massive amount of clinical and basic research. When PUVA was
observed to apparently eliminate selectively the abnormal cutaneous
lymphocytic infiltrates of mycosis fungoides, lichen planus and other
diseases, a stimulus was provided to study the effects of PUVA and UV
radiation on immune cells and immune function of animals and humans.

Early and apparently obligatory events in the induction of skin
cancer by UVR are immunologic in nature. Probably the most interesting aspects of photoimmunology have been the observations of systemic
alterations in the whole host induced by local exposure of skin to
UVR. Unexposed skin, circulating cells, and cells of the spleen are
altered by exposure of skin to UVR. Subcarcinogenic exposure doses
alter mice, making them tolerant to transplanted, highly antigenic
UV-induced tumors. Allergic contact sensitization capacity is also

Table 1. In Vivo UVR Exposure of Skin Alters Immune System.

CELLS
  Lymphocytes
    ↓ viability and function (T, B)
    membrane alterations
  Langerhans cell
    morphologic, antigenic, enzymatic, functional changes
  Keratinocytes
    ↓ ETAF production

SKIN (local effect on exposed sites)
  ↓ allergic contact sensitization
    (induction > elicitation)

HOST (systemic effect on unexposed sites)
  ↑ susceptibility to skin cancer (suppressor T cell)
  ↓ allergic contact sensitization
    (single exposure, antigen specific, suppressor T cell,
    induction > elicitation)
  ↓ DH to injected antigen
  Altered spleen cells

---

diminished in unexposed skin. Quantitative work studying action spectra, dose response curves, relative thresholds, time-intensity reciprocity, and other exposure parameters have provided encouraging evidence of adequate selectivity or specificity to allow manipulation of certain photoimmunologic phenomena.

THERAPEUTIC USE OF LIGHT

One specific aspect of photomedicine is the use of non-ionizing EMR to treat disease by altering cells and extracellular metabolites. Table 2 lists possible mechanisms which have been suggested to explain beneficial effects. Of the diseases listed, only those involving photochemical alterations of metabolites or phototoxic cell injury have proved practical and successful. In the latter case, it is felt that cells involved in the pathophysiology of disease are quantitatively or qualitatively more susceptible to radiation and can therefore be modified by doses of ultraviolet radiation tolerable to normal skin.

One example of phototherapy by altering metabolites is "blue light" phototherapy of hyperbilirubinemia. Newborn infants, especially premature infants, may have inadequately active hepatic glucuronly transferase, the enzyme which converts bilirubin to a more water-soluble "conjugated" form. High blood and tissue levels of bilirubin result and may cause brain damage. Exposure of the skin to visible light results in lowering of bilirubin levels. The most likely mechanism of blue-light phototherapy is photoisomerization of the albumin-bound lipophilic chromophore to a more water-soluble form that can be excreted into the gut.

Table 2. Phototherapy and Photochemotherapy.

POSSIBLE MECHANISMS:	DISEASES TREATED:
1) Alteration of Metabolite	Hyperbilirubinemia
	Uremic Pruritus
	Rickets
2) Killing Organisms	Acne ($\pm$)
	Recurrent Herpes Simplex (-)
3) Phototoxicity of Abnormal Cells	Psoriasis
	Cancer
	Polymorphous Light Eruption
4) Induced Normal Protective Responses	Photodermatoses
	Vitiligo ($\pm$)
	($\pm$) questionable success
	(-) unsuccessful

A large proportion of patients with uremic pruritus have significant difficulty with itching which is not always relieved by
hemodialysis. It has been reported that six to eight erythemogenic
exposures to UVB result in less itching. The effect appears to be a
systemic one because ultraviolet exposures of one side of a patient
result in modest decrease in itching on both sides. It is hypo-
thesized that a photolabile metabolite, normally cleared by the
kidney, accumulates in skin to cause the itching.

The disease most often treated by phototherapy is psoriasis.
Repeated cell injury by photons from a variety of wavebands, with and
without photosensitizers, causes improvement in psoriasis. While the
mechanism of UVB phototherapy is most often assumed to be DNA photo-
damage with a resulting decrease in cell proliferation, other possible vascular, leukocyte or cell regulatory mechanisms may also be
playing some role. The gradual improvement caused by each treatment
and the induction of remission may be by different mechanisms. Crude
coal tar has been used as an adjunctive agent in phototherapy. In
the presence of repeated erythemogenic UVB exposures, the lubricant
vehicle used to apply tar is as therapeutic as the tar plus vehicle.
Topical agents such as mineral oil, hydrated petrolatum and hydro-
philic ointment seem to selectively increase the transmission of
ultraviolet radiation into psoriasis by altering the optical properties of the surface of the lesions.

Photochemotherapy describes the combination of a systemically
administered drug plus nonionizing electromagnetic radiation to treat
disease. The site and nature of in vivo photochemistry and subsequent photobiology are altered by supplying an exogenous chromophore.
In the doses usually used, the drug alone or the electromagnetic
radiation alone have no significant effect. Photoradiation therapy
with hematoporphyrin derivative is an example of photochemotherapy.
Early enthusiasm about dye-light treatment of recurrent herpes
simplex has diminished since double-blind studies failed to show
effectiveness. Recently, the known potent photosensitizing properties of psoralens have been better quantified and utilized to
treat a variety of diseases including psoriasis, mycosis fungoides,
vitiligo and some forms of eczema. The treatment, which utilizes
orally-administered methoxsalen and subsequent exposure of the skin
to UVA, has been called PUVA. PUVA is reported to be effective in a
growing list of skin diseases (see Table 3). The mechanism of therapeutic benefit is unknown and is most likely different for the various diseases.

As we learn more about action spectra and mechanisms of phototherapy and photochemotherapy, treatments may become more selective
by increasing efficacy and decreasing side effects. Unique properties of the laser, especially its monochromaticity, high intensity
and short pulse duration, will permit selective alteration of specific structures within living tissue. Expertise in skin optics,

Table 3. Diseases Reported to Respond to PUVA.

	NUMBER OF TREATMENTS TO CLEARING	CONTINUED TREATMENT REQUIREMENTS	RESPONSE RATE
Psoriasis	10-40	Yes, variable 0 to 60/year	90%
Mycosis fungoides	20-60	Yes, usually frequent; occasional remissions	80+% improvement; ? cures
Vitiligo	100-400	Not usually, if totally repigmented	70% improved
Eczema	20-60	Yes, frequent	<90% with aggressive therapy
Lichen planus	10-20	Yes, variable	80+%
Urticaria pigmentosa	15-30	Yes	80+%
Polymorphous light eruption	10-20	Probably, each spring	90+%
Actinic reticuloid	12 (?)	Yes	1 case
Alopecia areata	5-30	Unknown	50%
Palmar/plantar dermatitis	20-60	Usually	90+%

synthetic chemistry, photochemistry, photoaffinity labelling and pharmacology will permit design of medications which can be administered systemically and subsequently activated, deactivated, bound, released or biotransformed _in vivo_ by electromagnetic radiation.

EFFECTS OF LASERS ON BIOLOGIC TISSUE:

OPTIONS FOR SPECIFICITY

J. A. Parrish

Department of Dermatology, Harvard Medical School
Massachusetts General Hospital
Boston, MA 02114, USA

THE EFFECTS OF LASER RADIATION ON BIOLOGIC MATTER

In any tissue most biologic effects of optical radiation are usually caused by either or both of two basic mechanisms: (1) responses to photochemical reactions, and (2) responses to the heat caused by non-radiative deexcitation and dissipation of the absorbed energy. Host responses to extensive photochemical cell injury are not qualitatively very different from the inflammation and repair which occur from injury by many physical and chemical agents. However, in the case of more subtle cell alterations, the inherent specificity of photochemical reactions may lead to quantitative differences in host response to various wavelengths. In fact many biological responses to _in vivo_ photochemical reactions can be very wavelength-dependant and quite specific. _In vivo_ extracellular photochemical reactions may also have very specific effects on the host.

In contrast, the biologic effects of radiant heating are typically nonspecific. Thermal damage to tissue often follows a common mechanistic pathway, regardless of the wavelengths used to cause the heating. The various mechanisms by which heat damages tissues all derive from kinetic excitations, whereas photochemically induced damage derives from electronic excitations, often with little increase in temperature. Most cell types suffer irreversible thermal damage over a relatively narrow range of temperatures near 45°C; most proteins become unstable and tend to unfold, and membranes become more permeable. If one wishes to achieve a high degree of specificity for thermal damage to tissues, the specificity must be primarily spatial rather than mechanistic.

17

Because of their high power, coherence, monochromaticity and capability of very short pulses, lasers are particularly suited for spatial confinement of optically induced thermal damage. Lasers have been used primarily in medicine and surgery to induce thermal changes and have been used largely as a tool for controlled destruction of tissue. Laser radiation can also be used to induce photochemical reactions in vivo and thereby obtain some of the specificity inherent in photochemical injury.

The collective properties of lasers also make it possible to induce damage in tissues by additional mechanisms related to very high photon density. The unique properties of the laser leads to unique list of possibilities for photon-tissue interactions vased on the character of the initiating event at a molecular level. These include: (1) photochemical, (2) thermal, or (3) an interaction dependent upon the electric field strength of the optical radiation. Photochemical events are more likely to occur with wavelengths less than 500 nm and can be seen with either short or long exposures lasting from milliseconds to seconds. Damage mechanisms dependent upon electric field strength occur after very short exposures of very high intensity. Many of these effects are seen after pulses in the range of nanoseconds or shorter and require peak powers in excess of 10^8 watts per cm^2.

PHOTOPHYSICS AND PHOTOBIOLOGY

The effects of lasers on living tissue can be analyzed by considering two separate events: a photophysical insult and a photobiologic host response (see Table 1). All phases of the photophysical insult usually occur in less than 1 second while the host response may evolve over months. With continuous wave or long-pulse lasers, the photophysical event can be continuously induced and biologic responses may begin before radiation stops, but on a molecular level the photophysical interations are sub-second events. The photophysical event may take place in living or dead tissue and for the purpose of analyzing this event the tissue can be studied as inert material. There energy for the event is supplied by the laser. Expertise in engineering, material processing, physics, and chemistry are needed to study these events.

The biological response requires a living host and lasts for hours to weeks. The energy for the response is provided by the metabolism of the host. This aspect of laser-tissue effects is best studied by physiologists, biologists, biochemists, and physicians. Table 1a, 1b and 1c give an overview of the two components of laser biology.

The photophysical events can actually be considered as a sequence of interactions of energy with matter. Electromagnetic

Table 1a. Laser Biology.

	PHOTOPHYSICAL EVENTS →	BIOLOGIC RESPONSES
DURATION (SECONDS)	10^{-15} TO 1	1 TO 10^7
ENERGY SUPPLIED BY:	LASER	HOST METABOLISM
RESEARCH REQUIREMENTS:		
VIABLE TISSUE	NO	YES
TARGET CONSIDERED AS:	"INERT" MATERIAL	VIABLE HOST
EXPERTISE	ENGINEERING, PHYSICS	BIOLOGY, CHEMISTRY

Laser biology can be studied as two complex collections of events.
The photophysical alterations or injury occur in less than 1 second
and initiates host responses which may last for hours to weeks.
Aspects of either of these events may be utilized for therapeutic
purposes.

Table 1b.   Photophysical Sequence.

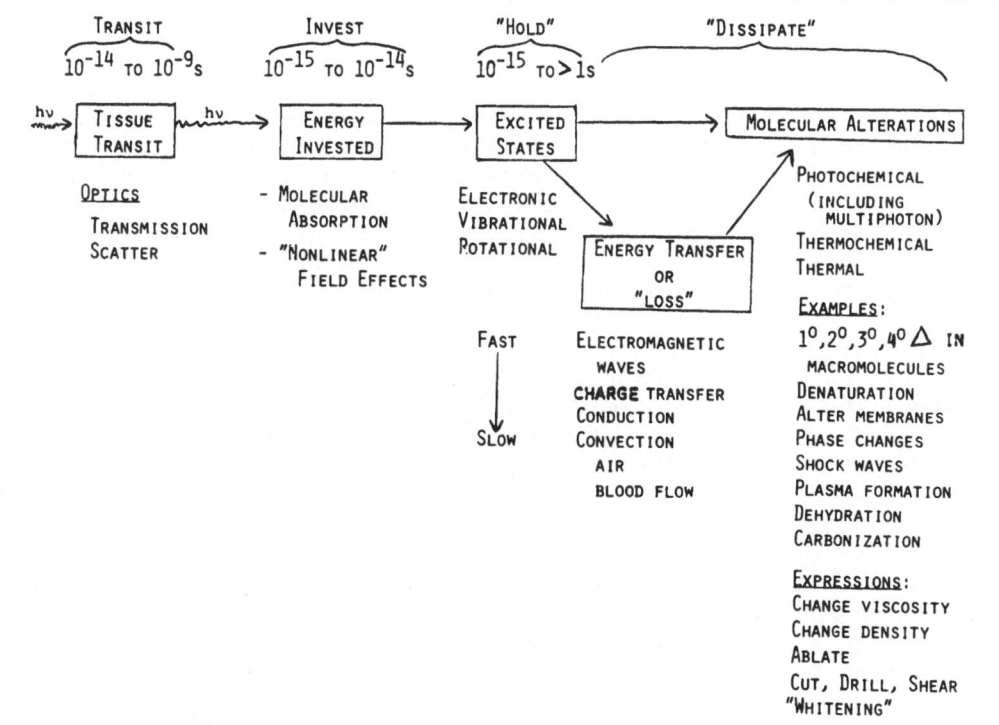

Electromagnetic energy travels through tissue until its energy is
invested in biomolecules. In the process of "dissipation" of that
energy molecules may be reversibly or irreversibly altered in ways
that cause measurable changes in tissue properties.

Table 1c.   Biologic Consequences (Host Responses).

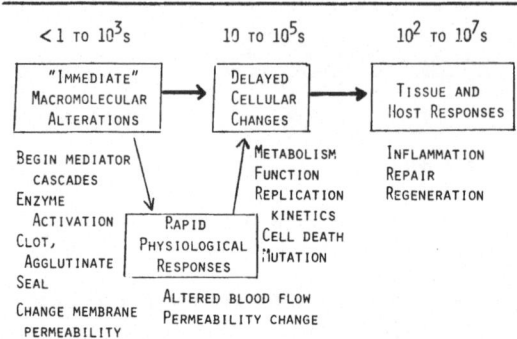

If enough macromolecules and cells are altered by the laser energy,
complex cascades of chemical mediators lead to further changes in
cell metabolism and function. If enough cells are injured, inflammation and repair are initiated in the host. These defenses may be
symptomatic. In some pathologic conditions segments of these events
may be therapeutic.

radiation is transferred through biologic materials until the energy
is invested in some way in the target tissue. With or without some
further energy transfer, the energy is eventually "dissipated" by
altering the structure of configuration of molecules within the
tissue. It is these molecular alterations which occur during the
energy dissipation which are the most important components of the
photophysical changes in the biologic target.

Electromagnetic radiation moves through tissue at the speed of
light decreased by a factor related to the refractive index of the
tissue. The refractive index of the stratum cornea of skin is about
1.55. Radiation transfer within tissue in complex. Essentially
every known photobiologic response following in vivo exposures of
humans depends in some way upon the optical properties of the exposed
tissue. The optical properties of tissue, viewed macroscopically as
a turbid medium, determine the optical radiation doses received by
various cell layers. In this sense the tissue's wavelength-dependent
absorption and scattering properties can greatly modify the ultimate
response to a given exposure, and often dictate as well the choice of
wavelengths feasible for treatment to a given cell.

Although the optical properties of connective tissue are very
complex, they can be measured and modeled. For example, spectral
scattering and absorption coefficient for human dermis can be calculated by application of a modified Kubelka-Munk theory. Dermal
scattering is markedly increased at shorter wavelengths, being
greater than 150 cm^{-1} at 400 nm, and less than 50 cm^{-1} at 800 nm and
beyond. The absorption coefficient, for bloodless dermis is smaller

than the scattering coefficient except at the prominent absorption
bands of water in the infrared region. Dermal scattering, therefore,
plays a major role in determining the depth to which radiation of
various wavelengths penetrates the dermis, and largely accounts for
previous observations that, in general, longer wavelengths across the
UV-visible-near infrared spectrum penetrate the dermis to a greater
extent than do shorter wavelengths. In vivo, the blood-borne pigments hemoglobin, oxyhemoglobin, carotene, and bilirubin are the
major absorbers of visible radiation in the dermis.

Photons traverse media, including tissue without effect until
they are absorbed. Upon absorption, photons cease to exist; their
energy is invested in the absorbing molecule or chromophore. The
excited molecule then dissipates the acquired energy by remission of
photons (fluorescence and phosphorescence), chemical reactions or
rearrangements (photochemistry), or nonradiative deexcitation (heat)
(Figure 1). In the vast majority of cases, most of the energy originally delivered by photons is expressed as heat, which dissipates,
mainly by diffusion, from those sites at which the absorption
occurred.

In most circumstances with most light sources, the amount of
light energy absorbed by a material is proportional to the amount
of light falling on that material. In this case absorption is a
specific molecular event which is wavelength dependant. Under
ordinary circumstances, this is also true for tissue. As radiant
exposure (light energy/area exposed) increased, enough light energy
may be deposited to increase tissue temperature in and near the
target molecules. Temperature increase is proportional to radiant
exposure after taking into consideration conduction, convection and
re-radiation. In this case because effect is proportional to cause,
both light absorption and temperature increase are called "linear"
processes. When irradiance is sufficiently high to alter the structural characteristics of the exposed material or the characteristics
of the incident light non-linear processes may occur. "Nonspecific"
or "bulk absorption", electron avalanche, electric field interaction
and ionization may occur.

Once energy has been invested in various biomolecules, it can be
transferred to other molecules by chemical energy transfer, conduction, transfer of particles, charge transfer, or by movement of
further electromagnetic waves. Eventually this energy is dissipated
and often this occurs in a way which changes molecules within the
tissue. The energy invested in a molecule may raise its electronic
of kinetic energy sufficient to cause unimolecular or biomolecular
reactions. Photochemical alterations with breaking or alteration of
bonds and multiphoton processes may occur. Thermochemical and
thermal alterations may result in changes in the primary, secondary,
tertiary, or quarternary configuration of macromolecules within the
tissue. This may result in alteration of membranes, denaturation,
dehydration, or carbonization of molecules. These changes may be

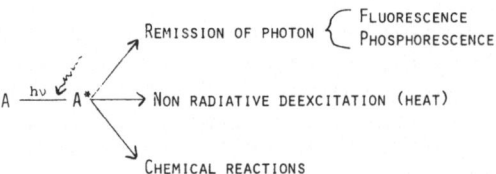

Fig. 1. Mechanisms of photon energy dissipation.

accompanied by breaks in tissue continuity. Both sonic and super-
sonic waves can be propagated within the tissue and when these exceed
the elastic limits of the tissue further breaks in continuity or
changes in membrane function may occur. These alterations may be
expressed microscopically and macroscopically as acute changes in the
physical properties or tissue such changes in viscosity, density,
optical properties and state of hydration. They may also be measured
as ablation, cutting, drilling, shearing, or discontinuous phases
within tissue. Tissue optics are acutely altered which leads to the
whitening of tissues sometime seen after laser impact.

BIOLOGICAL RESPONSES

     The photophysical events described above may lead to activation
or deactivation of enzymes or may alter the physical and chemical
properties of key macromolecules (See Table 1C). This may set off
complex chemical cascades which cause changes in viscosity, clotting,
sealing, and change in membrane potentials. Rapid physiologic
responses lead to changes in blood flow and vessel permeability.
These mechanisms may serve to amplify the impact of the photophysical
event on the tissue. Injured cells react with changes in function,
metabolism, and replication kinetics. Cell death may occur and
changes in DNA may lead to mutations in surviving cells.

     If enough cell injury occurs, chemical mediators are released
and primitive yet complex tissue and host responses are noticeable
histologically and grossly over the next few hours. Redness, swell-
ing, heat, pain and loss of function signify inflammation. Repair,
increased cell proliferation, regeneration, fibrosis and changes in
vasculature are long-term responses to cell injury manifested over
days to weeks. Certain components of these delayed responses may be
therapeutic in some pathologic conditions.

SPECIFICITY

     It is not very unique or exciting to be able to burn, boil,
explode or vaporize tissue. Therapeutic applications are possible
because unique collective properties of the laser make it possible

to have some control over degree and spatial extent of tissue injury. The most frequently used property is the capability of very high power density. In essence, the laser is a "light funnel" concentrating enormous light energy into a very small volume of tissue. This results in a kind of spatial specificity; a very small exposed field can be made very hot so that even though some surrounding tissue is affected, the affected target site is still so small that the sparing of normal tissue is maximal. The added laser property of collimation makes it possible to transport and focus the beam through a microscope and perform submicron microsurgery.

Absorption in very small tissue volume can lead to focal sites of phase changes which may lead to cutting, drilling, and ablation. Contiguous thermal burn injury without ablation can be confined to very small tissue volume (microscopic tissue layers to a few mm^3)

There are also ways to achieve specificity within the exposure field independent (or in addition to) spatial confinement by focusing (see Table 2). Lasers can be used to induce photochemical changes in tissue. Hematoporphyrin derivative photoradiation treatment is an example. Lasers make it possible to perform this treatment in any tumor accessible via endoscope, intravenous catheter, or hypodermic needle. Photochemotherapy[1] defines therapeutic application of light plus chemical in which neither alone has a significant effect on tissue. Examples include hematoporphyrin photoradiation therapy and psoralen photochemotherapy of skin diseases[2]. Because high power is not required to induce most photochemical reactions, conventional sources may sometimes be as effective in photochemotherapy. However, lasers may be used to induce multiphoton effects which could enhance all forms of photochemotherapy. In any case the inherent specificity of photochemical reactions provides a means of achieving selectivity of effects within the exposed field independent of focusing.

By controlling the rate of heating, exposure dose, and absolute temperature achieved within the exposed field it is also possible to obtain specificity dependent on differential thermal susceptibility of various biomolecules. New and separate events may be initiated as tissue temperature rises because different biomolecules have different temperature thresholds for permanent molecular rearrangements. Some biological thermal alterations are dependent also on the duration of temperature rise. By carefully controlling temperature, specific tissue alteration can be isolated.

Effects on blood vessels serve as an example of this kind of specificity[3,4]. The hamster cheek pouch has been used for direct microscopic observations of laser induced immediate changes in vessel integrity and morphology, and in the appearance and flow of blood, transient hemostasis, permanent hemostasis and hemorrhage were observed increasing exposure doses. In this case the photophysical

Table 2. Specificity in Photophysical Events.

	IMPORTANT LASER VARIABLE		
	POWER	λ	Υ
A. FOCUSING (SPATIAL CONFINEMENT OF EXPOSURE FIELD)			
"NONLINEAR EFFECTS"	✓	±	
CUT, DRILL, ABLATE	✓	±	
THERMAL BURN (MICROSCOPIC)	✓	✓	±
SUBCELLULAR MICROSURGERY*	✓	✓	
B. TARGET SPECIFICITY (WITHIN EXPOSED FIELD INDEPENDENT OF FOCUSING)			
PHOTOCHEMICAL (MOLECULAR)		✓	
DIFFERENTIAL THERMAL SUSCEPTIBILITY (MACROMOLECULE TO CELL)	✓	✓	±
SELECTIVE PHOTOTHERMOLYSIS (ORGANELLE TO MULTICELLULAR STRUCTURE)	✓	✓	✓

POWER = PHOTON DENSITY, CAPACITY FOR VERY HIGH POWER

λ = WAVELENGTH, CAPACITY FOR MONOCHROMATICITY

Υ = PULSEWIDTH, CAPACITY FOR VERY SHORT PULSES

* SUBMICRON MICROSURGERY ALSO DEPENDS ON COLLIMATION

event, the subsequent biological response and selection of the target itself is influenced by the temperature achieved in the tissue. Mechanistic and spatial specificity result.

Selective photothermolysis is a scheme for confining thermally mediated radiation damage to chosen pigmented targets at the ultra-structural, cellular, or tissue structural levels[5]. Experimental verification can be shown for two biologically interesting targets - blood vessels and melanocytes[6]. The confinement of damage can be as precise as with microbeam techniques, but millions of targeted structures are damaged simultaneously in vivo without precise aiming. This approach amy be particularly useful in turbid tissues, which unlike the eye, limit the precision with which isolated structures can be exposed. Tissues between targeted structures, including overlying or immediately neighboring cells, are spared, potentially reducing widespread destruction and nonspecific fibrosis.

Lasers are capable of very short pulses and this property is key to the spatial confinement of thermal change. Only certain targets within the exposed field are altered. This technique relies on selective absorption of a brief radiation pulse to generate and confine heat at certain pigmented targets. An absolute requirement is that the targets have greater optical absorption at some wave-

length than their surrounding tissues. This requirement can be met
either by choosing endogenously pigmented targets or by using staining or dye labeling techniques. Immunologic techniques (monoclonal
antibodies) can be used to deliver chromophore to specific targets
within the exposed field.

During laser exposure, absorption and radiationless deexcitation
convert radiant energy into heat within each target in the exposure
field. The targets begin to transfer this heat to their cooler
surroundings mainly by thermal diffusion, but this process takes some
time. Heat is initially confined to the targets during exposure. At
the end of an appropriately brief laser exposure, the temperature of
the target may have surpasses that required for thermal denaturation
(or other chemical or physical changes) while that of the surrounding
tissue remains well below this temperature. Immediately after the
exposure, thermal diffusion cools the targets and warms the tissue
between them but not necessarily to threshold temperatures for
observable biologic changes. The warmed tissue, with its specifically thermally damaged targets, then slowly cools.

Two examples of selective photothermolysis include specific
effects on targets as·large as blood vessels and as small as
melanosomes, one micron organelles within skin cells. On the basis
of models describing optical and thermal transfer in tissue, it was
predicted and has subsequently been shown that unfocussed, pulsed 577
nm laser irradiation of human skin causes widespread damage to micro-
vessels, with little or no direct damage to other tissue structures
such as the epidermis[7]. This is in contrast to the diffuse coagulation necrosis induced by continuous wave lasers, in which all
structures near the site of exposure are damaged. A dye laser with
0.3 sec pulse width (Candela model SLL-1100) was used to irradiate
uniform 3 mm sites with single pulses ranging from 0.5 to 5 J cm^{-2}
on the forearms of eight fair-skinned volunteers. Responses were
observed clinically and histologically immediately, 24, and 48 hours
after exposure. Striking vascular changes were noted with little or
no damage to the overlying epidermis or structures between vessels.
A dose of 1.5 to 2 J cm^{-2} consistently produced a pinprick sensation
and within a minute the forearm turned purple (purpura), indicating
superficial hemorrhage. Twenty-four hours after exposure, a necro-
tizing vasculitis without epidermal changes was seen histologically[8].

Targets within the unfocused field may be smaller than one
micron. Normal human volunteers exposed to brief (20 nanosecond)
351 nm wavelength pulses from a XeF excimer laser, showed whitening
of exposed skin. Transmission electron microscopy revealed the
earliest cellular alteration to be dose-dependent immediate disruption of melanosomes, both within melanocytes and basal keratinocytes.
Superficial keratinocytes and Langerhans cells were less affected
and it seemed that the XeF excimer laser was capable of organelle-
specific injury to melanosomes[9].

Table 3.   Manipulation of Biologic Variables
May Lead to Increasing Specificity.

A.	CHANGE PROPERTIES OF TARGET TISSUE
	TEMPERATURE
	OPTICAL PROPERTIES
	BLOOD FLOW
	PRESSURE
	SUCTION
B.	ALTER BIOLOGICAL RESPONSE
	STABILIZE MEMBRANES
	QUENCHERS
	ANTI-INFLAMMATORY AGENTS
	INDUCE OR INHIBIT ENZYMES
	ANTI-PROLIFERATIVE AGENTS

It is also possible to increase specificity or to augment or diminish selected aspects of the biologic response (see Table 3). This can be done to study mechanisms. For example, normal human volunteers were used to study the effects of heat, cold, pressure, suction and vasoactive pharmacologic agents on the threshold dose for laser-induced purpura. At the threshold doses the laser induced selective effects on vessels were found to be inconsistent with a shock wave mechanism for vessel rupture, and were more likely explained by microvaporization of red blood cells and blood[10]. By altering the physical properties of the tissue (see Table 3) the target is a different material and affected differently by photons.

By combining various techniques to obtain target specificity, selective responses can be achieved with less tissue destruction. Combinations of several different thermal and photochemical effects can be achieved by altering pulsewidth and intensity, by simultaneous exposure to more than one laser, or by combining laser exposure to more conventional sources. Photochemotherapy and supply of exogenous chromophores allow further manipulation of both the photophysics and biological response. By augmenting or diminishing selected components of the biologic response, it is also possible to specifically improve desired therapeutic response and minimize unwanted side effects.

REFERENCES

1. J. A. Parrish, T. B. Fitzpatrick, L. Tanenbaum, and M. A. Pathak, New Eng.J.Med., 291:1207-1212 (1974).
2. J. A. Parrish, R. S. Stern, and T. B. Fitzpatrick, "Dermatology Update: Reviews for Physicians," S. Moschella, ed., Elsevier-North Holland Inc., New York, 313-338 (1982).

3. R. R. Anderson, K. F. Jaenicke, and J. A. Parrish, Mechanisms of
    selective vascular changes by dye lasers, in: "Lasers in
    Surgery and Medicine," in press (1983).
4. R. W. Gange, K. D. Jaenicke, R. R. Anderson, and J. A. Parrish,
    Effect of pre-irradiation tissue target temperature upon
    selective vascular damage induced by 577 nm tunable dye laser
    pulses, in: "Microvascular Research," in press (1983).
5. R. R. Anderson and J. A. Parrish, Selective photothermolysis:
    Precise microsurgery by selective absorption of pulsed radi-
    ation, Science, 220:524-527 (1983).
6. J. A. Parrish, R. R. Anderson, T. Harrist, B. Paul, and G. F.
    Murphy, J.Invest.Dermatol., 80:75S-80S (1983).
7. R. R. Anderson and J. A. Parrish, Lasers in Surgery and Medi-
    cine, 1:263-276 (1981).
8. J. Greenwald, S. Rosen, R. R. Anderson, T. Harrist, F.
    MacFarland, J. Noe, and J. A. Parrish, J.Invest.Dermatol.,
    77:305-310 (1981).
9. G. F. Murphy, R. S. Shephard, B. S. Paul, A. Menkes, R. R.
    Anderson, and J. A. Parrish, Organelle-specific injury to
    melanin-containing cells in human skin by pulsed laser
    irradiation, Lab.Invest., in press (1983).
10. B. S. Paul, R. R. Anderson, J. Jarve, and J. A. Parrish, The
    effect of temperature and other factors on selective micro-
    vascular damage caused by pulsed dye laser, J.Invest.
    Dermatol., in press (1983).

INTERACTION OF LASER LIGHT WITH BIOLOGICAL TISSUE

A. J. Welch and M. Motamedi

Electrical Engineering and Biomedical Engineering Program
The University of Texas, Austin, Texas 78712, USA

INTRODUCTION

Various groups have reported photochemical, thermal and mechanical reactions as the result of irradiating tissue-with a laser. Photochemical reactions have included non-thermal retinal damage produced by exposure at blue wavelengths[1,2] and photoradiation therapy, in which hematoporphyrin derivative (HPD) is activated by visible light for the treatment of malignant tumors. Lasers provide the relatively high intensity narrow band light needed for the photochemical reaction to take place[3]. Primarily, lasers are widely used to generate heat through absorption of light by the different constituents of tissue. This process can produce photocoagulation or vaporization of tissue[4]. More recently, lasers have been used to induce optical breakdown and the creation of shock waves which disrupt the surrounding tissue. Plasma breakdown occurs when electric fields are strong enough to strip electrons from atoms, causing ionization of the medium and formation of a plasma. This phenomena is being used in ophthalmology to cut membranes in the eye[5].

A major concern during any laser irradiation is the distribution of the light in the tissue. In this chapter we examine laser light distribution in tissue and efforts to model this interaction. Since most of the current applications concern thermal reactions, a section of the chapter is devoted to heat deposition in tissue and the measurement of temperature in tissue.

LIGHT DISTRIBUTION IN TISSUE

When an incident wave strikes the interface between two media with different refractive indices, a portion of the incoming light is

reflected back into the incident medium. The magnitude of reflected
light depends on wavelength, polarization and incident angle as well
as the optical properties of the transmitting medium. A simple way
of presenting the interaction of light with matters is illustrated in
Figure 1.

Reflected light has two components, regular and diffused reflected light. Regular reflected light has an angle of reflectance
equal to the angle of incidence of the irradiation. The angles of
reflectance for diffuse reflected light are random and have a uniform
distribution. For tissue, diffuse reflectance occurs primarily at
interfaces within the tissue. For normal human skin, the regular
reflectance of a normally incident beam of light is always between
4-7% over the entire spectrum from 250 to 3000 nm, for both caucasian
and negroid skin[6]. Diffuse reflectance of human skin from 200 to
5000nm is shown in Figure 2[7].

When the transmitted portion is passed through matter in the
solid, liquid or gaseous state, its propagation depends of the
characteristics of the medium and is affected by two important mechanisms; namely scattering and absorption. Scattering is the removal
of energy from an incident wave because of inhomogeneities in a

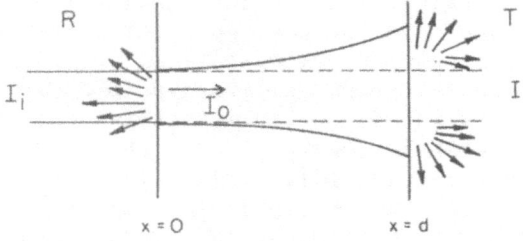

Fig. 1. The sample is illuminated with collimated beam I_o. R is the
reflected light, and T is the transmitted light.

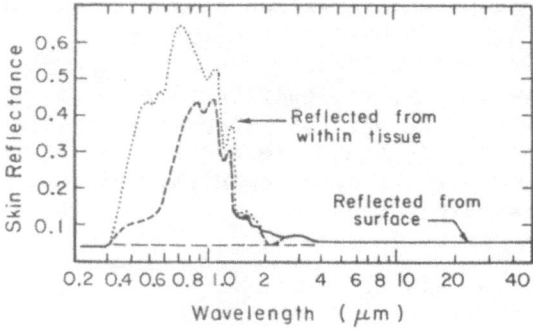

Fig. 2. Spectral reflectance of white and negro skin[7].

medium's refractive index that are related to physical inhomogeneities of the medium. The removed energy is subsequently reemitted in
the form of scattered light. It is the fundamental physical mechanism operative in reflection, refraction and diffraction. Absorption is the actual disappearance of light due to conversation of
energy into heat motion of the molecules of the absorbing material.

The medium in which a wave travels can be classified as being
either deterministic (such as guided wave) or random (such as atmos-
phere or ocean). Many biological media are randomly varying in time
and space, consequently, waves propagating in such media fluctuate
randomly in amplitude and phase. Statistical averages and probability densities are used to describe these random waves.

In general, wave propagation in random media can be grouped into
three categories[8]:

1.   "Wave random scatterers", random scatterers are random distribution of many particles such as ocean particles and red blood cells.
2.   "Wave in random continua", random continua are media whose
characteristics vary randomly and continuously in time and space such
as clear air and biological tissue.
3.   "Rough surface scattering" such as interfaces between different
biological media.

The spatial distribution and intensity of scattered light
depends on the size and shape of scatterers relative to the wavelength and the difference between refractive index of the medium and
the inhomogeneities. For particles with dimensions on the order of
about one tenth of the wavelength, scattering is generally weak and
isotropic and its intensity varies inversely with the fourth power of
wavelength. This is known as Rayleigh scattering. For particles
with dimensions on the same order as the wavelength, a much stronger
more forward-directed scattered intensity is present. If the dimensions of the particle are far greater than the wavelength, scattering
is weak but highly forward-directed. This is known as Mie
scattering.

Optical propagation in biological materials is dominated by
scattering because of the physical heterogeneities of biological
tissue at the cellular level which are of the order of an optical
wavelength. In order to understand the interaction of light with
biological tissue, it is important to know some of the characteristics of cells and tissues in the human body. Cells come in all
sizes and shapes and are usually several microns in diameter. Muscle
cells are a few millimeters long whereas some nerve cells are over a
meter long. Cells are grouped together and combined with other
material to form tissues. There are four basic tissue types, epithelial, connective, muscular, and nervous[9]. Each type of tissue has
specific optical properties which must be characterized individually.

DISTINCTION BETWEEN ABSORPTION AND SCATTERING

When light intensity I_0 enter a long glass cylinder filled with whole blood, the intensity I of the exit beam is less than I_0. For a given density of red blood cells, experiments show that I depends of the length L of the column according to Beer's exponential law.

$$I = I_0 e^{-\gamma L} \tag{1}$$

Where γ is usually called the extinction coefficient, since it is a measure of the loss of light from the direct beam. However, in this case, some of the decrease in intensity is not due to real absorption of the light but comes from the fact that some light is scattered to the sides by red blood cells and consequently is removed from the direct beam. (See Figure 3.)

Theoretical and experimental investigation in the darkly pigmented ocular media, and for any tissue irradiated with CO_2 laser indicate that γ is a close approximation of true absorption coefficient. In other cases, we can regard γ as made up of two parts, α due to true absorption and β due to scattering, where

$$\gamma = \alpha + \beta \tag{2}$$

Typically, attempts to model temperature rise and damage in laser irradiated tissue have not considered the scattering of light in the tissue. These models have assumed Beer's Law of absorption with β equal to zero as the basis of heat deposition in tissue. In a model for Nd-Yag irradiation of gastric tissue, Welch at al. tried a wide range of possible absorption coefficients[10]. Using reported absorption coefficients of 12 cm^{-1}, 5.7 cm^{-1}, 0.40 cm^{-1}[11,12,13,14] computed values of temperature rise and extent of damage were compared to measured temperature and the depth of damage in dog gastric tissue. Computed values far exceed the experimental data.

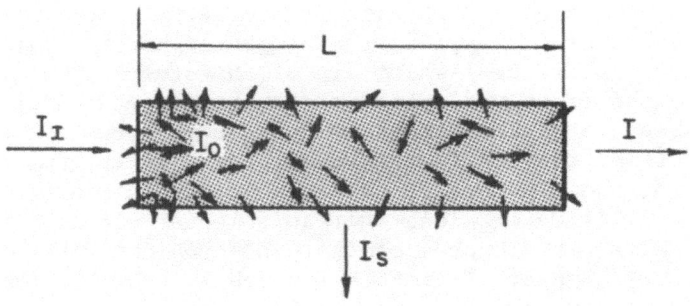

Fig. 3.   Scattering of light by red blood cells.

The model was not suitable for this application, because it did not take into account the scattering phenomenon[15]. Halldorson et al., proposed a three-dimensional time dependent temperature model based on single scattering approximation and eq. (2) for prediction of temperature rise in the laser irradiated tissue. Temperatures measured at the surface of exposed dog gastric tissue closely matched predicted temparature rises[14].

The significance of light scattering in biological tissue is observed in the optical density measurements of skin tissue. Using Figure 3 as an example, transmittance T and reflectance R are defined as the ratios of transmitted to incident intensity and reflected to incident intensity respectively. The optical density is defined as:

$$OD = \log (T^{-1}) \tag{3}$$

For a purely absorbing material described by Beer's law, a plot of OD versus L is a straight line, with a slope proportional to the absorption coefficient. When the radiation wavelength approaches the size of an object or inhomogeneity in the medium, scattering occurs and substantial deviations from Beer's law takes place. The effect of scattering on optical density as a function of thickness is shown in Figure 4 for different wavelengths for human skin[9]. At $\lambda = 2.2$ μm the optical density is due predominantly to absorption.

The analysis of wave propagation in random scatterer has been approached by using either "radiative transfer" theory (also called "transport theory") or "multiple scattering" theory ("analytical theory"). Radiative theory considers the propagation of intensities. It is based on heuristic observations of the transport if energy through a medium containing particles[16,17,18]. The basic differential equation in transport theory is called the "equation of transfer" and it is equivalent to Boltzmann's equation used in kinetic gas

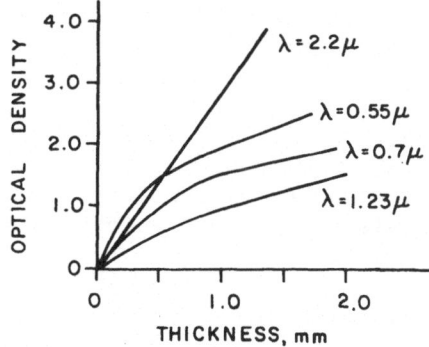

Fig. 4. Optical density plotted versus skin layer of white human skin at four wavelengths[9].

theory. Exact general solutions of the equations of transport theory have not been obtained; however, for some special cases, simple and practical approximate solutions have been derived. The most useful of these approximate solutions uses the diffusion equation.

In diffusion theory, photons are considered particles that are scattered and absorbed by the medium, their intensity is scattered isotropically (that is, the diffused intensity encounters many part-icles and is scattered almost uniformly in all directions). The probability density function for intensity as a function of angular distribution is almost uniform. Diffusion theory has been used to describe some of the observed phenomena in biological materials with a promising degree of quantitative precision and with analytical simplicity[9,19,20].

Reynolds used diffusion theory to model the scattering and absorption of light by randomly oriented red blood cells suspended in a homogeneous plasma medium. He derived equations for diffuse re-flectance defined for a finite size of a fiber optic catheter used for the spectrophotometric measurement of oxygen content in blood. Excellent agreement was seen between experimental data and the pre-dicted data by this model, verifying the theoretical approach used in describing light distribution in blood[19].

However, in very thin layers of biological tissue with less than 1mm thickness, or very near the surface, light scattering is not isotropic. Beyond a depth of 1mm light scatters isotropically and the diffusion equation may be used to describe the patterns of light in the tissue[20,21].

Diffusion theory considers the optical intensity or photon flux Γ at any given point (x,y,z) within a medium as given by:

$$\Gamma = -D\nabla\xi(x,y,z) \tag{4}$$

where $\xi(x,y,z)$ is the photon density and D is the diffusion coef-ficient for photons. (See reference 22 for details of the mathemat-ical development). Taking the divergence of both sides of (4) gives us:

$$\nabla\Gamma = -D\nabla^2\xi(x,y,z) \tag{5}$$

Because of the continuity of photons, the total flux of incident photons into a volume equals the rate at which photons are absorbed and scattered:

$$\frac{\partial\xi}{\partial t} = -\nabla\Gamma - \frac{\xi(x,y,z)}{\tau} + s(x,y,z) \tag{6}$$

where τ is the photon life time and $s(x,y,z)$ is the rate of genera-tion of diffusing photons per unit of volume (i.e., the rate of conversion of collimated photons to diffused photons per unit volume). In the steady state

$$\frac{\partial \xi(x,y,z)}{\partial t} = 0$$

so:

$$D\nabla^2 \xi(x,y,z) - \frac{\xi(x,y,z)}{\tau} = -s(x,y,z) \tag{7}$$

For the one-dimensional case (7) is reduced to:

$$\frac{d^2 \xi(z)}{dz^2} - \frac{\xi(z)}{D\tau} = - \frac{s(z)}{D} \tag{8}$$

and eq. (4) becomes:

$$\Gamma = -D \frac{d\xi(z)}{dz} \tag{9}$$

Depending of the boundary conditions of the problem, different approximations can be used to solve diffusion equations. For example, for a slab medium a green's function solution has been obtained using the simple boundary condition light flux $\xi = 0$ at $Z = 0$ and at $Z = d$[23].

More recently, Doiron et al.[24] used the diffusion theory approximation to calculate the space irradiance (the total radiation of a unit volume from all directions) in the tissue. They assumed spherical symmetry and monochromatic light source of wavelength λ. The space irradiance at distance r from the source was given by

$$\phi\ (r) = \phi_o\ (\tfrac{a}{r})e^{-\alpha\ \lambda} \tag{10}$$

where

ϕ_o = space irradiance at distance a=r
$\alpha(\lambda)$ = attenuation coefficient of the tissue (cm^{-1})
    = $\sqrt{\beta/\xi}$ (diffusion theory)
β = absorption coefficient
ξ = diffusion coefficient $\simeq 1/3\ \kappa$
κ = scattering coefficient
a = source radius.

Generally, diffusion theory is a reasonable approximation of light distribution in biological medium. However, care must be taken in defining boundary conditions and source distribution and selecting values for the coefficients of equation[10].

Kubelka and Munk[25] proposed another approximate solution of radiative transfer theory which is based on two fluxes[8]. They proposed a solution based on simple algebraic operations that gave results in agreement with experimental data. Historically, this theory was used in the paint and paper industry[26,27].

More recently the Kubelka and Munk method was applied to laser irradiation of skin[28,29]. Van Gemert et al. used this method in conjunction with reported experimental values of reflection and transmission of blood and skin to compute absorption and scattering coefficients of skin. Based on this analysis, they developed a model for laser coagulation of dermal vascular lesions[29].

Although the model applies only to diffuse light sources and presumes two light fluxes traveling in forward and backward directions, Von Gemert et al. believe it is a reasonable model for a collimated beam, such as laser. Ideally laser irradiation incident upon a medium should be described by four flux theory[8]. Unfortunately, no one has demonstrated the feasibility of determining all of the coefficients needed for the four flux theory.

The Kubelka and Munk theory is based on relations of diffuse reflectance and transmittance of a multi-layered system and the diffuse reflectance and transmittance of each component layer. Consider diffuse fluxes $I_+(x)$ and $I_-(x)$ travelling in the positive and negative x direction, respectively as illustrated in Figure 5. The magnitude of I_+ is a function of its absorption within dx and backward scattering which causes a decrease in magnitude, and backward scattering contribution form I_- which causes an increase in I_+. If A and S represent absorption and scattering coefficients respectively, then the change in I_+ in the interval dx is given by

$$+ \frac{dI_+}{dx} = -(A_+ + S_+)\, I_+ + S_- I_- \tag{11}$$

Similarly I_- decreases due to absorption and backward scattering but increases due to back scattering contribution from I_+.

$$- \frac{dI_-}{dx} = -(A_- + S_-)\, I_- + S_+ I_+ \tag{12}$$

Note that $A_+ \neq A_-$ and $S_+ \neq S_-$ unless the intensities are completely diffused and almost isotropic. Under these assumptions, A_+ and A_- are replaced by A and S_+ and S_- are replaced by S. Using differential equations (11) and (12), and appropriate boundary conditions and the values for total reflection coefficient R and total transmission coefficient T, a closed form solution can be derived, and the constants A and S can be expressed in terms of measurable boundary conditions[30],

$$a = \frac{1+R^2-T^2}{2R}$$

$$S = \frac{1}{xb} \frac{1-R/(a+b)}{T}$$

where

$$a = \frac{A+S}{S}$$

$$b = \sqrt{a^2-1}$$

Wan et al.[28] used this technique to predict light transmission through skin and estimate absorption and scattering coefficient of the skin. There was little practical difference in obtaining the epidermal transmittance using either collimated or diffused incident radiation. Transmittance of skin measured with an integrating sphere are compared in Figure 6 to values predicting using Kubelka-Munk theory. In this example the skin thickness was limited to the epidermis. The diffuse scattering (S) and absorption (A) coefficients for human dermis calculated from Kubelka-Munk theory of radiation transfer are shown in Figure 7[31].

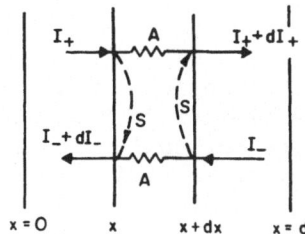

Fig. 5.   Positive and negative flowing fluxes I_+ and I_-. A and S represent absorption and scattering.

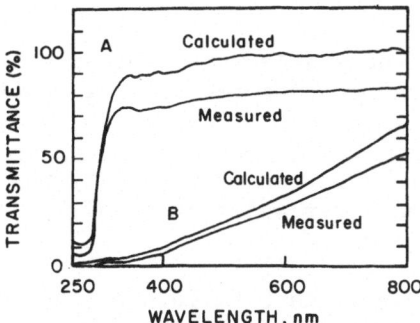

Fig. 6.   Comparison of measured epidermal transmittance with that calculated by Kubelka-Munk theory[28]. A.Caucasian, B. Dark.

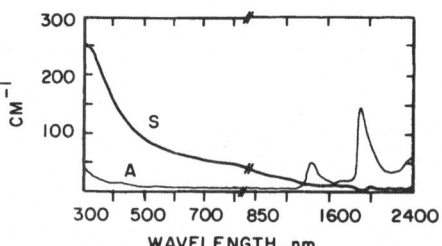

Fig. 7.   Diffuse scattering (S) and absorption (A) coefficients for
          human dermis in vitro, calculated from measurements of
          spectral remittance and transmittance of thin dermal
          sections under conditions appropriate to application of the
          Kubelka-Munk theory of radiation transfer[31].

Gijsbers[32] has recently derived the theoretical basis of
Kubelka-Munk theory from an equation of radiative transfer obtained
from phenomological considerations. He concluded that the Kubelka-
Munk model is correct for:

1.   perfect or nearly perfect diffused incident light,
2.   layers with infinite dimensions perpendicular to direction of
      light propagation, illuminated with an infinitely large and
      homogenous field of incident light,
3.   isotropic scattering.

He also derived a one-dimensional model from the equation of
radiative transfer that is similar to Kubelka-Munk theory. This
model can provide a solution for collimated incident light, whose
propagation is described by three differential equations.

In the same paper, modified versions of these models were developed to account for reflection at the boundary due to a change in
refractive index. Theoretical comparison showed that the reflection
at the boundary became important when scattering was much larger than
absorption.

He concluded that the models were theoretically valid only if
the light was isotropically scattered and the scattered light was
perfectly diffuse.

The most important deviation from reality was the assumption of
the infinitely stretched layer and flux in directions perpendicular
to the direction of incidence.

This suggested a significant failure of the models in the case
of a laser beam with a small diameter which could not be described
with a one-dimensional model. The one-dimensional models could be
used only if the diameter of incident light flux was large compared
to the thickness of the layer.

The other method for investigating wave propagation in random media is multiple scattering theory or analytical theory which is based on Maxwell's field equations and deals with averaged field quantities. Multiple scattering theory starts with a wave equation to obtain a solution for a single particle, introduces the inter-action effects of many particles, and then considers statistical averages.

One of the most useful multiple scattering theories has been developed by Twersky[33] which has been applied to optics in blood by Loewinger et al.[34] and Anderson and Sekleg[35]. They compared multiple scattering results and experimental optical density data for whole human blood.

HEAT DISTRIBUTION IN TISSUE

The production of heat can be represented by a source distri-bution function, which is equal to the rate of absorption of light by the medium. For a monochromatic beam of light that is passed through an absorbing medium, the absorbed portion of light in a given volume can be calculated from the equation of power conservation.

Basically

> outflow of flux of light   power generated per unit volume
>   per unit volume   = - power absorbed per unit volume

For the purpose of simplifying the derivation of the heat source, many investigators have assumed that the radiation travels only in one direction. This allows for the calculation of heat source based on Beer's exponential law of absorption.

Assuming that the incident irradiance is normally distributed, then

$$I(r,z=o) = I_o \, e^{-\frac{1}{2} \left(\frac{r}{\sigma_o}\right)^2} \tag{13}$$

where

> I_o = center irradiance (w/cm^2)
> σ_o = standard deviation of beam (cm)

The intensity of light reaching a depth of z in the tissue is described by Beer's law as

$$I(r,z) = I_o e^{-\alpha z} \, e^{-\frac{1}{2} \left(\frac{r}{\sigma_o}\right)^2} \tag{14}$$

Assuming that the absorption of energy takes place only in the z-direction, the heat source is given by

$$A(r,z,t) = \alpha I_o e^{-\alpha z} e^{-\frac{1}{2}(\frac{r}{\sigma_o})^2} \tag{15}$$

Use of eq. (15) as the heat source in the heat conduction equation provides a prediction of temperature rises in a highly absorbing medium ($\alpha/\beta \gg 1$). In these situations computed temperatures are highly correlated to measured temperatures[15]. However, for media in which absorption of light is not the dominating mechanism, Beer's law fails to predict laser induced temparature rises.

A modified version of Beer's law has been suggested that includes the contribution of scattering to the heat source[10]. The size of the laser image within the tissue is assumed to be a function of depth, i.e. the image standard deviation $\sigma(z)$ increases with depth in the media, as shown in Figure 8.

Typically, an extinction coefficient γ is defined as the sum of the absorption coefficient α, and the scattering coefficient β, as indicated previously in equation[2]. Then for a lossless medium ($\alpha = 0$), equation (14) becomes

$$I(r,z) = I_o e^{-\beta z} e^{-\frac{1}{2}(r/\sigma(z))^2} \tag{16}$$

Since the total power remains unchanged along the penetration depth, the standard deviation of the image is

$$\sigma(z) = \sigma_o e^{(\beta/2)z} \tag{17}$$

If total power is reduced as a function of depth by $e^{-\alpha z}$, then

$$I(r,z) = I_o e^{-\gamma z} e^{-\frac{1}{2}(r/\sigma(z))^2} \tag{18}$$

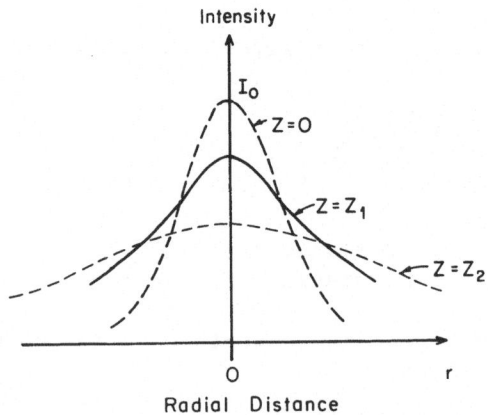

Fig. 8. Broadening of light in a lossless medium along the direction of penetration z.

and the heat source is given by

$$A\,(r,z,t) = \alpha I_o\; e^{-\gamma z}\; e^{-\frac{1}{2}(r/\sigma(z))^2} \tag{19}$$

The modified heat source was used for prediction of temperature rises in Nd-YAG laser irradiated dog stomach. Inclusion of spreading beam in the source term of the heat conduction equation produced a close match between measured and predicated temperatures as shown in Figure 9.

Our understanding of the interaction of laser light with tissue is often expressed in terms of models used to estimate light distribution, temperature and damage in tissue. Validation of the accuracy and credibility of our concepts and models requires experimental measurements of the transient temperature response, light intensity and histology of the irradiated tissue.

Neither a model or an experimental design should be considered independently. Models help identify critical exposure conditions, provide estimated results, and generally help fill gaps where experimental data is not available. When there are large differences between models and experimental data, the model, experimental design, and measurement system must be examined to locate the source of error. The model-experiment synergism (Figure 10) requires continual comparison and adjustment.

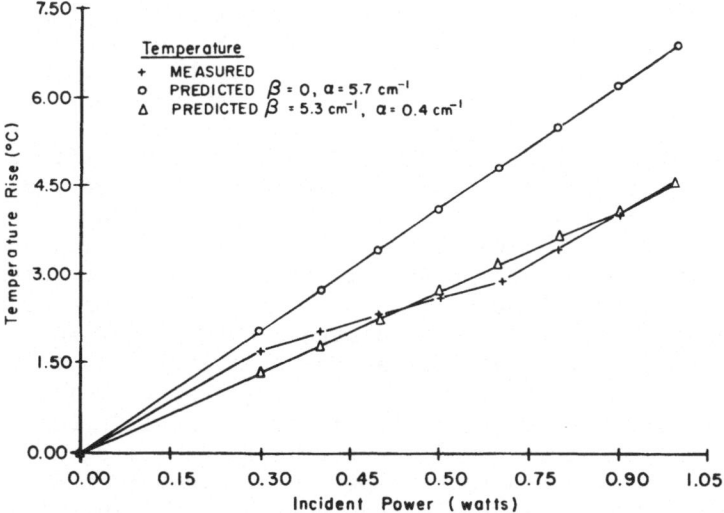

Fig. 9.   Measured and predicted temperatures for exposure time of 8 seconds for a sensor located at the center of the beam and depth of 1.2 mm in the tissue.

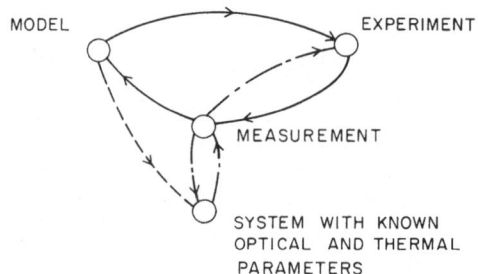

Fig. 10. Interaction of measurement, model, experimental systems.

     Light and temperature measurement in a laser irradiated biologic
medium are extremely difficult. It is difficult to establish the
precision of these measurements, especially if computed and measured
values do not agree. It is usually necessary to measure light intensity and temperature in a simple physical system with known optical
and thermal properties. When agreement is achieved between the model
and measurements from the known system, then there is a basis for
accepting measurements from the biological system. The inner relationship of measurement, model and biological systems is illustrated in Figure 10. Any experiment involving physical measurements
in a biological system should include some aspect of each component
represented in the figure.

TEMPERATURE MEASUREMENT IN TISSUE

     The determination of temperature at a particular point, p, in
a medium requires an invasive measurement with a sensor such as a
thermocouple. A calibrated characteristic (such as the junction
voltage for a thermocouple) reflects the temperature of the sensor
when thermal equilibrium is reached between the medium and the sensor. In the neighborhood of p the medium is replaced with the sensor
which introduces a material that may have thermal properties far
different than properties of the medium. Perturbation of the medium
by a metal cylinder is illustrated in Figure 11. A homogeneous
medium is placed between two parallel infinite surfaces that have
boundary temperatures, T_0 and T_1 respectively. The temperature of the
medium is linearly related to the distance between the surfaces and
the differences in boundary temperatures as illustrated in the
figure. The cylinder in the Figure could represent a wire thermocouple. Since the conductivity of a metal is usually much, much
larger than the conductivity of a water-based medium such as tissue,
the cylinder creates a constant temperature volume in the tissue.
The disruption of the original steady-steady temperature field is
illustrated in Figure 11. To minimize errors, the sensor should be
small with respect to temperature gradient and the structure of the
tissue; if possible the sensor should have thermal properties similar
to that of the tissue.

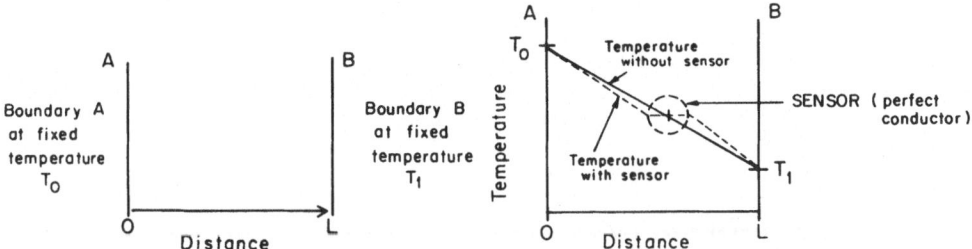

Fig. 11. Two infinite, parallel boundaries at fixed temperatures
          (left). Perturbation of temperature by sensor (right).

The measurement of transient temperatures in laser irradiated
tissue provides a challenging task. Problems that have been encountered in our laboratory include:

1. response time of the sensor;
2. absorption of laser radiation by the sensor;
3. high temperature gradients in the tissue; and
4. conduction of heat by the sensor.

RESPONSE TIME

        Response time is usually measured by plunging a sensor into a
heated bath of water. A typical response is illustrated in Figure
12. Unfortunately, the response contains two artifacts. The first
at the initial rise is caused by heat rising from the surface of the
heated bath. The second is an apparent overshoot in the response of
the sensor which may be due to a combination of temperature gradients
in the bath and recoil of the sensor plunger mechanism. Because of
these errors, 10-90% rise time indicated in the figure may be more
representative of response time than the traditional time constant,
also illustrated in the figure. The initial error can be eliminated
by driving the sensor into a cooled bath rather than a heated bath.
A fast but well-damped plunger and a stirred bath should minimize the
problem of overshoot.

        The response time of a sensor depends upon the mass of the
sensor junction and the thermal contact of the sensor and tissue. A
poor contact creates a thermal resistance between the surface of the
sensor and the tissue which decreases the response time of the sensor. The response of a probe to a step increase in temperature has
been discussed by Reed[36] and by Cain and Welch[37]. Reed numeric-
ally calculated the rise time of a cylinder in an infinite medium
based on a solution described by Carslaw and Jaegor[38] for the
following simplifying assumptions. First, the probe was entirely
embedded in the medium with no axial flow. Second, the heat transfer

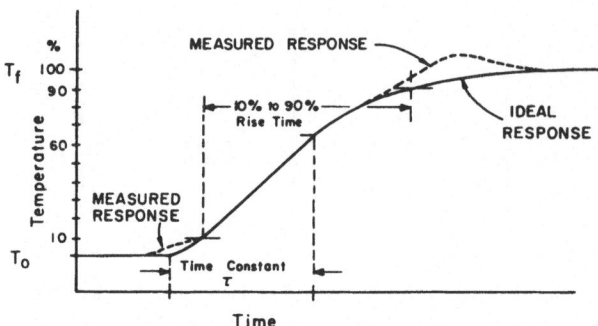

Fig. 12. Temperature response of sensor driven into a heat bath.

between the medium and the probe was by conduction with no thermal resistance at the interface. The probe initially at unit temperature and at T = 0 was placed in an infinite medium at zero temperature. The calculated fall time (cooling time) was equivalent to heating a probe in a medium at elevated temperature. Plots were obtained of rise time τ versus relative volumetric specific heats. The relative volumetric heat was defined as

$$\nu = \frac{\rho c \ (\text{medium})}{\rho c \ (\text{sensor})} \tag{20}$$

In Figure 13, the theoretical rise time (10-90%) response of quartz cylinders embedded in various mediums is compared with the rise time Cain and Welch measured by driving a micro thin film thermocouple into water. Thin films of nickel and copper were vapor deposited on a quartz substrate to form a 20 to 40 μm diameter thermocouple junction at the tip of the probe[38]. The experimental data confirmed Reed's analysis that the increase in rise time is a function of sensor radius. If a 1 msec rise time is required in a biomedium, the sensor radius should be of less than 10 microns.

When the thermal response time of a laser irradiated tissue is faster than the sensor, there is obviously a measurement error. The initial rate of temperature increase during the heating phase with a constant power of irradiation is given by:

$$\frac{\partial T}{\partial t} = \frac{\alpha I(r,z)}{\rho c} \tag{21}$$

where

α is the absorption coefficient of the tissue (cm^{-1})
I(r,z) is the intensity at a radial distance r from the center of the laser beam and depth z within the tissue (watts/cm^2)

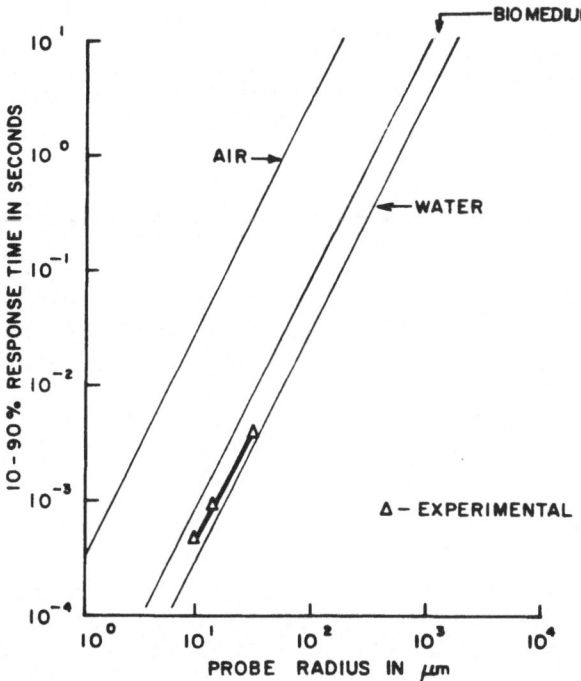

Fig. 13. Model and experimental rise time for Microthermocouples in water, air, and biomedium. The calculated values are from the model by Reed[36]. The experimental values are from Cain and Welch[37].

Typical values of the $\frac{\partial T}{\partial t}$ $(t = 0^{+})$ prior to the conduction of heat in the tissue may be of the order of 1000°/sec for therapeutic irradiations of stomach tissue and as high as 50,000°/sec during retinal surgery. Once the laser is turned off the thermal relaxation time depends upon the volume of tissue heated. The time constant for the stomach irradiation may be at least one second; whereas, the response time in the eye may be less than 20msec.

The error associated with response time with a sensor can be estimated if a few simplifying assumptions are made for both the sensor response (T_S) and the medium temperature response (T_m). The difference in these temperatures represents the measurement error, e, as shown in Figure 14.

Suppose the medium response to a constant power irradiation of 1 unit is:

$$T_m(t) = 1-e^{-t/k} \tag{22}$$

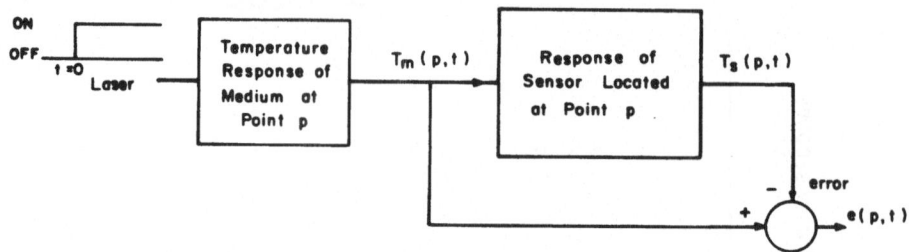

Fig. 14.   Model of medium and sensor response for temperature rise at
           point p.   The error represents the difference in actual
           rise of the tissue and the measured temperature.

where k is the time constant of the medium.   Suppose the sensor
response to a unit step change in temperature is

$$T_s(t) = 1 - e^{-t/a} \tag{23}$$

where a is the time constant of the sensor in the medium.   Using
Laplace transform techniques, the error between medium temperature
and measured temperature, as illustrated in Figure 14, is

$$e(t) = \frac{-r}{r-1} e^{-t/k} + \frac{r}{r-1} e^{-t/a} \tag{24}$$

where $r = a/k$.   Typical temperature responses are illustrated in
Figure 15.   The maximum error of the signal shown in Figure 15 occurs
at time t_{max}, and is given by

$$t_{max} = \frac{a}{r-1} \, Lnr \tag{25}$$

and the maximum error at t_{max} is

$$e(t_{max}) = (\tfrac{r}{r-1})\{r^{-(\frac{1}{r-1})} - r^{-(\frac{r}{r-1})}\} \tag{26}$$

     Maximum errors for various ratios of medium and sensor time
constants are shown in Figure 16.   Even using a time constant three
times faster than the medium response produces an 18% error with
respect to the steady-state temperature.   Precision measurement of
the entire transient requires a sensor with a time constant ten times
faster than the response time of the medium.   If the measurement is
limited by the response time of the sensor it may be possible to
compensate for sensor time constant using inverse Laplace transform
techniques[39].

     Large measurement errors occur any time the irradiation time is
less than or equal to the response time of the sensor.   Only during

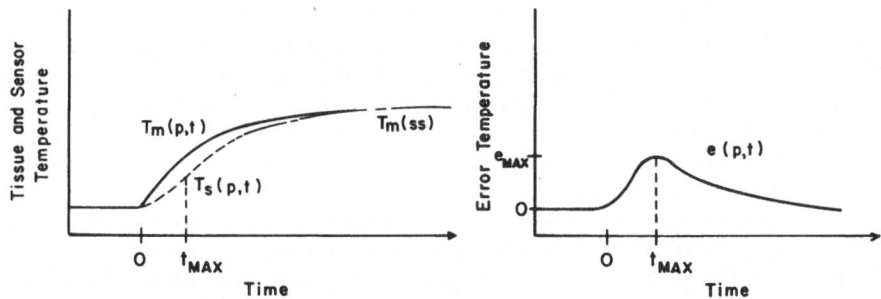

Fig. 15.   Tissue and sensor temperature responses (left). The
          difference of these temperatures is illustrated at right.
          T_{max} is defined as the time of maximum difference.

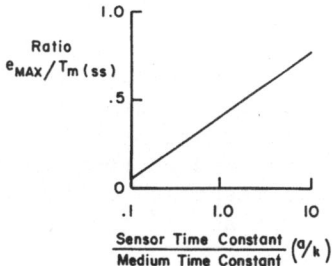

Fig. 16.   The ratio of maximum error(e_{max}) to medium (tissue) steady
          state temperature as a function of the ratio of sensor and
          medium time constants.

the relaxation phase of the temperature response does the sensor
catch up with the medium temperature. For example, if the irradiation duration is less than one millisecond, the rapid temperature rise is dominated by the magnitude of heat generated in the
tissue. Once the exposure ends, the volume of tissue heated and
conduction governs the relaxation time constant of the cooling
medium. At this point, the sensor should catch up with the medium
temperature with a time less than three time constants of the sensor.

DIRECT ABSORPTION

    Ideally, the sensor would not absorb laser irradiation. Un-
fortunately, irradiation striking the sensor is absorbed and the
light energy is converted to heat which is conducted to the medium or
along the sensor. The indicated temperature response of the sensor
depends upon the medium surrounding the sensor and energy absorbed by
the sensor. In water, the response time is rapid as equilibrium is

reached between heat generated in the sensor and heat conducted to
the water. However, irradiation of the sensor in air creates a high
temperature due to the poor transfer of heat between the sensor and
air (see Figure 17). The temperature rise in air may be ten times
larger than the steady-state temperature reached in water. Irradiation of a sensor in air can destroy the sensor unless caution is
taken to limit irradiation power.

The response time of the probe-medium system is of the same
order as the response times of Figure 13. The indicated sensor
measurement is a superposition of the temperature response created by
direct absorption of laser light and the actual medium temperature
response.

The direct absorption response is the major source of measurement error. It must be removed from the measured temperature response to obtain a reasonable indication of the temperature in the
medium. If the sensor response is ten times faster than the medium
response, the direct absorption response appears as a step increase
when the laser is turned on and a step decrease in temperature when
the laser is turned off. Both step responses should have the same
magnitude and can be averaged to create a correction factor for the
measured response[37]. When the medium response time has the same
order as the direct absorption response, it is impossible to differentiate between the medium response and the direct absorption
response.

SPATIAL TEMPERATURE GRADIENTS

The indicated temperature of a sensor represents the average
temperature over the volume of the active sensing element. When a

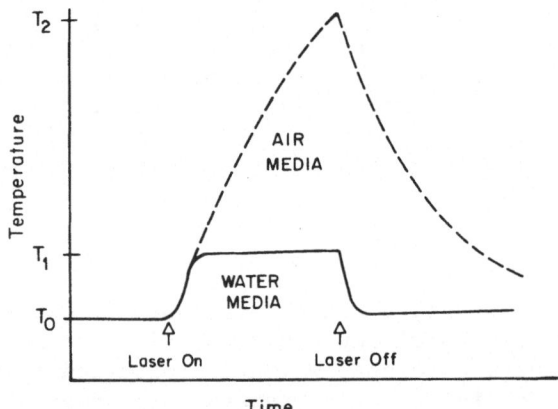

Fig. 17. Temperature response due to direct absorption of laser
          irradiation by sensor in water or air media.

large sensor with respect to the expected temperature gradients of
the medium is placed in the tissue, the sensor not only averages but
distorts the local temperature field. In effect, the sensor averages
tissue temperatures in the neighborhood of the sensor.

In the fundus of the eye, small images and the highly absorbing
pigment epithelium combine to produce extreme radial and axial gradients. The gradients are approximately Gaussian in shape in the
radial direction and exponential in the axial direction as illustrated in Figure 18. Irradiation of the eye with an 80 micron diameter retinal image that produces a 40° peak temperature rise can
produce maximum radial and axial gradients of the order of 300°/mm
prior to significant conduction. The maximum radial gradient occurs
at the image standard deviation and the maximum axial gradient occurs
at the surface of the absorbing layer ($z = 0$). At the center of the
image the radial gradient is zero.

The center of the beam and just below the surface of the absorbing layer of the pigment epithelial represents the point of maximum
temperature rise. Because the sensor averages temperatures across
its' surface, the indicated temperature of this point is less than
the peak medium temperature. If the radius of the sensor is equal to
the standard deviation of the Gaussian beam, then the indicated
temperature would be ten to fifteen percent less than the peak value
due to averaging of the radiant temperature profile. Further reduction of the indicated value would occur if there is a large axial
gradient across the sensor. Generally in laser irradiations of soft
tissue, the spatial gradients are rather small with respect to micro
thermocouple dimensions (50 μm or less) and do not present a major
problem in measurement.

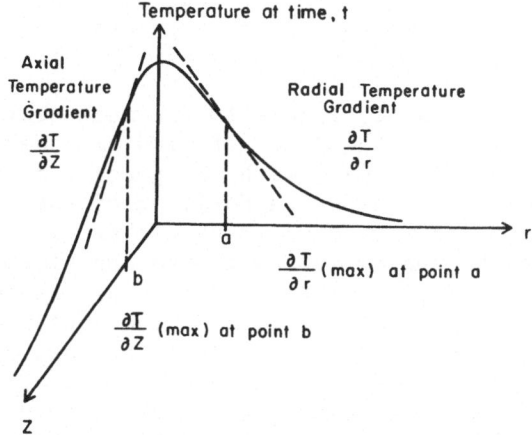

Fig. 18. Temperature gradients in tissue prior to significant
          conduction in the radial and axial directions.

AXIAL CONDUCTION

Ideally, thermal properties of the sensor should match those of
the tissue. Thus, any heat flow in the sensor would simply represent
the flow that would occur in the same region if the sensor was not
present. However, sensors with metal leads produce a region of high
conductivity in the tissue. If the sensor is measuring peak temperatures, the leads would be carrying heat away from the sensing element
and indicate a temperature below the actual temperature of the
medium.

In one experiment, a micro thermocouple was inserted through the
back of the eye until the sensing element was located in the retina.
Laser irradiation of the eye heated the pigment epithelium and the
conduction of heat in the tissue increased the temperature of the
retina. However heat transferred to the probe from the pigment
epithelium was rapidly conducted to the sensor tip, creating a large
artifact in the indicated retinal temperature.

LOCATION OF SENSOR

Even after solving all of the measurement problems, one sometimes overwhelming task remains: the placement of the sensor in the
tissue. Actually, placement is not so much a problem as knowing the
location of the sensor during laser irradiations. The precision
required for locating a sensor with respect to the surface of the
tissue and the center of the laser beam is a function of the temperature gradients created in the tissue.

For example, there is no way of a priori placing a sensor tip
10 μm ± 2 μm posterior to Brooks membrane in the eye. Yet, expected
temperature gradients in the pigment epithelium suggest this precision is needed if measurements are to be related to structural
regions of the retina.

In skin or stomach tissue, gradients are much smaller and the
required precision may be at the order of ±100 μm. Nevertheless,
subsurface placement of the sensor with the precision has never been
reported. One way of determining the position of the probe with
respect to a surface is to mark the tissue either with a stain on the
sensor or by leaving the sensor in the tissue and histologically
measuring the position of the mark. Even if depth is determined,
there is no certainty as to the position of the sensor with respect
to the center of the laser image. We have found that repeated exposures of low level laser irradiation and movement of the laser beam
can be used to find a maximum temperature rise. The point of maximum
temperature rise should theoretically be at the center of the laser
beam. Once again, the precision of locating the sensor with respect
to the center of the beam depends upon the size of the sensor and the
radial gradient of the temperature profile.

Reasonable measurements are possible, but care must be taken to assure accurate measurements and location of the sensor. Particularly, evidence should be presented that the reported temperatures actually represent the tissue temperature and are not artifacts.

SUMMARY

Temperature measurement in laser irradiated tissue is not impossible but it is an extremely difficult task. Reasonable measurements require an understanding of the above-mentioned problems and painstaking efforts to calibrate the dynamic response of the sensor in a laser irradiated medium.

Acknowledgements

This work was supported in part by a grant from Promed Technology.

REFERENCES

1. W. T. Ham, J. J. Ruffolo, H. A. Mueller, A. M. Clarke, and M. E. Moon, Investigative Ophthalmology and Visual Science, 17(10):1029-1035 (1978).
2. W. T. Ham, H. A. Mueller, and J. J. Ruffolo, SPIE, Ocular Effects of Non-ionizing Radiation, The Society of Photo-optical Instrumental Engineers, 229:46-50 (1980).
3. T. J. Doughtery and R. E. Thoma, "Lasers in Photomedicine and Photobiology," R. Pratesi and C.A. Sacchi, eds., Springer-Verlag, New York (1980).
4. J. A. Dixon, "Surgical Application of Lasers," Institute of Electrical and Electronics Engineers, Proceedings, 70 (1982).
5. D. Aron-Rose, J. J. Aron, M. Griesemann, et al., J.Am.Intraocul. Implant.Soc., 6:352-354 (1980).
6. J. A. Parrish, R. R. Anderson, F. Urbach, et al., "Biologic Effects of Ultraviolet Radiation with Emphasis on Human Responses to Longwave Ultraviolet," Plenum Press, New York (1978).
7. E. R. Hendler, J. Crosbie, and J. D. Hardy, "Measurement of Skin Heating During Exposure to Infrared Heating," Project NM 17-01-13-2 of Naval Air Material Center, Philadelphia, Pennsylvania, March (1957).
8. A. Ishimaru, "Wave Propogation in Random Media," Vol. 1 and Vol. 2, Academic Press, New York (1978).
9. C. C. Johnson and A. W. Guy, Proc.IEEE, 60(6):692-718 (1972).
10. M. Motamedi, A. Gonzales, and A. J. Welch, "Thermal Response of Gastro-intestinal Tissue to Nd-Yag Laser Irradiation: A Theoretical and Experimental Investigation," to be published

in Proceedings of 1983 International Congress of Lasers and Electro-optics, Laser Institute of America.

11. L. A. Priebe, L. E. Baker, and A. J. Welch, in: "Laser Surgery II," Isaac Kaplan, ed., 242-256, Jerusalem Academic Press.

12. P. G. Kiefhaber, Nath, Moritz, "Endoscopical Control of Massive Gastrointestinal Hemorrhage by Irradiation with a High-Power Nd-Yag Laser," Progressive Surgery, 15:140-155 (1977).

13. W. W. Rother, T. Halldorsson, J. Langerholc, and K. Schaffler, "Present Status of the Nd-YAG Laser in Endoscopy and Surgery," Laser Surgery II, 211-223, Isaac Kaplan, ed., Jerusalem Academic Press (1978).

14. T. Halldorsson, and J. Langerholc, "Thermodynamic Analysis of Laser Irradiation of Biologic Tissue," Applied Optics 17(24): 3948-3958 (1978).

15. A. N. Takata, L. Zaneveld, and W. Richter, "Laser-induced thermal damage of skin," Final Report SAM-TR-77-38, U.S. Air Force, Aerospace Medical Division, Brooks Air Force Base, Texas (1978).

16. S. Chandrasekhar, "Radiative Transfer," Oxford University Press, London and New York (1960).

17. V. V. Sobolev, "A Treatise on Radiative Transfer," Van Nostrand-Reinhold, Princeton, New Jersey, (1963).

18. D. H. Menzel, "Selected Papers on Transfer of Radiation," Dover, New York (1966).

19. L. O. Reynolds, "Optical Diffuse Reflectance and Transmittance from an Anisotropically Scattering Finite Blood Medium," Doctoral Dissertation, Electrical Engineering, University of Washington, Seattle.

20. L. O. Svaasand, D. R. Doizon, and A. E. Profio, "Light Distribution in Tissue During Photoradiation Therapy," Medical Imaging Science Group (MISG), University of Southern California (1981).

21. C. C. Johnson, IEEE Trans, BME 17(2):129-133 (1970).

22. R. J. Zdrojkowski and N. R. Pisharoty, IEEE Trans. BME 17(2): 122-128 (1970).

23. L. O. Reynold, "Three Dimensional Reflection and Transmission Equations for Optical Diffusion in Blood," M. S. Thesis, Electrical Engineering, University of Washington, Seattle (1970).

24. D. R. Doiron, L. O. Svaasand, and A. E. Profio, Lasers in medicine and surgery, SPIE 357:48-52 (1982).

25. P. Kubelka, "New Contributions to the optics of intensely Light Scattering Materials," Part I, J.Opt.Sci.Amer., 38:448-457 (1948).

26. P. S. Mudget and L. W. Richards, Appl.Opt.10:1485 (1971).

27. B. J. Brinkworth, Appl.Opt. 11:1434 (1972).

28. S. Wan R. R. Anderson and J. A. Parrish, Photochemistry and Photobiology 34:493-499 (1981).

29. M. J. Van Gemert and J. P. Hulsberger Henning, Arch.Dermotol. Res., 270:429-439 (1981).

30. F. Kotter, J.Opt.Soc.Amer., 50:483–490 (1960).
31. R. R. Anderson and J. A. Parrish, J.Of Invest.Derm. 77(1):13–19 (1981).
32. G. H. M. Gijsbers, "Optics of Tissue and In Vivo Fluorescence of Hematophophyzin-Derivative (HD)," Masters Thesis, Eindhoven University of Technology, The Netherlands (1983).
33. V. Twersky, J.Math.Phys., 3(4):724–734 (1962).
34. E. Lowinger, A. Gordon, A. Weinzeb and J. Gross, J.Appl. Physiol., 19(6):1179–1184 (1962).
35. N. Anderson and P. Sekely, Phys.Med.Biol., 12:185–192 (1967).
36. R. P. Reed, "Thin-Film Sensors of Micron-Size and Application in Biothermology," Doctoral Dissertation, University of Texas, Austin, Texas (1966).
37. C. P. Cain and A. J. Welch, IEEE Transactions on Biomedical Engineering, BME 21(4):421 (1974).
38. H. S. Carslaw and J. C. Jaeger, "Conduction of Heat in Solids," Second edition, Oxford Press, London (1959).
39. R. V. Churchill, "Operational Mathematics," Third Edition, McGraw-Hill, New York (1972).

BIOLOGICAL EFFECTS AND APPLICATIONS

BIOLOGICAL ACTION OF LOW-INTENSITY MONOCHROMATIC

LIGHT IN THE VISIBLE RANGE

T. I. Karu* and V. S. Letokhov**

*Laser Technology Center, Acad.Sci.USSR
Moscow Region, Academgorodok, 142092
**Institute of Spectroscopy, Acad.Sci.USSR
Moscow Region, Academgorodok, 142092

INTRODUCTION

Recently the low intensity (noncoagulative) visible laser radiation has been successfully used in some areas of medicine (photodynamic therapy of tumors, therapy of infant hyperbilirubinemia, some dermatological diseases, etc.)[1].

In addition the therapy with red (632,8nm) laser light (stimulation of tissue regeneration) used for irradiation of patients with trophic and indolent wounds has gained some acceptance in the clinical practice[2-4]. But there are still many doubts among scientists about the existence of that kind of stimulation.

Indeed, the healing effect of red light is often attributed to a psychological effect. From the physical standpoint the effect of biostimulation seems improbable since it is difficult to explain particular sensitivity of biological objects to the irradiation with red light as compared to the systematic irradiation with white light, the spectrum of which contains a red component approximately of the same power.

Clinicians often discuss a particular role of laser light coherence to explain its biological effects. This argument does not seem reasonable because the rate of molecular excitation with low-intensity light is many orders (10^{10} to 10^{14} times) lower than the coherence loss rate of the excited molecular states in condensed phase at normal temperature.

We have made an attempt to carry out quantitative studies with organisms of different degree of organization to prove or disprove the stimulating effect of low-intensity red light. Two types of eucariotic cells - mammalian cells grown in the culture, and yeast organisms as well as one type of procariotic cells - E.coli, have been used in this investigation. The effect of irradiation was evaluated by measuring the rate of the DNA and proteins synthesis and the rate of full reproduction.

MONOCHROMATIC LIGHT IN VISIBLE RANGE STIMULATES THE DNA SYNTHESIS IN MAMMALIAN CELLS

Since nucleic acids themselves do not absorb visible light, the changes in the synthesis rate of DNA after its irradiation with red light indicate deep changes in the cell activity after absorption of light by any chromophore.

At first we investigated the dose-response relationship after the irradiation of the culture of HeLa cells with coherent and incoherent red light (Figure 1). We found that in our experimental conditions both types of red light stimulate DNA synthesis from the energy flux of 10 Jm^{-2} with maximum at the level of 100 Jm^{-2}. No major differences were observed between the results obtained by coherent and inchoherent light sources.

So, it is evident a stimulation of DNA synthesis in cellular culture after the irradiation with monochromatic red light at 633 nm. Properties of light such as coherence and polarization do not play a role for this effect.

If there is stimulation at 633 nm, is it possible to stimulate DNA synthesis with light of other wavelengths in the visible spectrum? To clarify this question we measured the action spectrum of monochromatic light on DNA synthesis over the entire visible region using incandescent filament lamp or mercury lamp and monochromator as a source of tunable light.

DNA synthesis was stimulated in three spectral regions with maxima at about 400 nm, 620 nm and 760 nm (Figure 2a). It means that "red light syndrome" i.e. stimulation with red light, can be accomplished with relatively monochromatic irradiation at other wavelengths too. On the other hand, such a character of action spectrum points to the fact that cell has one or, most likely, several acceptors of light in the visible range. One of them may be any porphyrin-containing enzyme, like catalase, having absorption bands in these regions[5].

Another possibility is that there exist a special light absorbing chromophore (chromophores) in eucariotic cell, like phytochrome

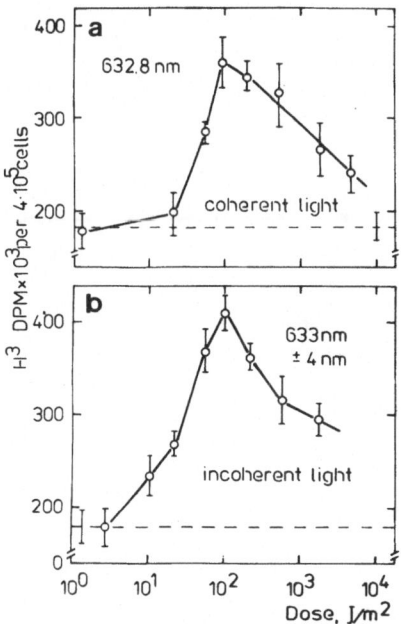

Fig. 1.   The stimulation of DNA synthesis in HeLa cells after
irradiation with coherent (He-Ne Laser, λ = 632,8 nm, Figure
a) and incoherent light (incandescent filament lamp with
monochromator, λ = 633 ± 8 nm, Figure b). The dashed lines
denote the control level (nonirradiated cells). HeLa cells
at logarithmic stage of growth were cultivated in scintill-
ation vials in 2 ml of medium 199, supplemented with 15%
bovine serum and antibiotic (100 units/ml canamycine). The
light beam was expanded by a set of lenses to the diameter
of the flask bottom (2.4 cm) covered with a monolayer of
cells (4.10^5 cells 72 h after transplantation). The irradiation was performed in darkness and the temperature was
kept all time at 37°C. The output power of He-Ne laser was
40 mW, the light intensity was reduced from 0,6 to 90 W/m².
The dose was reached by irradiating the cells during 2 to
60 s. The power density of lamp was 1.5 and 25 W/m², the
dose was reached by irradiating the cells during 2 s to
60 s. The incorporation of H^3-thymidine (concentration of
5 µCi/ml, Sp.act. 26 Ci/mM) was investigated with
radiometric technique by pulse labeling for 20 min at 37°C.
The experimental technique is described in[11].

for red light in plants[6] or so called "kryptochrome" for blue light
in microorganisms[9]. So far the existence of such specialized
chromophore in the mammalian cells is not confirmed, although such an
assumption has been made[7]. In any case, the primary photochemistry
and photophysics including the question about the chromophore, as

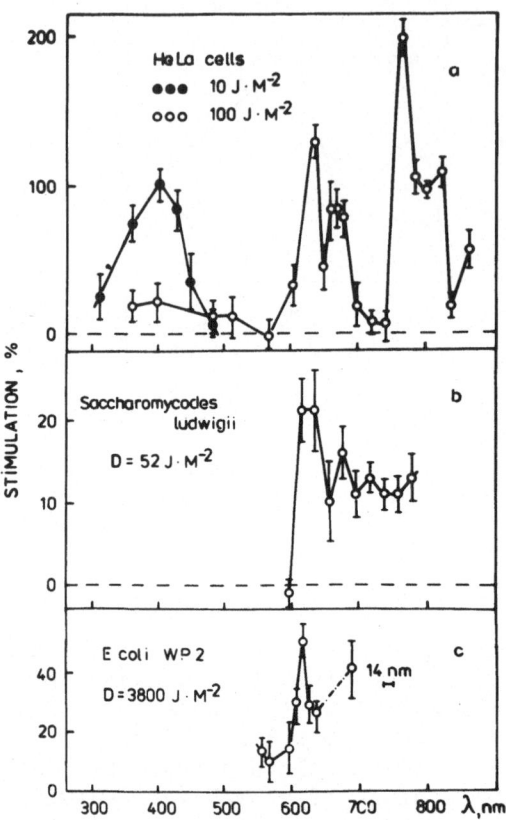

Fig. 2.   The action spectrum of visible light on DNA synthesis in
          HeLa cells (a), on yeast growth (b) and E.coli growth (c).
          The dashed lines denote the control level (without ir-
          radiation).   The radiation wavelengths were obtained by
          using monochromator with incandenscent (400-860 nm) lamp.
          Yeast cells were grown for 5 hours in wort (7 Bal) in
          rockermounted flasks, then separated from the nutrient
          medium by centrifugation and washed twice with sterile tap
          water.   A number of 5.10^7 cells were placed in Petri dishes
          with agar-wort and 2 hours later cells were irradiated
          during 10 s at room temperature.   Then yeast were grown
          during 14-16 hours at 28°C.   The suspension of three times
          washed cells was hydrolised with NaOH (0.1 M) in a boiled
          water bath for 10 min.   The content of protein in hydro-
          lasate thus obtained was determined by the Lowry method[10].

          Logarithmic phase cultures of E.coli WP2 were irradiated in
          0.05 M phosphate buffer, then the rich medium was added and
          cells were incubated at 37°C during 1.5 hours as described
          in /12/.   After the incubation the samples were taken,
          diluted and spread on plates of rich medium solidified with
          Bacto agar.   The visual count of the colonies was made after
          24 hrs of incubation.

well as the mechanisms of realization (metabolic, conformational etc. changes in the cell after the primary act of absorption) are not clear.

The finding, that the effect of DNA synthesis stimulation depends on the intensity of light (Figure 3) shows the nostationary or transient nature of this effect.

The question about the dynamics of the DNA synthesis stimulation effect is not fully clear. We have found, that 15 minutes after the irradiation the effect is almost maximal, remaining on that level for some hours, and then decreases slowly. 24 hours after irradiation with dose 100 $J \cdot m^{-2}$ it is approximately 40% up the control level. So the dose-response experiment (Figure 1), done 1,5 hours after the irradiation, reflects the maximal value of the stimulation effect.

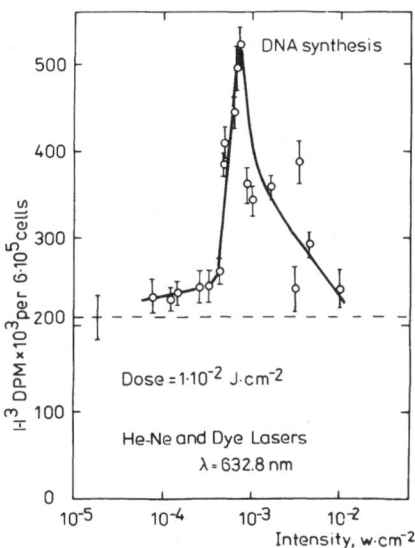

Fig. 3. The dependence of DNA synthesis' stimulation in HeLa cells on the intensity of light at $\lambda=633$ nm at constant dose 100 J/m^2. The dashed line denote the control level. The He-Ne laser with output of 4 mW and a dye laser pumped by Ar laser (output 24 mW) were used for the irradiation. The irradiation as well as the radiometric technique is the same as described in Figure 1.

Thus it can be concluded that the effect of DNA synthesis stimulation by low intensity visible light in the mammalian cellular culture exists and depends on wavelength, dose and intensity of monochromatic light, but not on the coherence of light.

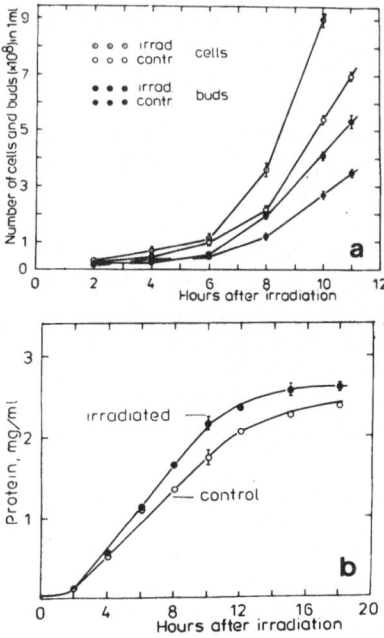

Fig. 4.   The effect of He–Ne laser irradiation on the reproduction
of Torulopsis sphaerica (Figure a) and the synthesis of
protein in Endomyces magnusii yeast (Figure b). The experimental technique is the same described in Figure 2.

RED LIGHT CAN STIMULATE THE REPRODUCTION RATE AND
SYNTEHSIS OF PROTEIN IN YEAST ORGANISMS

    The next natural step of our research was to understand the
ability of the red light to stimulate the vital activity of primitive
eucariotic cells when there is no regulation at organism or population level, i.e. each cell is independent. In our experiments we
investigated two types of yeast organisms with a He–Ne radiation
(λ=632,8 nm). They were the funguslike Endomyces magnusii reproducing by division and the Torulopsis sphaerica yeast reproducing by
budding.

    The growth curves show that the number of cells and buds in the
exposed culture is larger than in the control one in the exponential
phase of growth. The duration of lag-period remains almost constant
(Figure 4a). In the exponential phase the rate of protein synthesis
is increased which results in an active cell accumulation (Figure 4b)
and in a reduction of generation time by 1,5 to 1,8 times. The cell
accumulation in the case of irradiated yeast was maximal in the
second half of exponential growth phase as compared to the control
culture (Figure 5). The accelerated accumulation of cell by the
irradiated yeast is followed by a proportional increase in the number

of cells and buds at the same time in the exponential growth phase. It means that the size of the cells and the quantity of the protein in the exposed and control cultures are practically the same.

When the culture is in the stationary phase of growth the difference in the quantity of protein between the irradiated and control cultures gradually decreases. This is probably due to the decrease of the growth in a limited volume. The results of our experiments show that the response of various yeasts to irradiation, being similar qualitatively, may have great quantitative differences. The amount of protein in the irradiated culture of T.sphaerica at the end of the exponential phase, for example, is almost twice as much than in the unirradiated culture, and in E.magnussii the excess over the control is just 30% (Figure 5).

The stimulation of protein synthesis and reproduction rate begins from 2.10^3 J/m^2 and is maximal at $6.3.10^3$ J/m^2 for E.magnussii and at $4.2.10^3$ J/m^2 to T.sphaerica. An increase in dose causes the suppression of vital activity of the yeast. In T.sphaerica, for example, with a dose of $2.5.10^4$ J/m^2 only 27% protein, as compared to the control culture (100%), can be synthesized.

Using the same experimental conditions as earlier, we have measured the action spectrum of the stimulation of Saccharomycodes ludwigii yeast growth. The energy flux was optimal for this yeast on the dose curve. In this action spectrum there are two bands where the stimulation of growth was observed – near 640 and 680 nm.

RED LIGHT STIMULATES BACTERIA GROWTH

Since it has been found that the vital activity of eucariotic cells is stimulated under some wavelengths of visible light ir-

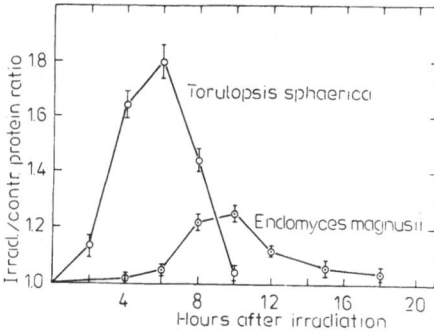

Fig. 5. The ratio between the rates of reproduction of exposed and control yeasts measured from the quantity of synthesised protein. The experimental technique is the same described in Figure 2.

radiation, it is of great interest to know how procariotic cells respond to this type of irradiation. It should be noted that there is almost no data in literature on the light sensitivity of micro-organisms in the visible spectrum. Most of the data are concerned with the suppression of vital activity under UV light[8]. The action of blue light is described in[9] but the green and red regions of visible light are still not investigated.

The growth curves of the exposed and unexposed cultures of wild type (WP) E.coli (Figure 6) show that the growth's stimulation is maximal in the first hours after irradiation. With further incubation the difference decreases and almost disappears when the culture passes to the stationary growth phase. As compared to yeast, the basic difference here is that the lag-period is reduced (from 1.6 hrs in the control culture to 0.2 hrs in the exposed one). This means that irradiated cells after being placed into a nutrient medium begin to reproduce with shortened delay. This effect is independent of light coherence, as shown on Figure 6.

The irradiation with red light at 633 nm stimulates the growth of E.coli in a definite dose range, from 5.10^2 J/m^2 to 5 or 6.10^4 J/m^2, with its maximum at 4 or 5.10^3 J/m^2[12].

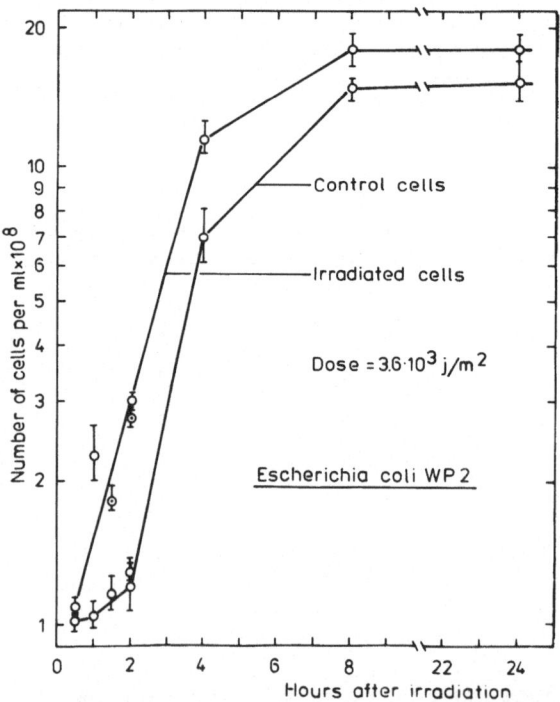

Fig. 6. Growth curves of control and irradiated by He-Ne laser (λ = 632,8 nm-(-o-o-) or incoherent red light λ = 633±4 nm (-⊙-⊙-)) cultures of E.coli.

An increase of the number of E.coli cells is followed by a stimulation of DNA synthesis. The measurement of the quantity of DNA synthesised within the first 10 minutes of cell's incubation shows that for irradiated cells the DNA synthesis level is 1.5 to 1.8 times higher than for unexposed cells. However 30 minutes after irradiation the DNA radioactivity does not differ anyhow from the control one.

We have also studied the action spectrum of light on the growth of E.coli in the red region of visible light (550 to 650 nm). In this region the action spectrum measured (Figure 2c) resembles very much the action spectrum of light on DNA synthesis for mammalian cells (Figure 2a, b). Growth stimulation can be observed in a range of 610 to 640 nm with a maximum at 620 ± 7 nm.

CONCLUSION - APPLICATIONS

In our experiments we have found that low-intensity red light, both coherent and incoherent, under certain conditions (dose, intensity, wavelength) can stimulate the vital activity of both procariotic and eucariotic cells. For bacteria and yeast the observed effect is maximal at a dose $\sim$ 50 times higher (5.10^3 J/m^2) than for human cells (100 J/m^2). We should note that the sensitivity differs even in one class of organisms as it has been observed in case of E.magnusii and T.sphaerica yeast.

All the observed effects are final macroeffects. The processes of primary photochemistry (light quantum absorption and signal transfer) and the mechanisms of realization (modification of metabolic pathways in cells) are still not clear. The presence of a characteristic spectrum of action qualitatively similar for procariotic and eucariotic cells in a region of 550 to 650 nm could indicate that there should exist a common way of low-intensity red light action.

In our experiments we have estimated that the biostimulating action of visible light is possible only in rather narrow interval of parameters (wavelength, dose, intensity). It seems to be one of the reasons why some times the clinical experiments have not been successful.

Proceeding from results of our experiments with mammalian cells we gave the recommendations for light-therapy into two clinics.

The clinical experiments started in 1981 in Gorki Medical Institute (Dr. V.F. Novikov, L.W. Paramonov) and in 1982 in Moscow District Central Clinical Institute (MONIKI) (Dr. G.A. Romanov) for treatment the stomach ulcers. By now in Gorki there have undergone a course of treatment approximately 100 patients and in Moscow almost 200, most of them having the ulcer with the diameter more than 2 cm and suffering from it (and treated with classical methods) for years. He-Ne laser (633 nm) or filament lamp with filters (narrow-band red light 640 ± 10 nm or wide band red light 600-800 nm) have been used as the light sources, introduced endoscopically. There are two new points comparing with earlier clinical experiments—noncoherent light was used and an attempt was made to treat the inner ulcers with low-power red light. The patients have been checked by double blind tasting. In most cases the light-therapy was used with combination of standard chemotherapeutical treatment, because the ulcer is only one symptom of the disease of the whole organism. In experimental groups more than 90% of patients have been found wholly recovered. Best results have been got using the wideband red light (600-800 nm).

The stimulation on wound healing with low-intensity red light is rather close to laser-mystery, as told Dr. R. Kaufmann some years ago[13], and as it was mentioned in our introduction. We do hope that with our experiments with organisms of different degree of organization, as well as with recommendations for clinic, we have, if not cleared away the mystery, at least explained it a little bit.

REFERENCES

1. R. Pratesi, C. A. Sacchi, (eds.) Laser in Photomedicine and Photobiology," in: "Springer Series in Optical Sciences," vol.22, Springer-Verlag, Berlin-Heidelberg, New York (1980).
2. N. F. Gamaleya, in: "Laser Applications in Medicine and Biology," vol. 3, M. Wohlbarsht, (ed.), Plenum Press, N. Y., London (1977), p. 54.
3. E. Mester, E. J. Nagy, E. Bascy, et al., in: "Medizinishe Physik in Forschung und Praxis," A. Kaul, Walter de Gruyter, (eds.) Berlin (1976), p. 117.
4. J. S. Kana, G. Hutschenreiter, D. Haina, and W. Waidelich, Arch-Surg. 116:293 (1981).
5. "Atlas of Protein Spectra in the Ultraviolet and Visible regions," D. M. Kirschenbaum, (ed.) IFI Plenum, New York, Washington, London (1972).
6. H. Smith, "Phytochrome and Photomorphogenesis," McGraw-Hill Book Company, London (1975).
7. S. A. Gordon and K. Surrey, Rad. Research, 12:325 (1960).
8. "Photobiology of Microorganisms," P. Halldal, (ed.) Wiley Interscience, N. Y. London, Sydney, Toronto (1970).
9. "The Blue Light Syndrome," H. Senger, (ed.) Springer-Verlag, Berlin-Heidelberg, New York, (1980).
10. O. H. Lowry, N. J. Rosebrough, A. L. Farr and R. J. Randall, J.Biol.Chem., 193:265 (1951).
11. T. I. Karu, G. S. Kalendo, V. S. Letokhov and V. V. Lobko, Ii Nuovo Ciments 1D:828 (1982).
12. T. I. Karu, O. A. Tiphlova, V. S. Letokhov and V. V. Lobko, Ii Nuovo Cimento 2D: 1138 (1983).
13. R. Kaufmann, in: "Lasers in Medicine and Biology," F. Hillenkamp, R. Pratesi, C. A. Sacchi, (eds.), Plenum Press, New York, London (1980), p.69.

CERTAIN ASPECTS OF HELIUM-NEON LASER IRRADIATION

ON BIOLOGICAL SYSTEMS IN VITRO

S. Passarella*, E. Casamassima*, E. Quagliariello*,
I.M. Catalano** and A. Cingolani**

*Istituto di Chimica Biologica and Centro di Studio
C.N.R. sui Mitocondri e Matabolismo Energetico
Università di Bari, Italy
**Dipartimento di Fisica, Università di Bari and Unità
GNEQP C.N.R., Italy

Living organisms create and maintain their essential orderliness
at the expense of their environment which they cause to become more
disordered and random; moreover the specificity of molecular interactions in cells results from the structural complementary of the
interacting molecules, so that linked sequences of enzyme-catalyzed
reactions provide the means for transferring chemical energy from
energy-yielding to energy-requiring processes.

The molecular basis of life is founded on the activity of
enzymes, proteins specialized to catalyze biological reactions, which
allow metabolism to be carried out in the different cellular compartments. Most enzymes require one or more non-protein components, such
as a metal ion or an organic molecule called a coenzyme for their
activity.

Thus the energy supply to a living system given by irradiation
of laser light can influence cellular metabolism of biomolecules and
consequently more complex systems in which they are contained absorb
laser light and therefore change their structural and biochemical
characteristics.

The use of laser in medicine and biology has become increasingly
widespread in the past few years. However, even if a great deal of
interest has been devoted to the interaction of laser irradiation
with tissue and with biological molecules, the effect of laser irradiation on biomolecules and the mechanism by which their behavior
is changed as well as the effect of laser irradiation on the differ-

67

ent components of the living cell is to date rather obscure on the molecular level.

This paper deals with experimental investigations made on the effect of laser irradiation on biological systems.

Initial work was carried out using high-peak power Q-switched Ruby laser and the reduced form of nicotinamide adenine dinucleotide (NADH). NAD and NADH act as electron acceptors and donors respectively, during the enzymatic transfer of hydrogen atoms from one specific substrate molecule to another and play a basic role in energetic metabolism. Two-photon absorption process by NADH has been demonstrated for Ruby laser irradiation[1] and the cross-section measured[2]. Low power c.w. Helium-Neon laser was subsequently used by us, owing to its application in phototherapy and biostimulation. Helium-Neon laser irradiation (4-5 J/cm^2) was found to cause changes in both absorbance and fluorescence spectra of NADH[3].

When irradiated NADH was tested in a typical reaction catalyzed by pig heart lactate dehydrogenase with pyruvate as substrate, irradiation of the coenzyme was found to cause uncompetitive inhibition of the reaction[3].

It should be noted that the effect of NADH irradiation on enzyme catalyzed reactions is dependent on the nature of the enzyme and on the energy dose supplied to NADH by Helium-Neon laser. Inhibition was found if lactate dehydrognases from pigs heart or rabbit muscle were used, while an increase in the NADH oxidation rate was found in the presence of lactate dehydrogenase from hog muscle. The effects found appear to be constant in time[4].

In order to assess the generality of these effects under different experimental conditions, investigation was made of the dependence of NADH oxidation rate on external pH. As reported in a typical experiment (Fig. 1), stimulation was found to occur in the oxidation rate of laser-irradiated NADH induced by hog muscle lactate dehydrogenase, in the pH range between 6.3 and 7.4. No distinction was found at higher pH values, suggesting that molecular interactions between irradiated NADH and lactate dehydrogenase may depend on the pH-dependent structure of the enzyme.

Investigation of the effect of laser irradiation on biomolecules was extended to ribonucleic acid (RNA) isolated from rat liver. RNA plays a basic role in the cell genetic system, since it is synthesized according to DNA structure (transcription) and directly dictates the amino acid sequence of proteins (translation). Thus the possible effect of laser irradiation in biostimulation and phototherapy might be related to change in the ability of cellular components to carry out protein synthesis.

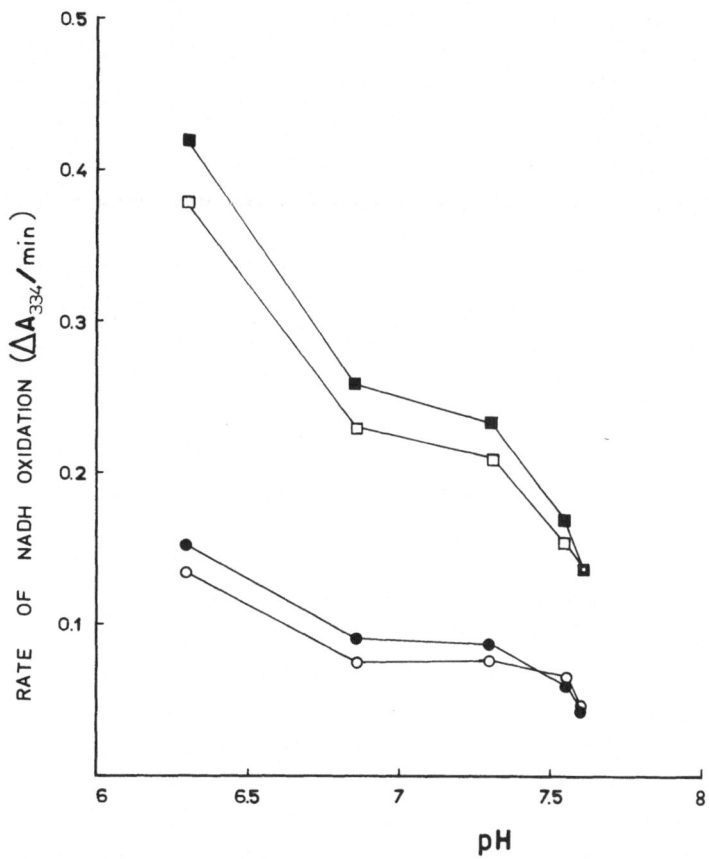

Fig. 1. pH dependence of the increase in the NADH oxidation rate
induced by Helium-Neon laser irradiation.
Two 1 ml aliquots were taken up from a freshly prepared
solution of NADH (10 mM), one of which was irradiated at 2°C
by a Helium-Neon continuous-wave single mode laser (13.3 mW,
4.87 J/cm^2) as previously described[4], while the other was
kept under the same experimental conditions without irradiating.
Measurements were carried out after addition of either
irradiated (●,■) or non irradiated NADH (○,□) (0.2 mM each)
to 2 ml of a reaction mixture containing 20 mM Tris-HCl at
pH values indicated and 0.3 mM sodium-pyruvate. The rate
to NADH oxidation was measured at 23°C by the absorbance
decrease at 334 nm following the rapid addition of either
0.19 μg (○, ●) or 0.58 μg (□,■) hog muscle lactate dehydrogenase to the reaction samples, by means of a Beckman DU-8
Spectrophotometer equipped with the Sypper System Compuset.
Rate measurements are expressed as ΔA_{334}/min. For other
details see ref. 4.

Preliminary investigation of the effect of laser irradiation on
RNA properties suggests that Helium-Neon laser light (10 J/cm²)
changes the specific extinction coefficient of isolated rat liver
RNA, which is transparent to the primary frequency of Helium-Neon
laser (λ = 632.8 nm) (Fig. 2). This effect was constant in time up
to at least 24 hours. No effect was found if energy irradiation was
lower (1 or 5 J/cm²). The mechanism by which the change in optical
properties of RNA occurs is not understood at present.

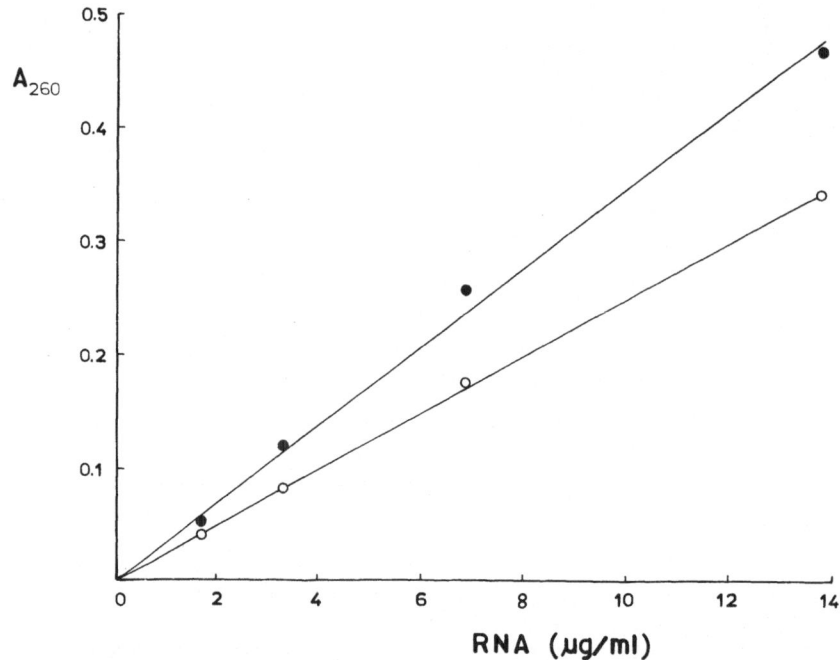

Fig. 2. Determination of specific extinction coefficient in L-RNA
and in N-RNA.
Polysomal RNA was isolated from rat liver essentially
according to Maccecchini et al.[7]. and then suspended in a
buffer containing 10 mM Tris-HCl pH 7.5, 1mM EDTA pH 7.6,
0.2% SDS. Two 1 ml aliquots were taken up from the stock
solution. One was irradiated at room temperature by Helium-
Neon laser (13.8 mW, 10 J/cm²), (L-RNA,●—●), while the
other was kept under the same experimental conditions without the irradiation (N-RNA,○—○). Different aliquots of
both samples were utilized to give the final concentrations
used in the measurements. Absorbance at 260 nm was measured
10 min after the irradiation time by means of a Beckman DU-8
Spectrophotometer. Values were respectively 23 mg/ml for
N-RNA and 32 mg/ml for L-RNA.

In order to gain further insight into the problem of laser-
biosystems interaction, mitochondria were used to investigate whether
laser induced effects could be maintained when biomolecules were
irradiated in a more physiological situation.

It should be noted that since mitochondria absorb light at
different wavelengths, unknown receptors other than NADH or RNA could
be responsible for the structural and functional changes in mitochondria induced by irradiation of Helium-Neon laser light.

Photoacoustic spectroscopy measurements reveal that Helium-Neon
low-power irradiation causes changes in mitochondrial optical properties. However, this process has not been found to damage either the
membrane of the mitochondria or any of their enzymatic activities.
The Helium-Neon laser irradiation effect has been investigated both
with regard to the redox state change of intramitochondrial pyridine
nucleotides, induced by addition to the mitochondria of oxaloacetate,
oxoglutarate plus ammonia and cis-aconitate, as well as to the permeability of rat liver mitochondria in relation to the same substrates. Laser irradiation was found to modify certain NADH-linked
dehydrogenase reactions of the mitochondria[5].

The chemiosmotic-coupling hypothesis of oxidative phosphorylation claims that electron carriers of respiratory chain act as an
active transport system or pump to transport H^+ ions from the mitochondrial matrix outside the inner membrane, thus generating an H^+
gradient across the inner membrane, which is impermeable to H^+ ions.
The electrochemical gradient or protonmotive force generated across
the membrane of respiration-energized mitochondria consists of a
gradient of H^+ (ΔpH) plus a transmembrane potential negative inside
mitochondria ($\Delta\Psi$).

Use was made of safranine, a photometric probe capable of
measuring mitochondria $\Delta\Psi$,[6] with the safranine response measured
due to fresh or irradiated mitochondria (Fig. 3). As shown by the
decrease of all safranine-mitochondrial suspension spectrum (3A),
mitochondria energization was found after irradiation was carried
out in the presence of the inhibitor of oxidative phosphorylation
olygomycin (3B), but completely blocked if irradiation was carried
out in the presence of electron-transfer chain inhibitors as rotenone
and antimycin (3C). It should be noted that the spectra of different
mitochondrial preparations added with safranine in the same experiment were perfectly overlapping, thereby confirming our results.
These findings clearly show that the irradiation of mitochondria
produce a remarkable change in mitochondrial energy metabolism, once
again in an incomprehensible way.

As a further investigation, the ability of mitochondria to
oxidize succinate was tested both in the absence and presence of
laser irradiation (Fig. 4). A small but significantly reproducible

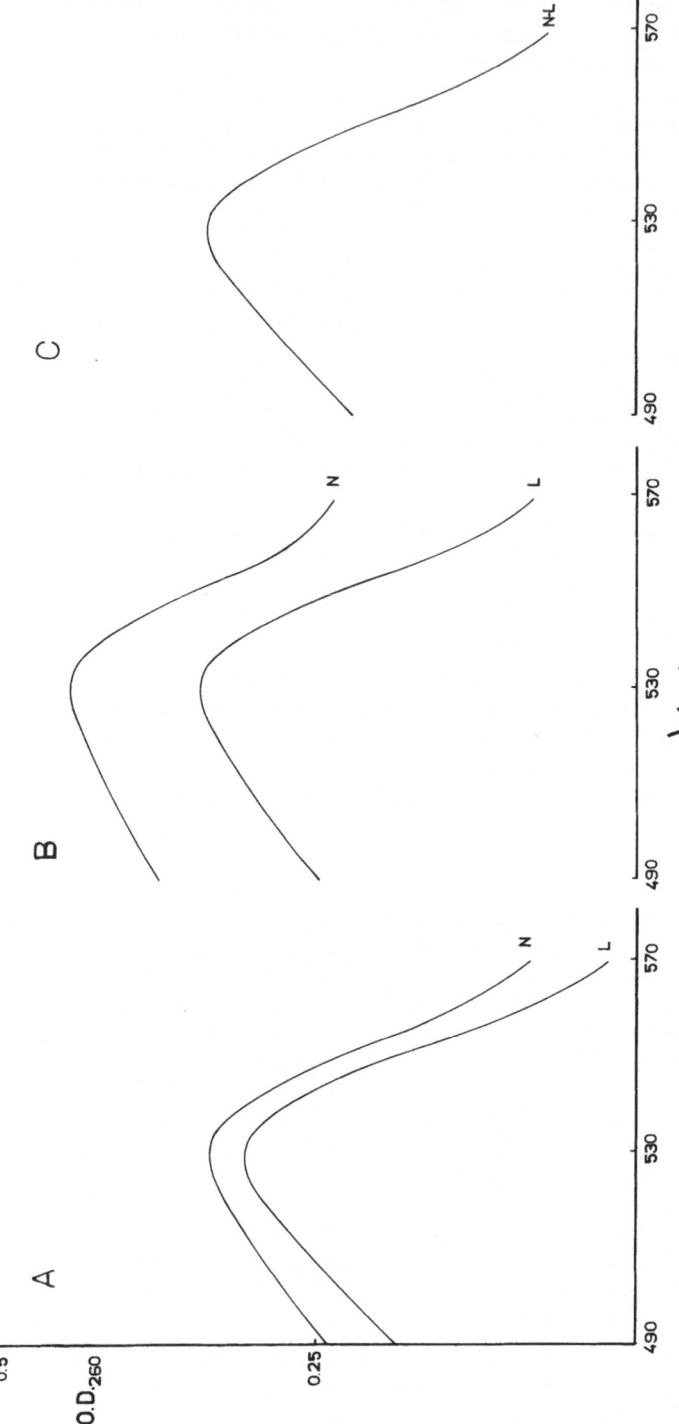

Fig. 3. Safranine response of either fresh mitochondria or mitochondria irradiated by Helium–Neon laser. Rat liver mitochondria isolated as previously described[5] were incubated at 2°C in a medium standard consisting of 250 mM sucrose, 20 mM Tris–HCl pH 7.25, 1 mM EGTA at final concentration of 1 mg/ml. In a typical experiment, two 1 ml aliquots were taken up from the suspension, one irradiated at 2°C by Helium–Neon laser (11 mW, 5J/cm^2), (L) and the other kept under the same experimental conditions with no irradiation (N). After irradiation, 1 ml of the standard medium was added to the samples to give a final concentration of 0.5 mg protein/ml. Safranine (final concentration of 12.5 μM) was added to both samples and absorbance spectra were measured by means of a Beckman DU-7 Spectrophotometer. Further additions prior to irradiation were as follows: A – 10 μl ethanol. B – 2 μg oligomycin dissolved in 10 μl ethanol. C – 2 μg rotenone and 2 μg antimycin A dissolved in 10 μl ethanol.

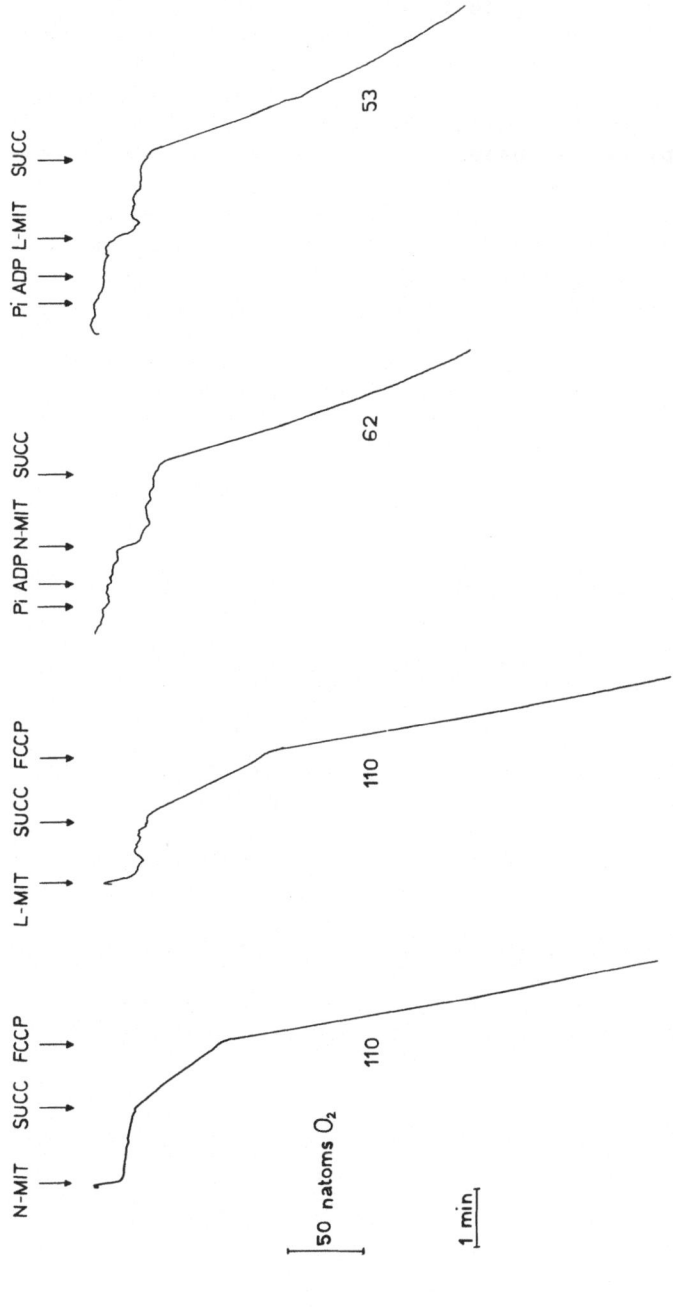

Fig. 4. Oxygen uptake induced by addition of succinate to either fresh or Helium-Neon laser irradiated rat liver mitochondria.
Mitochondria (2.8 mg of protein), isolated as previously reported[5], non-irradiated (N-MIT) or irradiated by Helium-Neon laser (13 mW, 5 J/cm²), (L-MIT), were incubated in 1.4 ml of a reaction mixture containing 250 mM sucrose, 20 mM Tris-HCl pH 7.25, 1 mM EDTA-Tris, 2 µg rotenone at 25°C in a thermostated cell of a GILSON Model 5/6H Oxygraph. At times indicated by the arrows, the following additions were made: succinate-Tris (SUCC, 1 mM); carbonyl cyanide p-trifluoromethoxy-phenyl-hydrazone (FCCP, 1 µM); adenosine 5'-diphosphate (ADP, 1mM); phosphate-Tris (Pi, 1mM), with the resulting oxygen consumption in the medium followed in the function of time. The rate of oxygen consumption was then recorded and calculated as tangent to the initial (linear) part of the experimental traces. It is expressed as nanoatoms O₂/min x mg protein taken up.

inhibition was found in the rate of oxygen uptake by the addition of 1 mM succinate (SUCC) in the presence of adenosine 5'-diphosphate (ADP) and inorganic phosphate (Pi), (both of 1 mM) while no effect was found following stimulation of oxygen uptake by addition of the uncoupler carbonyl cyanide p-trifluoromethoxy-phenylhydrazone (FCCP, 1μM). This experiment suggests that the effect of laser irradiation is in some way related to oxidative-phosphorylation rather than electron flow in the respiratory chain.

It appears obvious that a great deal of work is required to clarify the reported phenomena both with regard to the mechanism of the found effects and to the possibility of changing conditions of irradiation or the laser source.

Nevertheless the fact that clear evidence is now available to claim that laser light directly influences biological systems strongly suggests that investigation of this so far incomprehensible influence should be the aim of the scientific community.

REFERENCES

1. D. E. Round, R. S. Olson, and F. M. Johnson, NEREM REC., 158 (1966).
2. I. M. Catalano and A. Cingolani, Opt.Comm., 32:156 (1980).
3. S. Passarella, M. C. Dechecchi, E. Quagliariello, I. M. Catalano, and A. Cingolani, Bioelectrochem.Bioenerg., 8:315 (1981).
4. S. Passarella, E. Casamassima, M. Padolecchia, E. Quagliariello, I. M. Catalano, and A. Cingolani, Bull.Mol.Biol.Med., 7:25 (1982).
5. S. Passarella, E. Perlino, E. Quagliariello, L. Baldassarre, I. M. Catalano, and A. Cingolani, Bioelectrochem.Bioenerg., 10:185 (1983).
6. A. Zanotti and G. F. Azzone, Arch.Biochem.Biophys., 201:255 (1980).
7. M. L. Maccecchini, Y. Rudin, G. Blobel, and G. Schatz, Proc. Natl.Acad.Sci.USA, 76:343 (1979).

A BRIEF DISCUSSION ON PROBLEMS ARISING IN THE IRRADIATION

OF WHOLE BIOLOGICAL SPECIMENS BY LASER BEAMS

G. Delfino[1], S. Martellucci[2], J. Quartieri[3],
and E. Quarto[4]

[1]Istituto di Cibernetica del CNR, 80072 Arco Felice
Napoli, Italy
[2]Facoltà di Ingegneria della II Università Tor Vergata
00173 Roma, Italy
[3]Istituto di Fisica della Facoltà di Ingegneria
Università Piazzale Tecchio, 80125 Napoli, Italy
[4]Dipartimento di Biologia Cellulare – Universita della
Calabria 87030 Rende (Cosenza), Italy

INTRODUCTION

High power CW lasers such as CO_2, Nd:YAG and Argon are becoming
increasingly important tools in human surgery and operative endoscopy, finding applications as light scapels and coagulators to stop
bleeding or to coagulate tumor tissue. The major effects of interest
in medical applications are large penetration depth of laser radiation into biological tissue and the redistributive scattering inside
the tissue, which results in the heating of comparatively large
volumes through the thermal conversion of absorbed light energy.
This volume heating, coupled with the possibility of hot contacting
the tissue during an operation, and with the possibility of coupling
to the microscope, makes the laser an interesting alternative to
electrical cauterization or diathermal coagulation commonly used in
surgery and operative endoscopy.

At the impact area of a laser beam, the biological and physical
changes in tissue depend upon the temperature attained through its
heating effect; at temperatures above 58°C a denaturation of organic
matter (coagulation) occurs, and at 100°C, vaporization of cells
occurs. At still higher temperature there is further decomposition
of the organic molecules resulting in carbonization and burning of
the tissue. The speed of this process and the volume of tissue
affected depends, for a given power imput, on the parameters influ-

encing the penetration depth and internal redistribution of the light energy, namely, the absorption coefficient and the internal scattering coefficient. Because of absorptivity of tissue, the effect of scattered light is dependent on wavelength of radiation. In fact the absorptivity varies from a maximum value with CO_2 laser irradiation to a minimum value with Nd-Yag laser.

This effect explaining the anomaly between the absorption coefficient obtained from the literature and the temperature rises actually measured.

Therefore it is necessary to develop theoretical and experimental methods to clarify the discordance between expected and experimental results.

In this work we will show, as examples, experimental work and two theoretical models developed in order to study the thermal interaction of laser radiation with biological tissues. The accuracy of the model is tested by experimental measurements of temperature rises and extent of damage. Generally, the development of a mathematical model of temperature rise in laser irradiated tissue, is not difficult because heat conduction is a well studied phenomenon and various methods are available to solve conduction problems in steady state and transient. But the results depend upon hypothesis and boundary conditions, therefore it is essential that the models are built with great accuracy and in such a way important factors are not neglected, e.g. light scattering, phase change blood flow, constancy of parameters, etc. Various models are available using several analytic and finite difference solutions to the heat conduction equation to evaluate temperatures in laser irradiated tissue.

These models can be experimentally verified. We can measure the temperature of the irradiated tissue for single and multiple layer situations. The results of theoretical models and experiments are not competitive, but they are complementary. Results of a model can indicate important parameters and conditions for use in the design of experiments. Also analysis of experimental data provides an indication of modifications for improvement of the models. If a model is sufficiently accurate, it can be used to predict results of many experimental conditions. The comparison if calculated temperatures and measured temperatures can demonstrate of there are significant differences in presumed radiation conditions and resulting heat source term. Furthermore we can demonstrate systematic errors in measured temperatures which may due to factors such as (1) dimensions of sensor, (2) time response of sensor, (3) direct absorption of laser light by sensor and (4) uncertainty in position of sensor in the tissue.

In the last years, various approaches have been proposed to solve problems of heat flow in biological tissues[1,2,3,4,5].

It is known that when the laser beam strikes the tissue, a portion of it is reflected, a portion may be transmitted through the tissue and the remainder is absorbed by the tissue. The temperature field in the tissue is a function of

- the rate at which the beam energy is deposited;
- the time lag of irradiation;
- the interaction volume;
- the transport losses of the heat; and
- the heat capacity of the tissue.

In order to gain some insight in the interaction, both experimental and theoretical approaches related to whole laser-irradiated frogs, have been considered[6,7,8].

In the present lecture, we report comparatively some results of theoretical and experimental study of the irradiation of frog skin by the argon laser beams.

In particular, the mathematical model which assumes spherical scattering of light in tissue[4] and theoretical gaussian model[2], generally used in the heating of solids by laser beams, are considered. Also, some experimental results are compared to theoretical values in order to verify both theoretical hypothesis and possible alternative choice of the values of the involved physical parameters of biological specimens.

MATHEMATICAL MODELS

The first model is that reported in Ref. 4, were a cylindrical symmetry for the direct radiation is assumed. The Lambert-Beer law for absorption can be written as follows:

$$I^B(r,z) = I_0(r) \, e^{-\gamma \cdot z}$$

where $\gamma = \alpha + \beta$ is the absorption coefficient of radiation dependent on frequency, β the coefficient which takes into account scattering of radiation out of the beam, and I^B indicates the direct beam intensity. Omitting details, for the sake of simplicity and conciseness, we report just a few formulas in order to give an insight of the mathematical hypotheses of the model.

If we assume the scattered beam to be spherically distributed and unpolarized we have:

$$I^S(r,z) = \frac{\beta}{4\pi} \int\limits_0^\infty \int\limits_0^D \int\limits_0^{2\pi} \frac{r'}{\rho^2} I_0(r') \, e^{-[\gamma z' + \gamma \rho]} \, dr' \, d\theta' \, dz'$$

where I^S indicates the intensity of the radiation arriving at $P(r,\theta,z)$ having been scattered at $P'(r',\theta',z')$. So the heat produced at P is given by

$$S(r,z,t) = \alpha \{I^B + I^S\}$$

In order to have four-dimensional (r,θ,z,t) maps of temperature we solve the heat transport equation:

$$\rho c \frac{\partial}{\partial t} T - K \nabla^2 T = S \qquad (1)$$

The equation (1) is subjected to the boundary conditions:

a) $\frac{\partial}{\partial z} T(r, o, t) = \frac{\partial}{\partial z} T(r, D, t) = 0$

b) $T(r, z, t) = T_o$ \qquad\qquad $t < 0$

c) $\lim_{r \to \infty} T(r, z, t) = T_o$

Relation (a) is the condition for no radiation and as will be pointed out in discussing results seems to necessitate some modification if experimental results are to be recovered in all conditions. We implicitly took into account the third relation letting $S(r,z,t)$ to have a gaussian or rectangular profile on r, and the second relation writing:

$$S(r,z,t) = S (r,z) \cdot 1(t - \tau)$$

where $1(t-\tau)$ is the unitary Heaviside step function.

With the above mentioned limitation it is possible to exactly solve the heat transport equation by a double transformation (i.e., a Fourier transformation for z and a Hankel one for r), getting:

$$T(r,z,t) = 1(t-\tau) \mathop{S}_{t',\xi,m} \xi \, \tilde{S}_m(\xi) \, e^{-k(\xi^2 + \frac{m\pi^2}{D^2})t'} \times$$
$$\times J (\xi r) \cos (\frac{m\pi' z}{D}) \qquad (2)$$

where S indicates integration over continuous variables and sum over discrete ones, and $\tilde{S}$ is the source term doubly transformed. Solution of the right hand side of this equation have been obtained by an iterative, multi-dimensional Simpson scheme for numerical integration, paying particular attention to the optimization of computer time consumption. Our numerical program, written in Fortran language, have been independently implemented both in the University of Naples, Central Computer (Univac 1110/80) and in a professional computer, and required about 50 $\check{K}$ bytes of C.P.U. RAM without segmentation. We chose to use two computers of such different type and speed because in focalized and scattering conditions computer time consumption greatly increased.

The second mathematical model we want to discuss is that introduced in Ref. 2. We have considered this model in order to value its suitableness in the irradiation of thermocouple-biological tissue systems.

Now, when the intensity distribution in the focal area can be described by a Gaussian function, we can write the dimensionless temperature Φ, for a constant irradiation in 0-t, is[9]:

$$\Phi(\xi,\zeta,\theta) = \int_{0}^{\theta} \frac{\exp[-\xi^2/(\theta+1)]\cdot\exp(-\zeta^2/\theta)}{\theta^{\frac{1}{2}}(\theta+1)} d\theta \tag{3}$$

where $\theta = 4kt/d^2$; $\theta = r/d$; $\xi = z/d$; and $\Phi(\xi,\zeta,\theta) = 2hTd/\epsilon P_0\pi^{\frac{1}{2}}$, k being the thermal diffusivity; h the thermal conductivity; d the radius of the spot; z the distance along the propagation direct of the laser beam of power P_0; r the considered distance from the beam center; T the temperature produced in r and z; and ϵ the fraction of the incident intensity absorbed by the target. We recall that equation (1) assumes that the physical parameters involved are independent of temperature or it considers the relative average values in a given range ΔT. Equation (1) also introduces the thermal penetration depth z_0, the radial propagation length r_0 as $z_0 = (4kt)^{1/2}$, and

$$r_0 = (4kt + d^2)^{1/2} \tag{4}$$

This latter relation defines the irradiation time-lag necessary to produce temperature variations at   distances from the beam center. In addition, when $\xi = \zeta = 0$, equation (1) gives

$$T(0, 0, t) = \frac{\epsilon P_0}{\pi^{3/2} hd} \tan^{-1}(4kt/d^2)^{1/2} \tag{5}$$

in which the d parameter is particularly related to the generated Gaussian heat source.

We also note that all physical parameters involved in the previous relations are considered here as referring to a homogeneous target. For $t\rightarrow\infty$ and $\xi = \zeta = 0$, Eq. (1) becomes

$$T(0,0,\infty) = \epsilon P_0/2\pi^{\frac{1}{2}} hd \tag{6}$$

where h = K.ρ.c, being ρ and c density and specific heat, respectively.

EXPERIMENTAL CONDITIONS

The experimental set-up to determine temperature profiles rising
in the irradiation of frog skin by laser beam is reported in Fig. 1.

A Pockels cell was used to get an optical rectangular signal of
about 4 seconds long. The cell was gated using a square voltage
signal supplied by a Grass-48 generator. TC is a stainless steel
encapsulated thermocouple having a 250 μm external diameter. The
thermocouple is used to measure the temperature on the skin and under
it (0.2 mm depth) in various regions. The stereo-microscope aids the
positioning of the spot with respect to the thermocouple. The positioning of the TC junction with respect to the laser spot is
obtained by performing micrometric displacements of the frog and
thermocouple with respect to the laser beam. To make under-skin
measurements we cut the skin 2 cm far from the position of the spot
and let the thermocouple run through.

The selected area to be irradiated is the epiphysis area of <u>rana
esculenta</u>. The Fig. 2 shows the 120 μm diameter hole obtained by a
300 sec irradiation and a 100 mW power beam and the Fig. 3 shows
characteristic thermal responses for unfocalized beam of a dye laser
at 5900 Å and 80 mW in power.

In Fig. 4 are reported the temperature rise of the thermocouple-
frog system when repetitive laser pulses of 200 msec at 5145 Å
impinge on the skin.

Finally, Fig 5 shows temperature rises of a focused (f ≈ 8.33cm)
beam of 5145 Å at 0.5 mm from the beam center and in the skin for
several power levels of the laser beam.

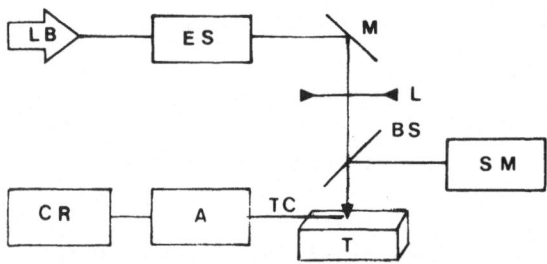

Fig. 1.   Experimental set up: LB, laser beam; ES electrooptic
          shutter; M, mirror; L, lens; SM Stereomicroscope; T
          biological target; TC, thermocouple; A amplifier and CR,
          chart recorder.

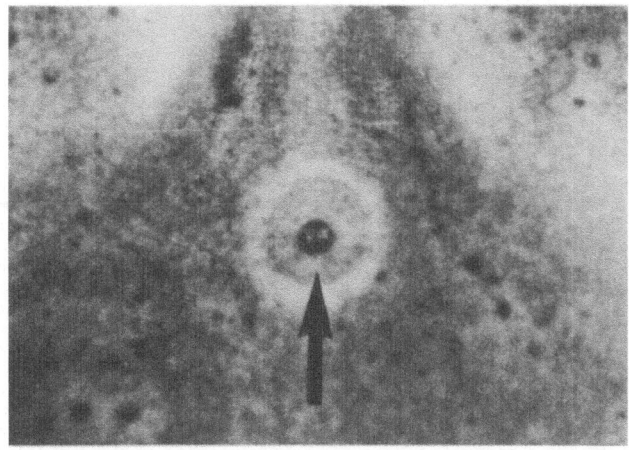

Fig. 2. Picture of the hole produced in 300 sec by 5145 Å, 100 mW,
multimode, multifrequency operating conditions (10x).

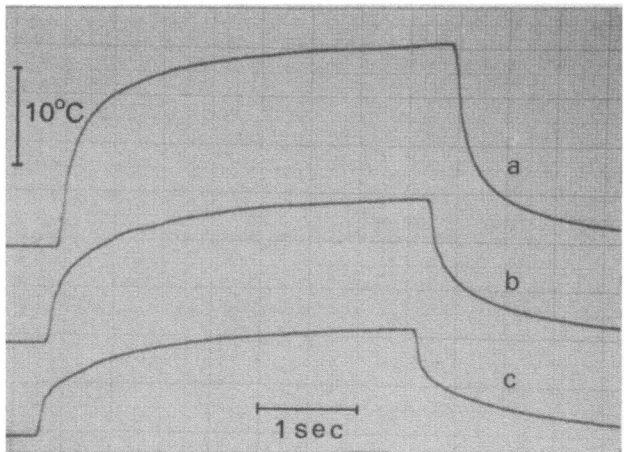

Fig. 3. Characteristic temperature rises for unfocalized dye laser
beam at 5900 A and 80 mW in power for some distances from
the beam center: a) 0 mm; b) 0.2mm; c) 0.4mm.

CONCLUSIVE REMARKS

    In our comparisons between numerical and experimental data,
temperatures of Eq. 2 fits sufficiently experimental results, particularly for beam powers lower than 40 mW.

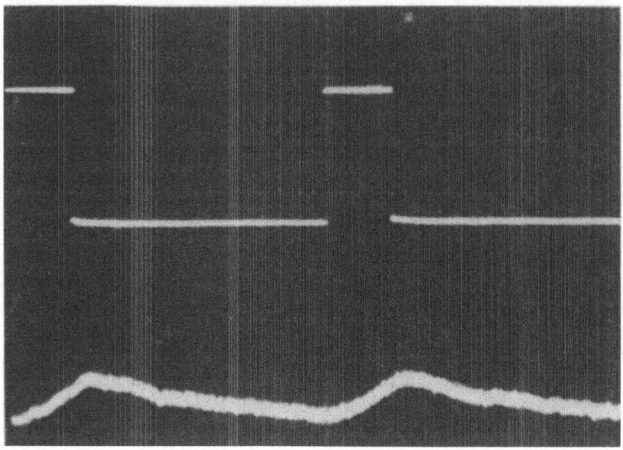

Fig. 4. Response of the thermocouple - frog system for repetitive
         laser pulses (upper) at 5145 A. Horizontal: 200 msec/div.
         Vertical: upper, 5V/Div; lower, 0.1 mV/Div.

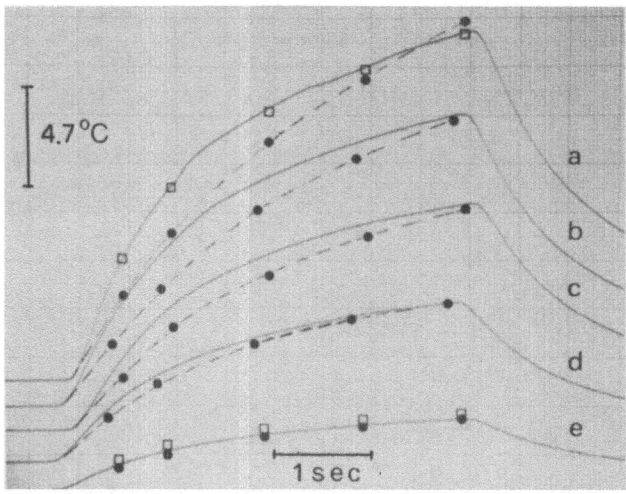

Fig. 5. Experimental temperature rises on the skin of a focused beam
         center and theoretical values of Eq. 2: □, $k/\rho c \simeq 2.57 \cdot$
         $10^{-3} cm^2/sec$; ●, $k/\rho c \simeq 1.55 \cdot 10^{-3} cm^2/sec$. a) 102,5 mW;
         b) 82.0 mW; c) 61.5 mW; d) 41.0 mW; e) 20.5 mW.

    For power values higher of about 40 mW, some differences exist
between them. These differences can be eliminated when in Eq.2 are
used diffusivity values different from the water one. In fact, in
Figure 5 are reported theoretical temperatures obtained from Eq.2

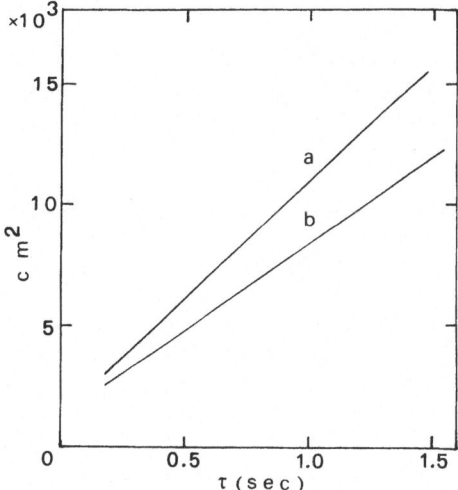

Fig. 6. Square distances from the beam center under the skin vs. τ
for a) 0.3 J and b) 0.065 J.

both for the water diffusivity value ($1.55 \cdot 10^{-3} cm^2/sec$) and for
$2.57 \cdot 10^{-3} cm^2/sec$ value.

This picture allows to conceive that theoretical and experimental temperatures are similar for power levels higher than 40 mW,
only when the diffusivity of the system is of about $2.57 \cdot 10^{-3} cm^2/$
sec.

This fact could be due to an effect of relatively high power
levels on the thermodynamic properties of the biological specimens.

In our case, if we measure the delay times τ related to the
maximum rise temperatures at r-distances, we can obtain, by means of
Eq. 4, the diffusivity values of the considered specimen. This
situation is reported in Figure 6 for two different levels of pulse
energies.

As we can see, these results give indications that the diffusivity coefficient is not constant for very different energies of
optical pulse and it is in particular higher than the water one.

We also remark that Eq.6 permits to obtain thermal conductivity
when the system has established the equilibrium for continuous radiation, once the ε value is known.

Acknowledgements

The authors are indebted with Prof. A. J. Welch, Prof. G.
Salvatore and Prof. M. Kemali for aids and discussions and with Mr.
E. Casale for his valid assistance in the experimental approaches.

REFERENCES

1. A. M. Clarke, W. J. Geeraets, and W. T. Ham, Jr. Appl.Opt.,
    8:1051 (1968).
2. J. Eichler, J. Knof, H. Lanz, J. Salk, and G. Schäfer
    Radiat.Environ.Biophys., 15:279 (1978).
3. L. A. Priebe and A. J. Welch, IEEE Trans., Biomed., BME-26, 244
    (1979).
4. T. Halldorsson and J. Langerholc, Appl.Opt., 17:3948 (1978).
5. G. Delfino, S. Martellucci, J. Quartieri, and E. Quarto, Lett.
    Nuovo Cimento, 38:353 (1983).
6. G. Delfino and E. Casale, Appl.Opt., 20:989 (1981).
7. M. Kemali, G. Delfino, and E. Casale, Z. Mikrosk.anat.Forsch.,
    95:321 (1981).
8. G. Delfino and E. Casale, Appl.Opt., 22:2712 (1983).
9. W. W. Duley in: "CO_2 Lasers: Effects and Applications," Academic
    Press, New York, p. 150 (1976).

TIME-RESOLVED FLUORESCENCE MICROSCOPY:

EXAMPLES OF APPLICATIONS TO BIOLOGY

F. Docchio*, R. Ramponi*, C.A. Sacchi*,
G. Bottiroli** and I. Freitas**

*Centro di Elettronica Quantistica e Strumentazione
Elettronica del CNR, Istituto di Fisica del
Politecnico, Milano, Italy
**Centro di Istochimica del CNR, Istituto di Anatomia
Comparata dell 'Università, Pavia, Italy

INTRODUCTION

Fluorescence microscopy (microfluorimetry) is an established technique that provides useful information about the biomolecules in a cellular environment[1,2]. The technique consists in exciting the samples (cells or part of them) with a light source of suitable wavelength, and in detecting and processing the fluorescence emission.

This is given, in some cases, directly from the biomolecules ("primary fluorescence"), usually through UV excitation. However, in most cases, information about the cells is obtained by labelling them with suitable fluorescent molecules ("secondary fluorescence").

Among the applications of this technique we mention: the evaluation of the cellular content of specific constituents (DNA, RNA, proteins)[3] and of the activity of several enzymes[4]; and the studies done to distinguish between differential structural or functional situations of biomolecules in a cellular environment[5,6]. In the latter case, the fluorescence molecule acts as a probe since its emission is modified as a consequence of the interaction of the molecule with the cellular structures[7].

Several accurate methods for the measurement of the fluorescence parameters (intensity, excitation and emission spectra, lifetimes, and polarization) at microscopic level have been described[8,9,10].

With continuous wave (cw) excitation, fluorescence intensity is commonly measured. This parameter is proportional to the product $N_0 \cdot q_0$, where N_0 is the number of fluorescing molecules, $q_0 = \tau/\tau_{sp}$ is the quantum yield, τ is the fluorescence decay time, and τ_{sp} is the spontaneous decay time of the fluorescent substance. Therefore a separate measurement of one of these two quantities (N_0, q_0), in addition to the intensity measurement, is required to evaluate both of them. Furthermore the quantum yield cannot be measured at microscopic level. On the other hand, if the fluorescence is excited by a pulse much shorter than the decay time, the fluorescent emission has a time-dependence expressed by

$$S(t) = S_o \exp(- t/\tau)$$

where $S_o \propto N_0/\tau_{sp}$. In this case the analysis of the waveform allows to evaluate both the decay time τ, and the peak amplitude S_o, and therefore, assuming that τ_{sp} is known, or that only a relative comparison is made between samples with the same τ_{sp}, to evaluate q_0 and N_o. It should be stressed that τ_{sp} is not a critical parameter, since it is a characteristic of each substance and is not affected by the environment.

The measurement of the decay time is usually considered a very promising approach since the lifetime of the excited state, which normally depends on the immediate surrounding of the fluorescent probe, is determined by the characteristics of the substrate. It should, therefore, be a valuable tool for the study of a large variety of biological processes[7]. The use of pulsed lasers in connection with the techniques of fluorescence microscopy has also made possible, in recent years, to measure this parameter at the cellular level with a high temporal and spatial resolution[11].

In general, the use of lasers, instead of conventional (Mercury or Xenon) flashlamps, to excite the fluorescence, offers several advantages:

i)   The possibility if obtaining pulses with a duration (10^{-9}–10^{-13}s) much shorter than those obtained with flashlamps (10^{-6}–10^{-9}s).

ii)   The possibility of focusing the excitation beam down to dimensions of the order of the wavelength, thus making possible "in situ" excitation of small portions of single cells.

iii) The possibility of making selective spectral excitation due to laser monochromaticity, and of scanning the excitation spectrum due to laser tunability.

iv)   The possibility of detecting fluorescing substances, at very low concentrations due to the high peak-intensity of the excitation beam.

A typical apparatus for microfluorimetry consists of (i) a light source for the excitation of fluorescence, (ii) a microscope for

fluorescence microscopy, (iii) filters or monochromators to select
input and output wavelengths, (iv) a light-detector, and (v) a signal
processor.

    In our laboratories, in connection with time-resolved fluor-
escence microscopy studies[12,13], a pulsed-laser microfluorometer
has been developed, which is briefly described in the next section.

THE LASER MICROFLUOROMETER

    The block diagram of the last version of the laser microfluoro-
meter is shown in Fig. 1. While a complete description of this
instrument is given in[14], we recall here its main properties.

    A pulsed nitrogen laser-pumped dye-laser is used as the exci-
tation source. It produces pulses with a duration of about 100 ps.
This rather inexpensive source[15], broadly tunable over the whole
spectrum from near UV to near IR, is well-suited for fluorescence
studies of molecules of biological interest.

    After collimation, the output beam is sent, via a beam-splitter,
to the microscope and to a reference photodiode, which provides a
signal proportional to the energy of the excitation pulse, suitable
for normalization purposes. A light chopper in the beam path has the

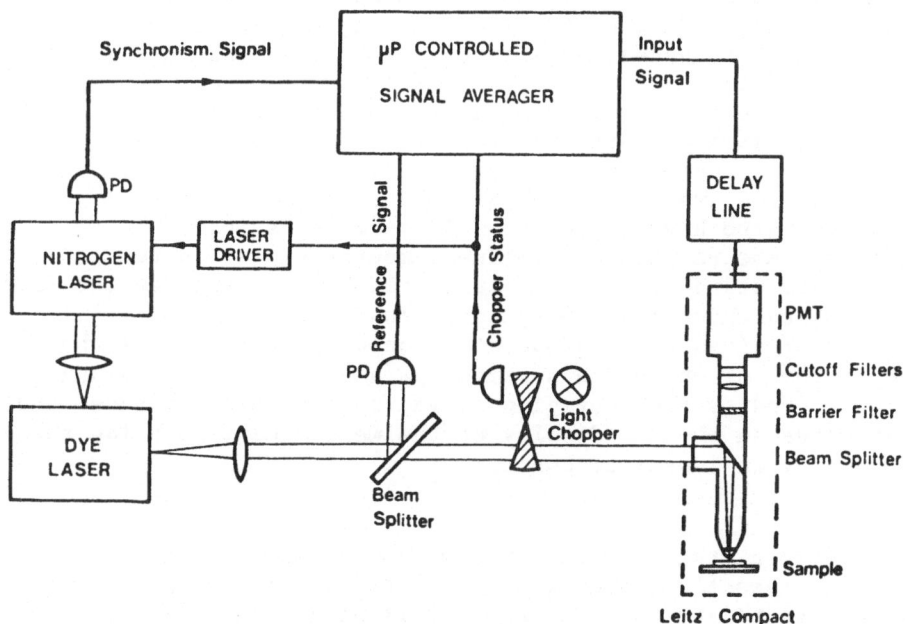

Fig. 1. Block diagram of the pulsed-laser microfluorometer.

function of transmitting and stopping the pulses alternately, in order to perform pulse-to-pulse background subtraction from the fluorescence curve.

The beam enters the side window of a Leitz Compact microscope and is focused on the sample through the objective. Since the beam is diffraction-limited, the minimum spot size is of the order of the laser wavelength ($\sim$0.3 µm). The fluorescence signal, collected by the microscope objective, passes through filters that select the spectral region of interest, and is detected by a photomultiplier tube (PMT-XP 1210, risetime $\sim$0.5 ns). Its electrical signal is sent to a microprocessor-controlled signal averager, developed in our laboratory[16] for the acquisition and processing of the fluorescence waveform. The averager has the following main properties:

- It performs the acquisition in the interrupt mode, in order to minimize the acquisition time, and thus optimize the overall procedure.
- It makes use of a sampling oscilloscope as a fast acquisition unit, in order to create a "slowed-down" replica of the waveform by sampling successive repetitions at different time-instants, i.e. in different positions of the waveform. The sampling instant is directly controlled by the microprocessor unit with great advantages in terms of system versatility, and in particular (i) it permits multiple time-scale operations, useful when both fast transients and long trails are present in waveforms; (ii) it makes for random sampling operation, thus avoiding deterministic errors in decay-time measurements when the fluorescent substances examined exhibit decomposition or bleaching.
- It converts the acquired data from analog to digital, and thus allows numerical data elaboration, resulting in an efficient background subtraction, normalization to the excitation pulse energy, and averaging process; the averaging process is necessary to have a good signal-to-noise ratio.
- It is connected to an off-line computer via a bidirectional IEEE 488 standard parallel interface-bus to transfer the data for higher numerical and graphic analysis.
- It is provided with an operator console, which allows the operator to set the measurement parameters at the beginning and to modify the averaging cycle via suitable commands, and it gives an on-line display of the measurement to allow the operator an immediate control.

It is evident that the use of a microprocessor acquisition and control unit gives the system much greater versatility, data pre-elaboration capability, and therefore reliability and ease of operation.

As a general remark, it is worth noting that the light-flux
technique, which is considered in this work, has some drawbacks
compared with other well-known Single Photon (SP) techniques in terms
of noise,e.m.i., and background immunity, independence of the waveform from the photomultiplier response, etc. However, our choice is
valid for three main reasons: (i) the system developed has the
maximum tunability and subnanosecond pulses (two requirements not yet
jointly met by lamps or mode-locked lasers) at the expense of a
repetition rate not yet suitable for SP; (ii) the low cost of N_2-
pumped dye-lasers as compared with mode-locked dye-lasers makes the
instrument suitable for many laboratories; and (iii) the number of
pulses required to perform a full measurement is much lower than that
needed for SP techniques. This is of great importance with regard to
the photodecomposition of the biological material under study.

The microfluorometer here described has an overall temporal
resolution of about 200 psec, mainly due to the sampling head of the
sampling oscilloscope. The spatial resolution is $\sim$0.3 µm, which
corresponds to the case of a diffraction-limited beam.

APPLICATIONS TO BIOLOGY

In this section, we report two examples of applications to
biology in which time-resolved fluorescence microscopy has proved its
validity and its potential usefulness.

The first example concerns the possibility of discerning,
directly at a cellular level, fractions of chromatin in different
functional states. This subject is presently investigated in several
laboratories: however, the most significant quantitative results have
been obtained on isolated and purified chromatin fractions rather
than through "in situ" measurements, as in our case.

The second example is a study of the fluorescence properties of
Hematoporphyrin-Derivative (HpD) both in solution and in single
cells. This drug is now extensively investigated since it is prefer-
entially retained by tumor tissues as compared with normal tissues.
This provides a powerful diagnostic and therapeutic aid, through
excitation at a suitable wavelength.

In Situ Evaluation of the Functional State of Chromatin.

The laser microfluorometer described in the previous section has
been used to study the functional state of chromatin at cellular
level.

Chromatin is a cellular component distributed throughout the
nucleus of the cell. It consists of DNA, histone proteins, non-

histone proteins, RNA, and enzymes and exhibits structural and com-
positional differences depending on its functional engagement. In
order to verify the possibility of evaluating these differences, the
biological samples were stained with the fluorescent probe Quinacrine
Mustard (QM). This dye is an intercalating and alkylating acridine
derivative, known as a chromosome Q-banding agent, which absorbs in
the violet-blu and emits in the green region of the spectrum.

Over the past few years, research has been performed in our
laboratories on the fluorescence behavior of both extracted and
purified DNA and of relatively simple biological structures, such as
bacteria, stained with QM[12,13]. The following results have been
obtained so far. (1) The interaction of QM can be made specific to
double-stranded DNA by controlling the staining conditions. (2) The
average decay time of the total fluorescent emission depends on the
base composition of the bound DNA. (3) The average decay time of the
DNA-QM complex depends (the base composition being the same) on the
spatial distribution of the QM-intercalated molecules which, in
solution, can be expressed at the stoichiometric ratio ℓ between the
number of QM-intercalculated molecules and the number of DNA base
pairs.

A non-radiative energy transfer mechanism in the Forster kinetics has been proposed as a possible explanation for the dependence
of the average fluorescence decay time on the two aforementioned
parameters[12]. In fact, molecular complexes formed by an acridine
molecule intercalated in a sequence of two Adenine-Thymine (AT) base
pairs fluorescence with a long decay time and act as energy donors.
On the other hand, when at least one of the base pairs of the
sequence is Guanine-Cytosine (GC), the fluorescence decay time is
much shorter and the molecular complexes act as energy acceptors.
Thus the DNA composition determines the donors (and acceptors) population, whereas its stainability (i.e. the parameter ℓ) determines
the average distance between donors and acceptors, which strongly
affects the energy-transfer probability per unit time.

To verify whether fluorescence decay time analysis makes possible an "in situ" evaluation of the functional state of chromatin, we
made an experiment[17] with two samples which are particularly suitable: (a) Polytene chromosomes of the embryo suspensor cells of
Phaseolus Coccineus and (b) Nuclei of parenchyma cells from dormant
tubers of Heliantus Tuberosus. In fact, in both these samples,
chromatin fractions with different degrees of activity can be singled
out easily due to their different morphology. In spite of this
difference, their DNA composition is the same. A difference in the
fluorescence decay time should therefore be ascribed to a different
chromatin stainability.

To perform the experiment, the laser wavelength was tuned to the
peak of the QM absorption spectrum (~420 nm).

Phaseolus Coccineus. The chromosomes to be investigated were
obtained from Phaseolus Coccineus developing seeds, and the measurements were made on the nucleus organizing region of S_1 chromosomes,
with a laser spot size of ∿0.5 µm on the object plane.

Fluorescence measurements performed on the corresponding segments of homologous chromosomes with active and inactive chromatin
gave the following results: the decay time for active chromatin
turned out to be 12.6 ± 0.2 ns, while for inactive chromatin it was
9.7 ± 0.2 ns, i.e. significantly shorter than the previous one.

Heliantus Tuberosus. Similar results have been obtained with
parenchima cells of the dormant tuber of Heliantus Tuberosus in the
G^0 phase of the cell cycle and, after stimulation with 2,4-dichloro-
phenoxyacetic acid, also in the G_1 and G_2 phases. As in the previous
case, slices of the sample have been stained with QM. In order to
illuminate the whole nucleus a laser spot size of ∿10 µm has been
used. The following mean values of the fluorescence decay time have
been obtained: 12.8 ± 0.2 ns for the G_0 phase, 13.9 ± 0.5 ns for G_1
and and 14.8 ± 0.6 ns for G_2, respectively.

As in the previous case (a), these results show that the decay
times obtained with active chromatin are longer than with inactive
chromatin.

Human Lymphocytes. On the basis of the previous results, a
similar experiment has been done with human lymphocytes, both in
quiescent condition (G_0 phase) and after cell cycle activation stimulated by Phytohemagglutinin (PHA) up to reaching the G_1 phase, and
stained with the fluorescent probe QM. The degree of template
activity of chromatin has been determined by means of cytoautoradio-
graphy and graded by a label index. This procedure was requested by
the heterogeneity of the induction period of lymphocyte response to
PHA.

The following results have been obtained: (1) Cells with the
highest label index (i.e. with template activity corresponding to the
G_1 phase) present a fluorescence decay time of 10.5 ± 0.3 ns. (ii)
Cells taken from the culture in absence of PHA, with a label index
"zero" (i.e. quiescent cells corresponding to the G_0 phase) present
a decay time of 8.5 ± 0.2 ns. For the cells with an intermediate
label index, the decay times are included between the above mentioned
values.

A reasonable interpretation of these results lies in the model
proposed for the chromatin structure in a different functional state.
It is well known that active chromatin differs from inactive chromatin in a number of chemical and physical properties, such as melting
temperature, packing ratio, and availability to DNAase attack, that
can be related to the structure of the biopolymer. Moreover, active

chromatin presents a higher percentage of non-histone proteins than
does inactive chromatin. It is known that non-histone or acidic
proteins are more tightly bound to DNA than are histones, and that
the binding of non-histone proteins involves the small groove of DNA,
i.e. the portion of the biomolecule in which the electrostatic interactions that stabilize the intercalated QM molecules take place[18].

   Moreover, these data agree with the evidence, obtained by spec-
tropolarimetric techniques in isolated chromatin titrated with
ethidium bromide, that the higher DNA superpacking of chromatin in
the quiescent phase produces a more favorable thermodynamic condition
for the intercalation process[19]. These data suggest a lower stainability of the active chromatin thus explaining the longer decay time
found in the corresponding fluorescence waveforms.

   The results presented in this section show that the fluorescence
decay time analysis makes possible and "in situ" evaluation of different chromatin functional states. This can be done not only in
chromosomes, where a well-defined deoxyribonucleoprotein (DNP) pattern exists, but also in interphasic nuclei in which the DNP organization is much more complicated. These results, interpreted on the
basis of the study performed "in vitro" on purified DNA[13], can give
an insight on the bond stoichiometry of the fluorescent probe and
DNA, directly "in situ", despite the great complexity of the biological substrate.

   It is worth noting that the laser microfluorometric technique,
based on the fluorescence decay time analysis, allows to obtain
results which are independent of the absolute amount of dye or DNA
present in the irradiated sample, opposite to what happens when
intensity measurements are performed with cw illumination.

   Important prospects can be opened by the possibility of ident-
ifying the functional state of cells when no variations of the amount
of DNA are involved. The possibility of distinguishing between
quiescent and cycling cells can have a far-reaching implication when
the presence of cycling cells, or any abnormal increase in their
amount, can be related to pathological conditions: for instance, this
would be the case of lymphocytes in relation to leukemia.

Fluorometric Studies of Hematoporphyrin-Derivative
in Solution and in Cells

   The low toxicity, the relative specific retention in malignant
tissues as compared with surrounding normal tissue, and the capability of photo-activation by red light make Hematoporphyrin-
Derivative (HpD) one of the most commonly-used photosensitizers in
the diagnosis and treatment of malignant tumors. The literature
describes an increasing number of applications of this drug for both

the location and the destruction, by photoactivation, of solid
tumors[20,21].

HpD, first introduced by Lipson et al.[22] in 1961, is a very
complex mixture of porphyrins. The literature includes several
studies on the chemical composition of HpD with the object of ident-
ifying the active ingredients, as a first approach to the definition
of the specificity of the drug towards tumors[23-25]. In fact, many
questions about the interaction of this drug with the cellular structures have not yet been clarified.

To investigate this interaction we performed an experiment of
time-resolved fluorescence microscopy to study the dependence of the
HpD fluorescence on the cell functional state[26]. This experiment
was however preceded by a careful study of the fluorescence properties of HpD in saline, in different environmental conditions[27].
In this case the laser dye was a 10^{-3}M solution of α-NPO in ethanol
and the laser was tuned at 400 nm.

As reported in the literature[28,29], the emission spectrum of
HpD in polar medium is characterized by two emission peaks at 610 and
670 nm. The corresponding excitation spectrum shows, in the Soret
region, a narrow band at about 395 nm, ascribed to monomeric forms,
which exhibit a fluorescence decay time of about 16 ns. The absorption spectrum, however, shows a second large band at about 365 nm,
ascribed to non-fluorescent dimeric forms. The spectral characteristics of fresh HpD solution are shown in Fig. 2 (dashed curves).
Nevertheless, an emission band around 580 nm has been observed in the
"in vivo" spectra of HpD in mice with mamma tumor[30], and in the
spectra of HpD components extracted SDS micelles from tumors (MBL-2
Lymphoma and Yoshida Hepatoma), but not from normal tissues[31].
Finally, it has been observed in the spectra of suspension of HpD-
treated mouse 3T3 cells[32]. In the latter case, the band is
centered at 590 nm, and its relative importance, in the fluorescence
spectrum, increases with time after HpD treatment. A similar
emission band in cells had already been observed in some of our
previous works[33,34]. Therefore a study of the conditions for the
appearance of this fluorescence peak should be useful for a better
understanding of the interaction of HpD with cellular structures and,
therefore, to clarify its tumor specificity.

We found that a new porphyrin species (NPS) emitting the yellow
region of the spectrum is formed in HpD in saline, after aging of the
solution. In the Soret region, it is characterized by a narrow
absorption peak centered at ∿405 nm, and two emission peaks around
575 nm and 630 nm (Fig. 2 unbroken curves), and a typical fluorescence decay time of about 3.5 ns.

The dependence of this NPS formation kinetics on several environmental partners (gas diluted in the solution, concentration, pH,
and temperature) has then been studied. The results are here summarized.

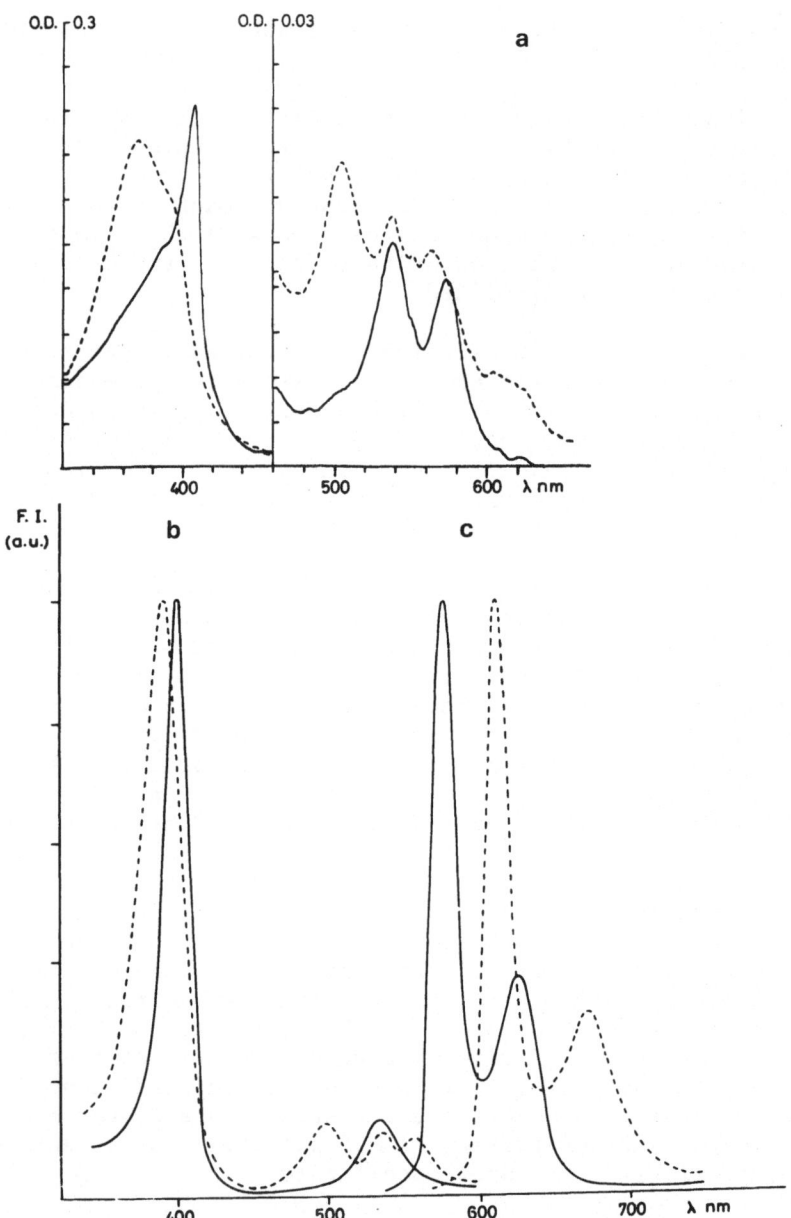

Fig. 2. Absorption (a), excitation (b) and emission (c) spectra of
HpD solution in saline. Dashed-line curves refer to air-
equilibrated solution immediately after preparation.
Unbroken-line curves refer to N_2-equilibrated solution 24 h
after preparation (HpD concentration: 3.10^{-6} M). In figures
(b) and (c), dashed-line curves refer to $\lambda_{emission}$ = 610 nm
and $\lambda_{excitation}$ = 395 nm, respectively; unbroken-line curves
refer to $\lambda_{emission}$ = 575 nm and $\lambda_{excitation}$ = 406 nm, respectively.

   The dependence of this NPS formation on several environmental
parameters (gas diluted in the solution, concentration, pH, and
temperature) has then been studied. The results are here summarized.

   The absorption spectra in the Soret region (Fig 3), where the
differences in the component absorption features are more evident,
were recorded at fixed time intervals, using fresh solution as a
reference. The measurements were done in different solution aeration
conditions. It has been found that the presence of Oxygen is an
inhibiting or, at least, a strongly delaying, factor for the formation of the species absorbing at a 405 nm. In fact, the rate of
formation in air-equilibrated solution is 8-10 times slower than in
de-aerated solution. The effect of concentration has been studies by
measuring the extinction coefficient at 405 nm (absorption peak of
the NPS) and 365 nm (absorption peak of the dimer) normalized to that
at 395 nm (absorption peak of the monomer) as a function of HpD
concentration. It has been found that the formation of the NPS
induces a dissociation of the dimer, and occurs at a rate that decreased as the HpD concentration increases.

   The effect of hydrogenionic concentration in the spectral
characteristics of the NPS has been evaluated at different pH, starting from an aged N_2-equilibrated solution of HpD, containing only
NPS. To vary the solution pH, concentrated HCl or NaOH were added.

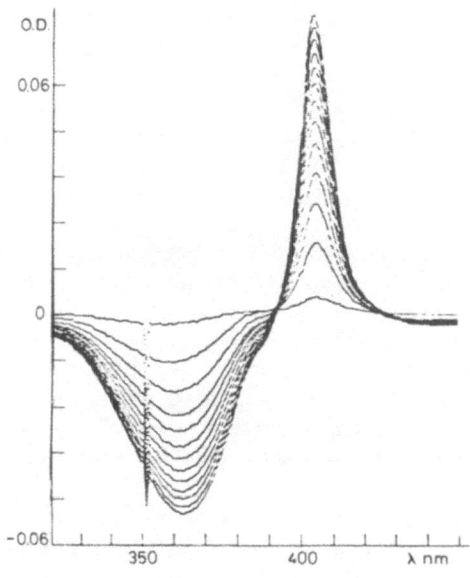

Fig. 3. Kinetics of the formation of the porphyrin species absorbing
         at 405 nm in N_2-equilibrated solution: the absorption
         spectra are taken using fresh solution as a reference. Time
         interval between consecutive curves: 15 min.

It has been found that acid pH's inhibit the appearance of the 405 nm peak. In fact, in the case of acidification, the absorption peak at 405 nm initially diminishes, but without any evident spectral shift. The absorption peak shifts to 400 nm at a pH value of about 3. In a more acid environment, there is only an increase in absorption. The same behavior is found for the excitation spectrum.

In a basic environment, inhibition occurs for pH values greater than 8. At higher pH's, with aging, a non-fluorescent species with an absorption peak at 408 nm appears. Finally, the NPS formation kinetics is also temperature-dependent. In particular, the temperature-increase is a factor favoring the appearance of NPS up to about 80°C. At higher temperatures, a gradual decrease of the absorption at 405 nm can be seen.

In conclusion, these results indicate the existence of a new porphyrin species, initially no present in the fresh HpD solution. The process of formation of this new species is strongly affected by the components if the air diluted in the solution, and by the pH of the solution itself. The NPS, in fact, forms in a practically neutral environment, and in anaerobic conditions.

From the results obtained, the NPS is likely to form starting from the monomer initially present. The transformation if the monomer alters the monomer-dimer equilibrium, with consequence dissociation of the latter. The NPS formation thus causes a decrease in the absorption peak in the 365 nm region, which corresponds to the dimer.

As to the chemical nature of the NPS, the results obtained do not enable any definite assumption to be formed. The possibility that this NPS may be a monomeric one, as in the case of free-base Hematoporphyrin[35], seems, however, not to be ruled out.

With this background we performed the experiment[26] of time-resolved fluorescence microscopy to study the dependence of the HpD fluorescence on the cellular functional state. As a biological model, we chose normal human lymphocytes, both in the quiescent state (G_0 phase) and in the pre-replicative (G_1) phase, after stimulation with Phytohemagglutinin (PHA) and treated with HpD. The fluorescence has been studied in two spectral regions: one in the yellow (521-581) and one in the red (604-649 nm) part of the spectrum.

First of all, the primary fluorescence of both stimulated and unstimulated lymphocytes was studies. In both cases, primary fluorescence was practically absent in the freshly-prepared smears in the cells in the red part of the spectrum. On the other hand, primary fluorescence was present in the yellow region. We observed, however, that repeated measurements in different lymphocytes systematically gave a peak intensity smaller by about a factor 10 than that given by

HpD in lymphocytes. It is worth remembering that, as already reported[36], primary fluorescence is substantially stabilized in cells
treated with HpD. This effect had also already been observed in
human lymphocytes[37].

    The fluorescence of HpD-treated lymphocytes was then analyzed.
Fig 4 shows an example of a fluorescence waveform of HpD in a stimulated lymphocyte. Both the experimental points and the fitting curve
are given. The method of analysis is also shown.

    Table I summarizes the results obtained with a set of measure-
ments on several samples, once the primary fluorescence contribution
has been subtracted. The fluorescence curve of both stimulated and
unstimulated lymphocytes, in the red region, is well fitted by the
sum of two exponential decays with similar decay times in the two
samples (∿3.5 and ∿15 ns). Going from the red to the yellow region

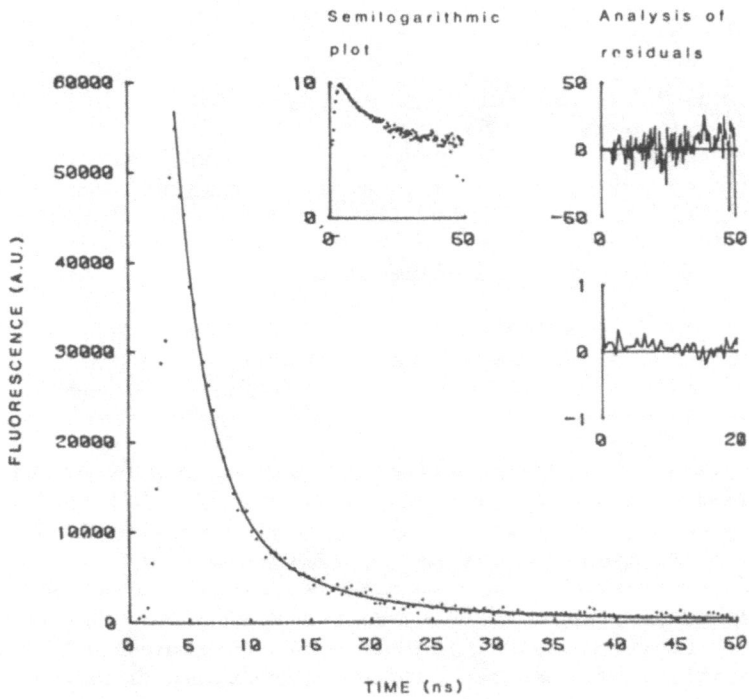

Fig. 4. Fluorescence waveform of an HpD-treated stimulated human
         lymphocyte in the spectral region 521-581 nm. The waveform
         has been obtained by averaging over 200 sweeps. The best
         fitting curve has been obtained after statistical analysis
         by the non-linear least-square method. The accuracy of the
         fit was evaluated through analysis of the residuals (b) and
         of their autocorrelation function (c).

Table 1. HpD Fluorescence in Unstimulated and Stimulated Lymphocytes
          Incubated (1 h) in Culture Medium after Addition of HpD
          (5 x 10^{-5} M).

	spectral range (nm)	peak intensity	decay times (ns)
Unstimulated lymphocytes	521–581	I_1	3.3
Stimulated lymphocytes	521–581	3.9 I_1	3.1
Unstimulated lymphocytes	604–649	I_2	3.7–14.9
Stimulated lymphocytes	604–649	2.1 I_2	3.5–15.2
		(I_2 = 2.5 I_1)	

of the spectrum the 15 ns component disappears and only the 3.5 ns
one remains. Moreover, the absolute peak amplitude is higher in
stimulated than in unstimulated lymphocytes, and this difference is
greater in the yellow part of the spectrum. Thus, we may consider
two emission bands, one at about 570–580 nm with a typical decay time
of ∿3.3 ns which can be attributed to the NPS, and one at about
610–620 nm, with a typical decay time of ∿15 ns which can be attributed to the monomer. To interpret these results, two hypothesis
may be advanced: 1) lymphocytes in a different functional state
exhibit a different uptake capability; 2) lymphocytes in a different
functional state favor, to a different extent, the formation of NPS,
due to their microenvironmental conditions.

    As to the first hypothesis, the different membrane conditions
surely justify a higher HpD uptake in stimulated lymphocytes.
Indeed, considering the whole HpD emission spectrum, the fluorescence
peak intensity resulting from time-resolved measurements is higher
in stimulated lymphocytes, thus indicating a greater overall HpD
content. However, the content differences are much more evident in
the yellow emission band, as if there were a preferential uptake for
the NPS. To see if this was reasonable, we studied the behavior of
HpD in the culture media used in the lymphocyte preparation. Only
48 h after adding the HpD to the culture media, it was possible to
observe the appearance of a small peak corresponding to the NPS (data
not given). It is worth noting that both the absorption/excitation
and the emission spectra are red-shifted by about 10 nm in culture
media, as compared with the saline solution. Since the incubation
time in our experiments was no longer than 1 h, we may assume that
the HpD absorbed by cells had no 585 nm fluorescence component in
it; thus, the new species should form directly inside the cells, as
stated by the second hypothesis. Indeed, in the light of the results
obtained in solution, the greater oxygen consumption in stimulated
lymphocytes, due to their higher metabolic activity, may be assumed
to favor NPS formation in these cells.

In conclusion, the finding of the aforementioned fluorescence differences in the yellow spectral region, related to a specific fluorescence compound, if widely confirmed in tumor cells, should be useful for two main purposes: (i) to clarify the interaction of HpD with tumors, the main aspects of which are not yet understood, and (ii) to increase sensitivity in the location of early-stage tumors, through the detection of fluorescence, a promising technique that is increasingly used.

REFERENCES

1. S. Undenfriend, "Fluorescence Assay in Biology and Medicine," Vol. II, Academic Press, New York (1969).
2. J. Duchesne, "Physicochemical Properties of Nucleic Acids," Vol. I, Academic Press, New York (1973).
3. F. Ruch and U. Leeman, Cytofluorometry, in: "Micromethods in Molecular Biology," V. Neuhoff, ed., Springer-Verlag, Berlin (1973).
4. G. Prenna, G. Bottiroli, and G. Mazzini, Histochem.J., 9:15 (1977).
5. M. Kapoor, Biol.Rev., 47:27 (1976).
6. S. A. Latt, Can.J.Genet.Cytol., 19:603 (1977).
7. L. Brand and J. R. Gohlke, Ann.Rev.Biochem., 41:843 (1972).
8. G. von Segenbusch and A. Thaer, "Fluorescence Techniques in Cell Biology," A. Thaer and M. Sernetz, eds., Springer-Verlag, Berlin (1973).
9. F. W. D. Rost, "Fluorescence Techniques in Cell Biology," A. Thaer and M. Sernetz, eds., Springer-Verlag, Berlin (1973).
10. S. Cova, G. Prenna, and G. Mazzini, Histochem.J., 6:279 (1974).
11. A. Andreoni, A. Longoni, C. A. Sacchi, O. Svelto, and G. Bottiroli, "Tunable Lasers and Applications," A. Mooradian, T. Jäger, and P. Stokseth, eds., Springer-Verlag, Heidelberg, New York (1976).
12. G. Bottiroli, G. Prenna, A. Andreoni, C. A. Sacchi, and O. Svelto, Photochem.Photobiol., 29:23 (1979).
13. A. Andreoni, S. Cova, G. Bottiroli, and G. Prenna, Photochem. Photobiol., 29:951 (1979).
14. F. Docchio, R. Ramponi, C. A. Sacchi, G. Bottiroli, and I. Freitas, An automatic pulsed laser microfluorometer with high spatial and temporal resolution, J.Microscopy, in press.
15. R. Cubeddu, S. De Silvestri, and O. Svelto, Opt.Comm., 34:460 (1980).
16. F. Docchio, A. Longoni, and F. Zaraga, Rev.Sci.Instrum., 52:1671 (1981).
17. G. Bottiroli, P. G. Cionini, F. Docchio, and C. A. Sacchi, In situ evaluation of the functional state of chromatin by means of quinacrine mustard staining and time-resolved fluorescence microscopy, Histochem.J., 16: (1984) in press.
18. D. E. Comings, B. W. Kovacs, E. Avelino, and D. C. Harris, Chromosoma, 50:111 (1975).

19. C. Nicolini, Basic Appl.Histochem., 25:319 (1981).
20. T. J. Dougherty, J. E. Kaufman, A. Goldfarb, K. R. Weishaupt, D. Boyle, and A. Mittleman, Cancer Res., 38:2628 (1978).
21. A. E. Profio, D. R. Doiron, and E. G. King, Med.Phys., 6:523 (1979).
22. R. L. Lipson, E. J. Baldes, and A. M. Olsen, J.Natl.Cancer Inst., 26:1 (1961).
23. J. Moan and S. Sommer, Photobiochem.Photobiophys., 3:93 (1981).
24. R. Bonnet, R. J. Ridge, P. A. Scourides, and M. C. Rosenbaum, J.Chem.Soc.Perkin T., 12:3135 (1981).
25. D. Kessel and T. Chow, Cancer Res., 43:1994 (1983).
26. F. Docchio, R. Ramponi, C. A. Sacchi, G. Bottiroli, and I. Freitas, Time-resolved fluorescence spectroscopy of Hemato-porphyrin-Derivative (HpD) in human lymphocytes, Chem.Biol. Inter., in press.
27. G. Bottiroli, I. Freitas, F. Docchio, R. Ramponi, and C. A. Sacchi, The time-dependent behavior of Hematoporphyrin-Derivative in saline: A study of spectral modifications, Chem.Biol.Inter., in press.
28. J. Moan and S. Sommer, Photobiochem.Photobiophys., 3:93 (1981).
29. S. B. Brown, H. Hatzikonstantinov, and D. G. Herries, Int.J. Biochem., 12:701 (1981).
30. W. J. M. Van der Putten and M. J. C. Van Gemert, Hematoporphy-rin-Derivative fluorescence spectra in vitro and an animal tumor, in: "Proc. Laser "81 Opto-Elektronik," München, West Germany (1981).
31. G. Jori - personal communication.
32. M. W. Berns, A. Dahlman, F. M. Johnson, R. Burns, D. Sperling, M. Guiltinan, A. Siemens, R. Walter, W. Wright, M. Hammer-Wilson, and A. Wile, Cancer Res., 42:2325 (1982).
33. F. Docchio, R. Ramponi, C. A. Sacchi, G. Bottiroli, and I. Freitas, Fluorescence studies of biological molecules by laser irradiation, in: "New Frontiers in Laser Medicine and Surgery," K. Atsumi, ed., Excerpta Medica, Amsterdam (1983).
34. G. Bottiroli, I. Freitas, F. Docchio, R. Ramponi, and C. A. Sacchi, Towards a better understanding of the mechanism of action of Hematoporphyrin-Derivative at the cellular level, in: "Proc. 13th Cancer Congress," Seattle, USA (1982).
35. A. Pasqua, A. Poletti, and S. M. Murgia, Med.Biol.Environ., 10:287 (1982).
36. F. Docchio, R. Ramponi, C. A. Sacchi, G. Bottiroli, and I. Freitas, Time-resolved fluorescence microscopy of Hemato-porphyrin-Derivative in cells, Lasers in Surgery and Medi-cine, 2:21 (1982).
37. F. Docchio, R. Ramponi, C. A. Sacchi, G. Bottiroli, and I. Freitas, Time-resolved fluorescence microscopy of Hemato-porphyrin-Derivative in tissue- and culture-cells, in: "Laser Tokyo '81," K. Atsumi and N. Nimsakul, eds., Inter Group Corp., Tokyo (1981).

TIME-RESOLVED SPECTROFLUOROMETRY OF

MELANIN IN HUMAN MELANOMA

M. Aricò*, M. Barcellona,** M. S. Giammarinaro**
and S. Micciancio

*Istituto di Dermatologia Sperimentale dell'Università
Via del Vespro, 129, I-90100 Palermo, Italy
**Istituto di Fisica dell'Università and CNR-GNSM
Via Archirafi, 36-I-90123 Palermo, Italy

INTRODUCTION

The intermediate steps of melanogenesis, the process of production of melanin, are not well known because of the difficulties
met in isolating the intermediate products of the process. In the
case of natural melanogenesis there are further difficulties because
of the inclusion of the substances of interest in a proteic matrix.
Usual methods for the extraction of melanin from organs or tissues
are very drastic (e.g. they may include boiling in HCl); very likely
they destroy intermediate products which are less inert than melanin
at the final melanogenesis stage.

It is known, on the other hand, that in histological specimen
the melanogenesis process is, in a sense, freezed, and visual observation of thin cuts of specimens rich in melanin often show a variety
of granules whose hue ranges from yellow to dark brown, or even
black. These granules have been identified as melanin at different
stages of the melanogenesis process, the darker hue corresponding to
a later melanogenesis stage.

We undertook a physical study of natural melanins without extracting them from the tissues and found indeed that melanin granules
at different evolution stages exhibit quantitatively different responses to certain stimuli, in particular to the irradiation with UV
light.

Previous work[1], essentially based on microphotometry and
visual inspection, showed that the melanin present in human melanoma,

exposed to a steady UV light intensity, undergoes complete bleaching; moreover the region which contains the melanin granule emits a yellow luminescence whose intensity, negligible at the turn-on of the UV light, begins to increase after some dead time reaches a maximum and then decreases. The maximum is attained in a time comparable with that required to bleach the granule. The characteristic times of the whole process are the order of several hours and depend on the initial hue of the melanin granule. In fact it was proved that the darker melanin granules exhibit a slower kinetics.

These results, showing the occurrence of a chain of reactions whose early steps are certainly photo-assisted, stimulated a more detailed investigation. We report here some preliminary data obtained by means of a time-resolved study of the spectrum of the yellow emission of melanin. Our results show that this emission actually is the superposition of two bands whose intensities follow different kinetics.

EXPERIMENTAL

Thin cuts ($\sim$10μm) of human melanoma specimens prepared as already described[2], were sealed between glass slide and cover and observed with a microscope (LEITZ DIALUZ 20) equipped for epifluorescence (100w Hg lamp. Excitation filter band: 355-425 nm). The microscope was also equipped with a computer controlled spectra acquisition system described elsewhere, which takes fluorescence spectra (2.5 nm spectral resolution) at 0.1 s interval, averages up to 65,000 consecutive scans for noise reduction purposes and stores the (averaged) spectra on a disk for off-line processing. In the experiments reported here we stored one spectrum every ten minutes. Each spectrum was the average of 600 scans (acquisition time of one averaged spectrum: one minute). The use of 100 x immersion objective together with a collimation of the exciting beam allowed to probe a specimen region of only $\sim$15μm of diameter.

RESULTS AND DISCUSSION

We chose regions of the specimen rich of melanin granules, then we turned on the UV light and started the automated collection of the emission spectra which exhibited, as a function of time, a continuous change of their shape as well as of their integral intensity. The time dependence of the latter was very similar to that reported in Ref.1, as expected. In Figure 1 we show two spectra of the same run taken respectively 60 min and 500 min after the turn-on of the UV light. The two spectra exhibit a remarkable difference of shape evidenced by the shift of the wavelength λ_{max} at which the spectra exhibit their maxima. In general λ_{max} shifts with time to longer wavelength, exhibiting a sudden jump a few hours after the turn-on of the UV light (see Figure 2).

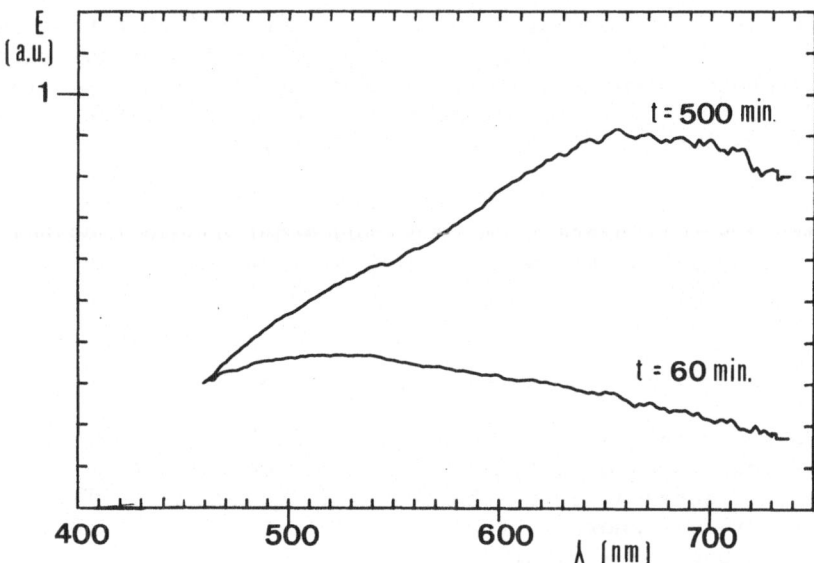

Fig. 1. Spectrum of the luminescence associated with melanin of
         human melanoma subjected to a steady UV light intensity.
         The two spectra have been recorded at different times after
         the turn-on of the UV light. The amplitudes of the two
         spectra are drawn to the same (arbitrary) scale. Remark the
         change of intensity as well as the shift of the maximum to
         longer wavelength.

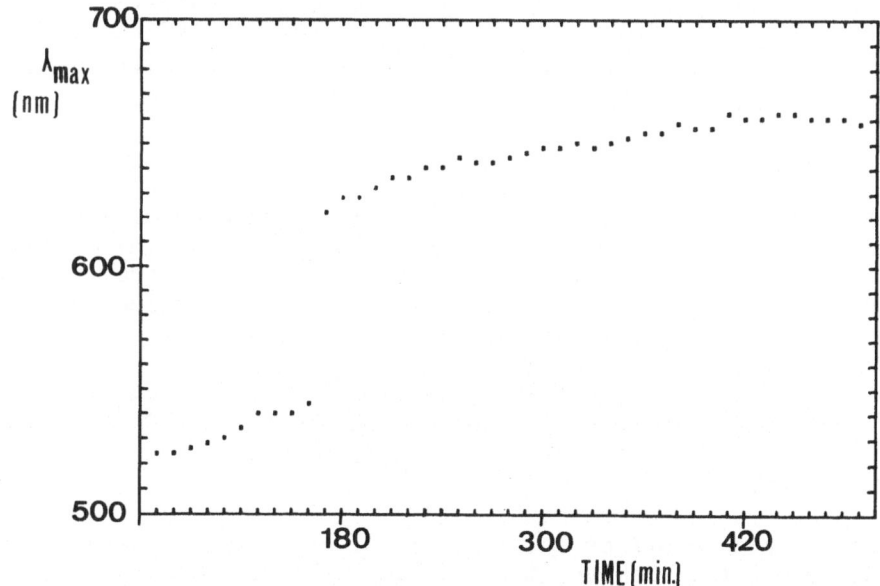

Fig. 2. Plot of λ_{max} (wavelength corresponding to the maximum
         emission amplitude) versus time. Remark the sudden shift of
         λ_{max}.

This kind of time dependence of λ_{max} suggests that the observed emission spectra E (λ, t) might be expressed as a superposition of two bands whose intensities follow different kinetics. According to Figure 2 the two bands are expected to be centered in the green and in the red regions of the spectrum, respectively.

Although these two hypnotized bands are not resolved (see Figure 1), we may try to isolate them from each other and to find how the two band intensities vary with time. Our argument is the following.

We expect that:

$$E(\lambda,t) = A_G(t) \, I_G(\lambda) + A_R(t) \, I_R(\lambda) \qquad (1)$$

where the subscripts G and R stand for "green" and "red", $I_G(\lambda)$ and $I_R(\lambda)$ are the two band profiles suitably normalized and $A_G(t)$, $A_R(t)$ give the time dependences of the two band intensities. The emission spectra $E(\lambda,t)$ are known in the interval $\lambda_s < \lambda < \lambda_\ell$ (in our experiments $\lambda_s = 460$nm, $\lambda_\ell = 750$nm).

We assume that

$$I_R \ (\lambda_s) = 0 \qquad (2)$$

and divide Eq.(I) by $E(\lambda_s,t)$. After some rearrangement we get the normalized spectra

$$E_N(\lambda,t) = E(\lambda,t)/E(\lambda_s,t) = \alpha I_G(\lambda) + B_R(t) \, I_R(\lambda) \qquad (3)$$

where $\alpha = 1/I_G(\lambda_s)$ and $B_R(t) = A_R(t)/A_G(t) \, I_G(\lambda_s)$ The difference between two normalized spectra taken at different times t, t':

$$D(\lambda;t,t') = E_N(\lambda,t) - E_N(\lambda,t) = [\alpha B_R(t) - B_R(t')] I_R(\lambda) \qquad (4)$$

does not contain the profile $I_G (\lambda)$, but gives a scaled replica of the band profile $I_R(\lambda)$. Of course this is true only if the assumption of Eq.(2) is correct. In other words, if Eq.(2) is true, the curves $D(\lambda;t,t')$, scaled so that they assume the same value in their maximum, are expected to superpose exactly to each other.

We have thus the possibility of extracting $I_R(\lambda)$ from $E(\lambda),t)$ as well as testing a posteriori if the starting hypotesis, Eq.(2), was reliable. A similar procedure yields $I_G(\lambda)$ together with a test for its reliability.

Once the band profiles $I_G(\lambda)$ and $I_R(\lambda)$ have been reliably obtained, we may fit them to $E(\lambda,t)$ to get $A_G(t)$ and $A_R(t)$.

Our experimental data on $E(\lambda,t)$ were processed with a computer program inspired to the algorithm sketched above. Figure 3a shows

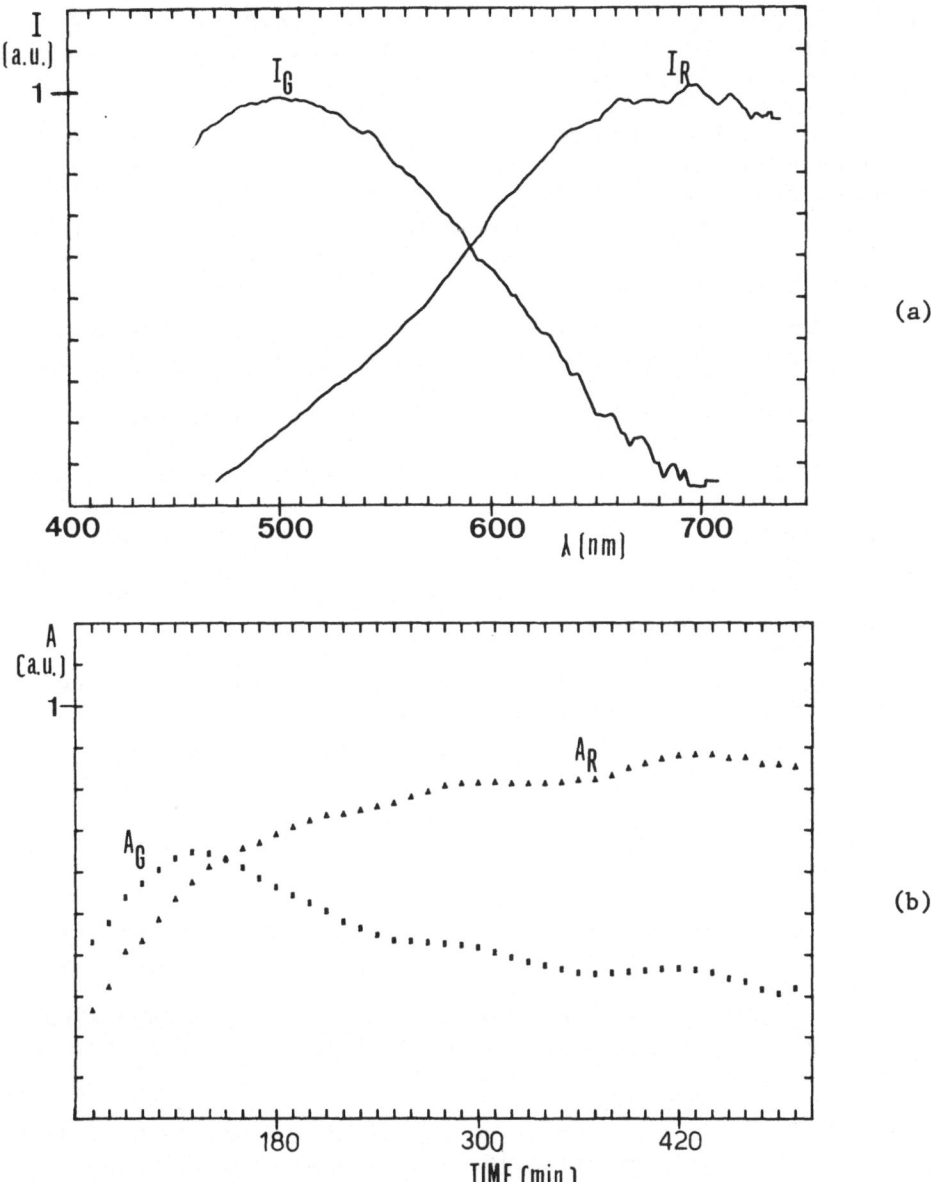

Fig. 3. (a) Normalized plot of the two bands present in the
         emission spectra of figure 1. (b) The relative weights of
         the two bands vary with time and follow different kinetics.

the normalized band profile $I_G(\lambda)$ and $I_R(\lambda)$ while Figure 3b shows how
the two band intensities A_G and A_R vary with time. The spread of the
band profiles $I_R(\lambda)$ obtained by means of Eq[4] for different couples
of times t,t' was about 5%. In the case of the band $I_G(\lambda)$ the spread
was somehow smaller. If we consider the amount of noise affecting
our experimental data (observe the non-smoothness of the spectra of

Figure 1) we were not authorized to expect a much smaller spread.
The relatively small spread indicates that the assumption that $I_R(\lambda_s)$
= $I_G(\lambda_\ell)$ = 0 is fairly reliable.

These results, together with those of Ref.[1], allow to draw a
preliminary phenomenological picture of the effects of the exposition
of melanoma melanin to UV light.

1.   There is evidence for the occurrence of a chain of processes.
      The steps identified up to now are:

      a) Transformation of melanin into a colorless substance. This
         first step certainly cannot take place without exposition to
         UV light.
      b) Production (and later destruction) of a substance emitting a
         fluorescence band centered at about 505nm.
      c) Production of a substance exhibiting a fluorescence band
         centered at about 700nm. This red-emitting substance is
         perhaps a modification of that produced in step b).

2.   The whole process is faster when the melanin is at an earlier
      step of melanogenesis.

According to the results reported by several workers and re-
viewed by Nicolaus[3], melanogenesis takes place in these main steps

      i)   oxidation of precursors and production of pigmented
            "building blocks"
     ii)   production of colorless oligomers
    iii)   gradual polymerization yielding the final black polymer

Very likely we probed some intermediate stages of step iii) and
produced, by means of the UV light, a fragmentation of melanin, i.e.
a process that, to some extent, is a regression of melanogenesis. In
particular the bleaching suggests that a stage similar to step ii) is
attained. The fact that the processes observed by us are slower in
the darker granules is consistent with the well known extreme stability of melanin in its final stage.

We believe that our experimental technique may be helpful for a
quantitative characterization of some phenomenological aspects of
natural melanogenesis.

REFERENCES

1. M. Aricó , M. S. Giammarinaro, A. Grana, S. Micciancio,
      Ann.It.Derm.Clin.Sper., 35:429 (1981).
2. M. Aricó, M. Bosco, A. Grana, Ann.It.Derm.Clin.Sper., 34:241
      (1980).
3. R. Nicolaus, "Melanin", Hermann, Paris (1968).

PHOTOCHEMOTHERAPY

PHOTOPHYSICAL PROPERTIES OF HEMATOPORPHYRIN

DERIVATIVE AND RELATED COMPONENTS

A. Andreoni and R. Cubeddu

Centro di Elettronica Quantistica e Strumentazione
Elettronica del C.N.R., Istituto di Fisica del
Politecnico, Milano, Italy

INTRODUCTION

Hematoporphyrin derivative (HpD), prepared from hematoporphyrin (Hp) by chemical modification[1], is increasingly used as a photo-sensitizer in phototherapy of several kinds of neoplasias[2]. Recent studies[3,4] indicate that a major reaction pathway in the photo-induced killing of tumor cells involves the intermediate of $^1\Delta$ O_2 generated by electronic energy transfer from triplet HpD. The reported greater phototherapeutic activity of HpD as compared with Hematoporphyrin (Hp)[5] is likely to arise from a higher affinity of the former porphyrin for tumor cells, resulting in specific greater accumulation and/or in longer retention times of the drug[6]. The biochemical bases of the higher affinity of HpD for tumor cells are still unknown, partly owing to the incomplete characterization of the HpD components. It must be noted that the differences in photosensi-tizing activity between HpD and Hp in vivo cannot be ascribed to large differences in the efficiency of $^1O_2$ production, since Hp is endowed with a higher quantum yield for singlet-triplet intersystem crossing and $^1O_2$ generation[7]. Although this dye has already been used in tumor therapy of human patients, many aspects are still unclear related both to the drug itself and to its interaction with the biological substrate once it is photo-activated by laser light.

In this paper we review the photophysical properties of HpD and its components. Spectroscopic studies of HpD in different solvent systems showed that one important feature of HpD, as compared with Hp, is the presence of large amounts of remarkably stable aggregated species that could be responsible for its strong photosensitizing activity in vivo. Results on a new blue-shifted emission of HpD are also reported. This new band peaked at 580 nm was first observed in

vivo[8] and seems to be important in the mechanism of biological
uptake of porphyrins[9].

EXPERIMENTAL PROCEDURES

The following substances were investigated: (i) Hematoporphyrin, (ii) Hematoporphyrin Derivative and (iii) Photofrin II (PII).
Hematoporphyrin, free base, from Porphyrin Products (Logan, Utah,
USA), was dissolved and characterized as described elsewhere[10].
HpD is the substance marketed by Oncology Research and Development
(Cheektowaga, N.Y., USA) as Photofrin. Photofrin II was obtained
from Oncology Research and Development. Being the chemical composition of most substances undetermined, all solutions to be compared were made at the same w/v concentration. Sodium dodecylsul-
phate (SDS) and cetyltrimethylammonium bromide (CTAB) were obtained
from Aldrich Chemical Co., Triton X-100 (alkylphenylpolyethylene
glycol) was obtained from Merck. The organic solvents used were
spectroscopic grade samples obtained from Merck. All other chemicals were commercially available reagent-grade products.

The absorption spectra were taken by a Perkin-Elmer 554 UV-VIS
spectrophotometer with 2 nm-slit using matched quartz cuvettes of 1
cm optical pathway. The emission and excitation spectra were
measured by a Perkin-Elmer 650-40 spectrofluorometer with 5 nm-slits
in both the excitation and observation monochromators. The spectra
were not corrected for either the lamp spectrum, nor monochromator
and photomultiplier responses. Fluorescence quantum yield values ϕ
were calculated by using a Hp solution in acetic acid as a standard
with $\phi = 0.15$ upon 400 nm-excitation[11].

The excitation source for the time-resolved fluorescence was a
dye laser pumped by an atmospheric-pressure nitrogen laser. A 10^{-3} M
solution of α-NPO [2-(1-Naphthyl)-5 phenyloxazole] in ethanol was
used as the lasing medium. The laser could be tuned either at 395 nm
or at 405 nm as required during the experiments. The dye laser
provided pulses of 150 ps duration (FWHM, full width at half maximum)
with peak power of $\sim$ 50 KW at a repetition rate of up to 100 Hz. The
solutions to be measured were contained in a 1 cm^2 cross-section cell
and their emission was observed at 90° through either a Kodak Written
nr. 22 cut-off filter or Corion SS 580-00 or Karl Zeiss IF 615 interference filters. The fluorescence decay was measured using two different techniques. In some cases the fluorescence signal was detected by a Varian 154 M photomultiplier with 400 ps FWHM response and
averaged over many repetitions of the excitation pulse by a digital
signal averager[12]. In other cases it was used the single photon
timing technique, where the fluorescence was detected by a single-
photon semiconductor photodiode operating in non-proportional
avalanche multiplication[13]. In both cases the experimental data
were transferred to a Tektronix 4051 graphic system for processing
and plotting.

RESULTS AND DISCUSSION

The absorption and emission properties of Hp and HpD were compared in the concentration range 2-100 μM either in pH 7.4 phosphate
buffer (0.1 M) or in aqueous solutions at pH 10 (0.1 M Tris) containing micellar dispersions of SDS and CTAB (20 mg/ml) or in absolute methanol. For instance the absorption spectra of Hp showed an
increase of the shoulder at 370 nm with the increasing concentration
in phosphate buffer. In fact while the absorption spectrum at 2 μM
is peaked at 395 nm (monomer peak), at 100 μM the 370 nm peak becomes
dominant. In the case of HpD the absorption peak in the same range
of concentrations is always at 370 nm. This fact suggests that HpD
contains some components with higher hydrophobicity than Hp, e.g.
some derivatives blocked at the level of the two hydroxyethyl groups
that are present in the side chains departing from the 2 and 4 positions of the tetrapyrrolic macrocycle[14].

The presence of aggregated species is evidenced also by the
emission quantum yield measurements. In fact while the shapes of the
emission spectra do not change in all the experimental conditions,
the quantum yield values decreases with increasing concentration for
Hp and HpD, being lower for HpD (0.24) respect to Hp (0.32) at 2 μM.
This is in agreement with the quenching of the fluorescent emission
induced by aggregation. A similar set of experiments was repeated
for Hp and HpD in methanol and in the micellar suspensions, where Hp
is known to exist entirely in monomeric form. The independence of
both the absorption spectra and the quantum yield from the HpD concentration in both SDS dispersions and methanol solutions indicates
that the equilibrium among the HpD components does not change under
these conditions.

All these results were confirmed by the time-resolved fluorescence measurements. The fluorescence decay, observed at λ > 565 nm
and measured by the Varian photomultiplier and the digital averager,
could be fitted in all the experimental conditions by two exponential
components whose decay times and relative amplitudes are listed in
Table 1. The lifetimes of the slow and fast-decaying components are
essentially identical for Hp and HpD.

In phosphate-buffered aqueous solutions, the relative weight of
the fast component increases for increasing HpD concentrations and
becomes similar to that of Hp at concentrations above 50 μM. On the
contrary, an essentially constant amount of fast-decaying HpD component is observed in all the methanol and micellar systems examined
by us; under the same conditions, Hp exhibits exclusively the slow-
decaying component. We therefore attributed the slow decay to monomeric Hp and the fast one to aggregated porphyrins. It must be
noted, however, that the presence of the latter component for HpD in
all solutions indicates that a more stable aggregated form is present
in this compound.

Table 1. Fluorescence decay times constants (τ_1 and τ_2) and relative initial amplitudes (A_1 and A_2) for different concentrations of HpD and Hp in various media

Medium	Porphyrin Concentration [μm]	HpD				Hp			
		τ_1(ns)	A_1(%)	τ_2(ns)	A_2(%)	τ_1(ns)	A_1(%)	τ_2(ns)	A_2(%)
Phosphate buffer pH = 7.4	2	16.50	85.8	4.10	14.2	15.40	93.5	3.94	6.5
	50	16.14	77.3	4.03	22.7	15.33	78.8	3.64	21.2
	100	16.67	69.3	3.69	30.7	15.52	70.3	4.07	29.7
Methanol	2	12.75	86.2	4.02	13.8	12.33	100	--	--
	100	13.30	87.5	6.44	12.5	11.95	100	--	--
SDS	2	17.57	82.2	3.83	17.8	16.61	100	--	--
CTAB	2	14.59	84.2	4.98	15.8	13.09	100	--	--

Recently it has been suggested that the stable aggregated species play a major role in the therapeutic activity of HpD. A new compound (Photofrin II) obtained by taking the most aggregated fraction of HpD after gel filtration, has been shown to be indeed more efficient in vivo than HpD. Our measurements of the optical properties of PII solutions in water confirmed the presence of a large amount of aggregates as indicated by the strong absorption at 370 nm, the low fluorescence quantum yield and the high percentage of the fast component in the fluorescence decay. In fact at a w/v concentration of 1.25 $\mu g/cm^3$, equivalent to 2 μM of Hp, a percentage of this decay of $\sim$ 26% was found. Moreover the emission spectrum revealed the presence of a new band peaked at 580 nm.

Other authors reported the existence in vivo of this band. It was found by observing the fluorescence of a solid tumor in mice previously injected intravenously with HpD[8]. Since the emission at 580 nm seems to be relevant in the biological interaction of porphyrins[9], we investigated the characteristics of this band under both cw and pulsed excitation in aqueous solutions of Hp, HpD and PII, that differ both in the amount of aggregated species and in their biological activity.

In order to simulate the temperature conditions in which the 580 nm band was reported to occur in vivo or in vitro[8,9], the solutions were kept at 37°C in the dark for 48 h. The resulting emission spectra measured at room temperature after the incubation period and excited at 405 nm showed a relevant increase of the 580 nm band in all solutions. For PII, this peak is greater than the one at 615 nm and the appearance of another band at 635 nm is noted. These modifications in the emission spectra were found to be irreversible. The excitation spectra of the 580 nm and 615 nm emissions were found to be peaked at 405 nm and 395 nm, respectively.

These findings are in agreement with the time resolved measurements on the PII solution under excitation at 405 nm and 395 nm using the single-photon timing technique. The fitting parameters of the decay curves are reported in Table 2. The data indicate that: (i) the relative amplitudes of the two components depend on the excitation wavelengths when the fluorescence is observed at λ > 565 nm. (ii) The amplitude of the fast component is always higher for excitation at 405 nm. (iii) Only this component is present for observation at λ = 580 nm. These results indicate the presence of two molecular species and it is possible, therefore, to attribute the fluorescence time constant of $\sim$ 2.6 ns to the molecular species with emission peaks at 580 nm and 635 nm and the time constant of $\sim$ 14.2 ns to the monomeric form of Hp, with emission peaks at 615 nm and 670 nm in agreement with our previous results. Moreover, for the PII solution, the 580 nm and 635 nm bands increase with time even at room temperature and, in a few days, the 615 nm and 670 nm disappear. Only the fast component could be detected in this case in agreement

Table 2. Fluorescence-decay time constants (τ_1 and τ_2) and relative
amplitudes (A_1 and A_2) for a 1.25 µg/ml water solution of
PII after 48 h incubation at 37°C.

Excitation	Observation	τ_1	A_1	τ_2	A_2
405 nm	$\lambda > 565$ nm	14.71	11.3	2.58	88.7
	IF 580	–	–	2.51	100.0
	IF 615	14.56	17.6	2.66	82.4
395 nm	$\lambda > 565$ nm	14.86	28.7	2.75	71.3
	IF 580	–	–	2.48	100.0
	IF 615	14.80	37.9	2.74	62.1

with our attribution. The formation of this new molecular species
seems to require the presence of both stable aggregates and monomers.
Actually the 580 nm band develops faster in PII than in HpD and at
low concentration values. According to this behavior, we suggest
that this species originates from monomers chemically bound to polymeric structures of porphyrins. Similar findings were reported for
uroporphyrin when covalently bound to amino terminal agarose gel[15].
On the basis of the above considerations the appearance of the 580 nm
band in vivo and in vitro could arise from the binding of monomers to
aggregates or other polymeric structures.

REFERENCES

1. T. J. Dougherty, J. E. Kaufman, K. R. Weishaupt, D. Boyle and
    A. Mittleman, Photoradiation therapy for the treatment of
    malignant tumors, Cancer Res., 38:2628 (1978).
2. T. J. Dougherty, G. Lawrence, J. Kaufman, D. G. Boyle, K. R.
    Weishaupt and A. Goldfarb, Photoradiation in the treatment
    of recurrent breast carcinoma, J.Natl.Cancer Inst., 62:231
    (1979).
3. T. J. Dougherty, C. J. Gomer and K. R. Weishaupt, Energetics
    and efficiency of photoinactivation of murine tumor cells
    containing hematoporphyrin, Cancer Res., 36:2330 (1976).
4. T. J. Dougherty, R. E. Thoma, D. G. Boyle and K. R. Weishaupt,
    Interstitial photoradiation therapy for primary solid tumors
    in pet cats and dogs, Cancer Res., 41:401 (1981).
5. T. J. Dougherty, K. R. Weishaupt, D. G. Boyle, J. Kaufman and R.
    Johnson, Phototherapy of human tumors, in: "Progress in
    Photobiology", A. Castellani, ed., Pergamon Press, London
    (1977).
6. R. L. Lipson, E. J. Baldes and A. M. Olsen, The use of
    derivative of hematoporphyrin in tumor detection, J.Natl.
    Cancer Inst., 26:1 (1961).

7. C. Lambert, R. S. Sinclair and T. G. Truscott, Photoprocesses involving-hematoporphyrins, in: "Abstracts 9th Annual Meeting", Am.Soc.for Photobiol., Williamsburg, USA (1981).

8. A. J. M. van der Putten and M. J. C. van Gemert, Hematoporphyrin derivative fluorescence spectra "in vitro" and in animal tumor, in: "Proceedings of Laser '81 Opto-Electronik", München, West Germany (1981).

9. M. W. Berns, M. Wilson, P. Rentzepis, R. Burns and A. Wile, Cell Biology and Hematoporphyrin Derivative (HpD), "Lasers in Surgery and Medicine", (1983).

10. S. Cannistraro, G. Jori and A. Van de Vorst, Photosensitization of amino acids by di-cyan-hemin: kinetic and EPR studies, Photochem.Photobiol., 27:517 (1978).

11. G. Cauzzo, G. Gennari, G. Jori and J. D. Spikes, The effect of chemical structure on the photosensitizing efficiencies of porphyrins, Photochem.Photobiol., 25:389 (1977).

12. A. Andreoni, A. Longoni, C. A. Sacchi and O. Svelto, Laser fluorescent microirradiation: a new technique, in: "The Biomedical Laser: Technology and Clinical Applications", L. Goldman, ed., Springer-Verlag (1981).

13. S. Cova, A. Longoni, A. Andreoni and R. Cubeddu, A semiconductor detector for measuring ultraweak fluorescence decays with 70 ps FWHM resolution, IEEE J.Quantum Electr., QE-19:630 (1983).

14. R. W. Henderson, G. S. Christie, P. S. Clezy and J. Lincheam, Hematoporphyrin diacetate: a probe to distinguish malignant tissue by selective fluorescence, Br.J.Exp.Phatol., 61:345 (1980).

15. J. D. Spikes, B. F. Burnham and J. C. Bonner, Photosensitizing properties of free and bound Uroporphyrin I, in: "Porphyrins in Tumor Phototherapy", A. Andreoni and R. Cubeddu, eds., Plenum Press, New York-London, in press (1984).

PHOTORADIATION THERAPY WITH HEMATOPORPHYRIN AS A SELECTIVE TECHNIQUE FOR THE TREATMENT OF MALIGNANT TUMORS

L. Tomio,* F. Calzavara* and G. Jori**

*Division of Radiotherapy, Civil Hospital of Padova
**Institute of Animal Biology, University of Padova
35100 Padova, Italy

INTRODUCTION

Photoradiation therapy (properly, the technique should be classified as photochemotherapy) is a new branch of oncology for the treatment of malignant tumors. The method is based on the ability of certain porphyrins[1-3] to be taken up and/or retained by tumor cells for longer periods of time than normal cells. Consequently, porphyrins, once photoexcited by absorption of visible light, promote a specific photosensitizing action on the tumor cells to which they are bound. Ultimately, the photosensitized processes can lead to cell death. In particular, upon irradiation with red light ($\lambda = 635 \pm 5$ nm), the porphyrins are selectively photoexcited[4], with no competing light absorption by other tissue components, so that any direct damage of the latter is avoided.

However, only a limited number of porphyrins are endowed with good tumor-localizing and -photosensitizing properties. In 1908 it was shown[5] that rabbit red blood cells hemolyze when illuminated in the presence of hematoporphyrin (Hp). About twenty years later, Policard[6] demonstrated that some tumors exhibited a red fluorescence emission upon illumination with a Wood lamp. This fact was ascribed to the presence of endogenous porphyrins in the tumor. The ability of Hp to accumulate in malignant tissues was shown to be a general property only in 1948 by Figge et al. who used different experimental tumors[1]. A derivative of Hp (HpD), obtained by treating commercial Hp with acetic-sulphuric acid[6], was reported to have a greater specificity than Hp toward a large variety of tumors in humans and animals[7-10].

117

The first regression of subcutaneously implanted glioma in rats was first reported by Diamond et al.,[11] who used photoactivated Hp. Later, Dougherty et al.[12] showed that the combination of HpD and red light is curative for a number of experimental tumors; the cyto-toxicity of photoexcited HpD appears to be related with the intra-cellular formation of singlet oxygen($^1O_2$). The destruction of a human bladder carcinoma transplanted in nude mice was obtained[13] by local exposure to light after HpD injection with no concomitant effect on normal bladder. The use of HpD and red light for the treatment of cutaneous and subcutaneous malignant lesions in humans was assessed in 1978[15]. Complete or partial responses were achieved in 111 out of 113 treated lesions. Since then, approximately 2,000 patients have been phototreated especially in USA and Japan[16].

Several unanswered questions still exist. Thus, very little information is available as regards the factors governing the tendency of porphyrin action at the cell level, and the precise mechanisms leading to photosensitized cell death. HpD, which is generally used in tumor phototherapy, is a complex mixture of aggre-gated porphyrins[17,18] which appear to play a major role as tumor localizers. HpD is able to enter all cell types being retained in reticuloendothelial and neoplastic cells for extended periods after the porphyrin levels in the serum have dropped[19]. On the contrary, Hp although forming aggregates in aqueous solutions does not enter the parenchymal and neoplastic cells and readily penetrates into the vascular stroma. This fact could explain the high Hp levels we found[3] in an experimental tumor system. As known, many tumors exhibit an increase of the interstitial tissue, especially in the case of some lines of hepatomas[20]. On the other hand, in vitro studies failed to shown any difference between normal and neoplastic cells as regards HpD uptake. The preferential accumulation of por-phyrins in tumor tissues thus seems to depend on the tumor morphology and architecture.

The lack of intracellular distribution of Hp may not be relevant for tumor photodestruction. In tumors treated with HpD and red light[19,21], one observes important damages of the micro-circulation which may then lead to cell death. However, our previous studies[22] showed that, at least for some types of neo-plastic cells, Hp binds with the cells the highest affinity binding sites being located at the level of the plasmatic membrane. Hp has also a larger ability than HpD to photogenerate $^1O_2$ at least in homogenous media[23]. On these bases, we began a clinical study for the treatment of cancerous patients with Hp and red light.

MATERIALS AND METHODS

Hp (Porphyrin Products, Logan, Utah) appeared to be 97% pure by HPLC. The sterile and apirogenic injectable solution of Hp was

prepared by Monico Farmaceutici (Venice, Italy) by mixing 1 part of
Hp and 50 parts of 0.1 M NaOH and stirring for 1 h at room temperature. The pH was brought to 7.2-7.4 by addition of HCl and the
solution was made isotonic by addition of NaCl. The final volume of
the solution was reached (200 parts) by adding a suitable amount of
0.9% NaCl. The Hp solution (5 mg/ml) was stored in the dark at -18°C
until used.

A standard Hp dose (5 mg/kg body weight) was used, while the
usual interval between Hp administration to the patient and phototreatment was 24 and 48 h. Two light sources were employed. First,
the 590-690 nm wavelength interval was isolated from the emission of
a 4,000 W high pressure Xenon arc-lamp by a combination of chemical
and heat-reflecting optical filters. The total light intensity at
the patient level was 25 mW/cm^2 within a circular light spot whose
area approached 20cm^2. In each session, the phototreatment was
prolonged up to 60 min. Practically, a total light dose of 22.5
J/cm^2 was delivered for each session of the phototreatment. As a
second light source, we used a group of five He/Ne laser sources
(Valfivre, Florence) each delivering ca. 20-25 mW at the distal end
of the optical fiber. The red light beam (632.8 nm) was focused
within a quartz fiber with a diameter of 1.1 mm. We used light
intensities in the 15-100 mW/cm^2 range, corresponding with a dose
range of 17-72 J/cm^2.

Recently, we obtained a 10W Argon laser, 457.9-514.9nm, (GSG
Laser, Torino, Italy) coupled with a dye laser utilizing Rhodamine B.
The 0.4 mm quartz fiber was coupled with the laser by means of alignment apparatus. The emission in the 620-640 nm region reaches a
maximum power of 2 W. Until now, we treated three patients in one
session at 24 h from Hp administration.

RESULTS

Our clinical experience with Hp-involving phototherapy presently
includes 17 patients (Table 1) and the following histological types
of cancer.

Basal Cell Carcinoma

Six lesions were treated, whose size ranged from a very superficial tumor with an approximate diameter of 1 cm to a lesion with a
5 mm deep central ulceration and a total diameter of about 2.5 cm.
All lesions were treated with the He/Ne láser. Those having smaller
dimensions received total light doses of 57 J/cm^2 and underwent a
complete regression. Only one lesion underwent recurrency after 3
months; the recurrency originated at the border of the original
lesion. Larger lesions received total light doses of 26 J/cm^2 to
avoid a too long exposure to irradiation; no response of the tumor
was detected.

Table 1. Cases and Physical Data**

Case No.	Age	Sex	Histology	Site	No. of Lesions	Dimension (cm)	Total dose (J/cm^2)
1	35	F	Malignant Melanoma	Shoulder	2	2	22.5
2	72	M	Squamous Carcinoma	Anus	1	6	22.5
3	79	F	Kaposi's Sarcoma	Foot	3	0,6-3	22.5
					4*	0,6-0,3	180-68
4	61	F	Breast Carcinoma	Chest Wall	2	0,5 (mult.)	22.5
5	62	F	Breast Carcinoma	Chest Wall	1	5	22.5
					2*	2	38
6	48	F	Breast Carcinoma	Chest Wall	1*	2	38
					3*	1	26
7	54	M	Squamous Carcinoma	Neck	1	5	22.5
8	49	M	Kaposi's Sarcoma	Arm	3*	0,5 (mult.)	35
				Oral ph.	1*	0,6	60
9	61	M	Basal Cell Carcinoma	Face	1*	2,5	26
10	75	F	Basal Cell Carcinoma	Face	2*	1	57
11	82	M	Basal Cell Carcinoma	Ear	1*	1	57
12	72	M	Basal Cell Carcinoma	Head	2*	1	57
13	80	F	Squamous Carcinoma	Head	1*	1	68
14	71	F	Malignant Melanoma	Foot	1	6	22.5

 * Lesions treated by He/Ne laser. Other lesions treated by Xenon
   lamp.
** Patients treated by dye laser are omitted.

Kaposi's Sarcoma

    Small (3-6 mm) cutaneous lesions were treated by the He/Ne
laser; using total light doses between 68-180 J/cm^2 we observed a
complete tumor regression, whereas only partial responses were
noticed after delivery of 35 J/cm^2. One patient was also affected
by an oripharingeal lesion. The latter gave a completely positive

response after 60 J/cm^2 irradiation from the He/Ne source. The Xenon arc-lamp was used to treat more advanced lesions involving confluent nodules with a diameter of about 3 cm. Surprisingly, light doses as low as 27.5 J/cm^2 were found to be sufficient for inducing a partial regression. The phototreatment was repeated after 20 days; the patient received a second administration of Hp. In this case, the regression of the tumor appeared to be complete with a one year follow-up.

Breast Carcinoma

Three patients with an overall number of 9 lesions were treated; in all cases, the metastatic nodules were subcutaneous and involved both small multiple lesions and large single lesions. No appreciable effect was observed after exposure of the patients to either the He/Ne laser of Xenon lamp, probably as a consequence of the infiltrating nature of the lesions and the low light doses (22.5-38 J/cm^2) delivered.

Melanoma

Our clinical data are in agreement with those obtained by other investigators: even relatively large nodules can be destroyed when the tissue displays a light pigmentation, whereas strongly pigmented nodules are insensitive to the phototreatment unless multiple optical fibers are available, which can be inserted into several positions of the tumor mass.

Squamous Carcinoma

Only two primitive cutaneous lesions have been treated: one lesions having small dimensions was exposed to the He/Ne laser, whereas one voluminous infiltrating lesion was irradiated by the Xenon lamp. The former lesion underwent complete regression, the latter gave only a partial response lasting for a short period of time. Another patient from this group exhibited cutaneous metastases at the neck level from an epidermoid carcinoma which was neatly differentiated from the oral pavement. After three months subsequent to the standard phototreatment with the Xenon lamp one can clearly observe a progression of the lesion, especially at the superior and lower border.

Phototreatment with the Argon-Dye Laser

Until now, three patients have been treated. The neoplastic lesions included a squamous carcinoma of the oral pavement, a squamous carcinoma of the tongue, and a basocellular carcinoma of the

tongue. Only the first patient can be presently evaluated owing to
the short interval of time from the phototreatment. The patient,
male, 60 year-old, exhibited one deeply infiltrated lesion into the
tongue; the dimensions were 3x2x3 cm. The lesion showed recurrency
subsequent to interstitial radiotherapy. The patient was subjected
to interstitial phototherapy through one optical fiber which was
repeatedly implanted into the tumor mass so that the whole area of
the target was certainly exposed to light. The phototreatment was
performed for 10 min. delivering 500mW at each insertion of the
fiber. At two days after the phototreatment an important edema of
the tongue and necrosis were apparent. After 15 days, the patient
reported a complete regression of pain while the tumor mass exhibited
an objective regression of about 30%.

DISCUSSION

Our initial data show that commercial Hp, at least in the form
supplied by Porphyrin Products, is uptaken by different tumor types,
which can be thus photosensitized to the action of red light; a
complete regression of the neoplastic lesion can be eventually
obtained.

The quality of the tumor response is strongly influenced by the
dimensions of the lesion and the total light dose. Actually, the 32
lesions treated in the initial 14 patients can be subdivided into two
groups. One group (Table 2) includes 14 superficial lesions either
single or multiple, whose diameter was equal or smaller than 1 cm;
the lesions have been treated by He/Ne laser. Specifically, 1 complete responses were obtained when the total light dose exceeded 50
J/cm^2. The second group (Table 3) includes single or multiple
lesions which were often exophytic or infiltrating with an external
diameter generally larger than 2 cm. Usually, the lesions of the
second group were treated with the Xenon lamp. A complete response
was only observed for the lesions from the Kaposi's sarcoma; however,
the latter were subjected to a second phototherapeutical treatment.
A partial (lower than 50%) response was observed for the single case
of melanoma lacking appreciable pigmentation, three metastatic
nodules from mammary carcinoma treated with the He/Ne laser, and one
case of squamous carcinoma. Finally, for the unique patient exposed
to argon-dye laser who can be evaluated until now, the response
appears to be similar with that observed by Dougherty et al. after
insertion of the optical fiber directly into the tumor mass and
delivery of very high total light doses.

Our results indicate that phototherapy with Hp shows a particular efficacy in the case of small superficial lesions; moreover, in
selected cases, the phototherapy is successful in the treatment of
large lesions provided adequate total light doses can be delivered.
In any case, the further improvement of the technique requires more
detailed investigations in order to determine:

a) the most suitable photosensitizing agent (Hp, HpD, other compounds);
b) the precise mechanisms leading to tumor regression;
c) the optimal doses of the photosensitizing dye;
d) the porphyrin levels in the tissue;
e) the light dosimetry and the occurrence of thermal effects at various depths in the tissues;
f) the most convenient time interval between porphyrin administration and exposure of the patient to light;
g) the advantages possibly arising from a fractionation of the light dose;
h) the importance and prevention of side effects.

Table 2. Details of First Group of Lesions.

Histology	No. of Lesions	Dimension (cm)	Light Source	No. of Treat-ment Courses	CR	PR	R	S
Basal Cell Carcinoma	5	1	He/Ne	1	5			
Kaposi's Sarcoma	8	0,5–0,6	He/Ne	1	5		3	
Squamous Cell Carcinoma	1	1	He/Ne	1	1			

Table 3. Details of Second Group of Lesions.

Histology	No. of Lesions	Dimension (cm)	Light Source	No. of Treat-ment Courses	CR	PR	R	S
Malignant Melanoma	3	2–6	Xenon	1			1	2
Kaposi's Sarcoma	3	0,6–3	Xenon	1–2	3			
Squamous Cell Carcinoma	2	5–6	Xenon	1–2			1	1
Breast Carcinoma	9	0,5–5	Xenon He/Ne	1 1			3	2
Basal Cell Carcinoma	1	2,5	Xenon	1				1

REFERENCES

1. F. H. J. Figge, G. S. Weiland, and L. O. J. Manganiello, Proc.
    Soc.Exper.Bid.Med., 68:640 (1948).
2. C. J. Gomer and T. J. Dougherty, Cancer Res., 39:146 (1979).
3. G. Jori, G. Reddi, L. Tomio, B. Salvato, P. L. Zorat, and F.
    Calzacara, Tumori, 65:43 (1979).
4. T. J. Dougherty, K. R. Weishaupt, and D. G. Boyle, Photo-
    radiation therapy of malignant tumors, in: "Principles and
    Practice of Oncology," V. De Vita, S. Hellman and S.
    Rosermberg, eds., J.B. Lippincott Publ. Co., New York (1981).
5. W. Hausmann, Biochem., 2, 14:275 (1908).
6. A. Policard, Compt.Rend.Soc.Biol., 91:1423 (1924).
7. R. L. Lipson and E. J. Baldes, Arch.Dermatol., 82:508 (1960).
8. H. B. Gregorie, E. O. Horger, J. L. Ward, J. F. Green, T.
    Richards, A. C. Robertson, and T. B. Stevenson, Ann.Surg.,
    167:820 (1968).
9. D. Kessel, Cancer Res., 42:1703 (1982).
10. M. C. Berembaum, R. Bonnett, and P. A. Scourides, Br.J.Cancer,
    45:571 (1982).
11. I. Diamond, S. G. Granelli, A. F. McDonagh, S. Nielson, C. B.
    Wilson, and R. Jaenicke, Lancet, 2:1175 (1972).
12. T. J. Dougherty, G. B. Grindey, R. Fiel, K. R. Weishaupt, and D.
    G. Boyle, J.Natl.Cancer Inst., 55:115 (1975).
13. K. R. Weishaupt, C. J. Gomer, and T. J. Dougherty, Cancer Res.,
    36:2326 (1976).
14. J. F. Kelly, M. E. Snell, and M. C. Berembaum, Br.J.Cancer,
    31:237 (1975).
15. T. J. Dougherty, J. E. Kaufman, A. Goldfarb, K. R. Weishaupt, D.
    G. Boyle, and A. Mittleman, Cancer Res., 38:2628 (1978).
16. T. J. Dougherty, An overview of the status of photoradiation
    therapy, in: "Porphyrin localization and treatment of
    tumors," D. Doiron, ed., Alan R. Liss Inc., New York (1983).
17. T. J. Dougherty, D. G. Boyle, K. R. Weishaupt, B. A. Henderson,
    W. R. Potter, D. A. Bellnier, and K. E. Wityk, Photoradiation
    therapy: Clinical and drug advances, in: "Porphyrin Photo-
    sensitization," D. Kessel and T.J. Dougherty, eds., Plenum
    Press, New York (1983).
18. T. Christensen, J. Moan, J. B. McGhie, H. Waksvik, and H.
    Stigum, Studies of HpD: Chemical composition and in vitro
    photosensitization, in: "Porphyrin Photosensitization," D.
    Kessel and T.J. Dougherty, eds., Plenum Press, New York
    (1983).
19. P. J. Bugelski, C. W. Porter, and T. J. Dougherty, Cancer Res.,
    41:4606 (1981).
20. P. M. Gullino, Extracellular compartments of solid tumors, in:
    "Cancer: A Comprehensive Treatise," Vol.3, F.F. Becker, ed.,
    Plenum Press, New York (1975).
21. W. M. Star, J. P. A. Marijnissen, A. van den Berg-Block, and H.
    S. Reinhold, Destructive effect of photoradiation on the

microcirculation of a rat mammary tumor growing in "sandwich" observation chambers, in: "Phorphyrin Localization and Treatment of Tumors," D. Doiron, ed., Alan R. Liss Inc., New York (1983).

22. I. Cozzani, G. Jori, E. Reddi, A. Fortunato, B. Granati, M. Felice, L. Tomio, and P. L. Zorat, Chem.Biol.Interactions, 37:67 (1981).

23. S. Cannistraro, A. Van de Corst, and G. Jori, Photochem. Photobiol., 28:257 (1979).

PHOTOBIOLOGY AND DERMATOLOGY

N_2 LASER RADIATION EFFECTS ON IODOAMINO ACIDS*

G. Palumbo[1], R. Massa[2], I. Vassallo[2], R. Bruzzese[2],
S. Solimeno[2], and S. Martellucci[3]

[1]Centro di Endocrinologia ed Oncologia Sperimentale del
CNR, c/o Istituto di Patologia Generale, Via Sergio
Pansini, 5-80131 Naples (Italy)
[2]Istituto di Fisica Sperimentale, Università di Napoli
PAD.20 Mostra d'Oltremare - 80125 Naples (Italy)
[3]Facoltà di Ingegneria della Seconda Università di Roma
Via Orazio Raimondo, 00173 Roma (Italy)

INTRODUCTION

In the last few years the interest in laser applications to
biology and medicine has grown up enormously (see reference 1 and 2
for a full review). This is due to peculiar properties of laser
radiation, as monochromaticity, high peak power, tunability, and
great variability of pulse temporal length (from CW emission to
picosecond pulses), which make the laser a powerful and versatile
tool in biological investigation.

In this paper we discuss the results obtained studying the
interaction of UV laser radiation with the four most commonly occurring iodoamino acids: 3 - iodotyrosine (MIT), 3,5 - diiodotyrosine
(DIT), 3,3',5, - triiodothyronine (T3), and thyroxine (T4).

The chemical and physical properties of these molecules have
been well established over the years[3]. Also the effects of X- and
γ-rays[4,5], and Raman spectra[6] have been investigated in full
details. The biological interest in these molecules arises from the
fact that T3 and T4 represent the thyroid hormones and the MIT and
DIT their precursors. The biosynthesis of hormones in the thyroid
gland requires complex mechanisms which involve several enzymatic
pathways, iodine and a large polypeptide as substrates.

---

* Work partially supported by the CNR "High Power Lasers National
  Project" and GNEQP.

The iodoamino acids and hormones formation takes place within the polypeptide matrix of thyroglobin, a large thyroid glycoprotein. It seems that the unique conformation of this protein gives the right environment necessary to capture the circulating iodine (to form MIT and DIT) and allows the coupling of two DIT to form T4[7,8].

The elucidation of the enormous relevance of possible defects in the hormone synthesis, release, and transport in generating a large number of diseases, is clearly out of our scope. An extensive and comprehensive review in this regard has been recently presented [9,10].

The iodoamino acids have interesting optical characteristics[3]. Namely, their absorption spectra (see Figure 1) show two major absorption peaks: a first one nearby 250 nm, and a second one, less intense, in the near U.V. region ($\sim$330 nm). As for the second absorption peak, the very interesting feature is that, although less intense, only a few biological substances absorb nearby that optical wavelength.

The above feature in the absorption spectra of iodoamino acids is due to the presence of iodine (I) atoms in these molecules. In fact, iodoamino acids differ from tyrosine and thyronine for the substitution of one, or more, hydrogen atoms with iodine atoms. The introduction of a single I atom ("MIT") produces the appearance of an absorption peak at 305 nm. Adding more I atoms to the molecule causes a more "red-shifted" absorption peak (325 nm for T4).

On the basis of the above considerations, it is evident that the N_2 pulsed laser system, which is characterized by a high peak power emission at 337 nm, lends itself as a unique mean of "selective" excitation of the system under investigation. Namely, laser ir-radiation at 337 nm should cause absorption, and, consequently, changes in the thyroid hormones only.

Thus, in this work we have aimed to study the photodeiodination of the iodoamino acids, produced by laser irradiation. In particular, we have studies absorbance changes in our samples after laser irradiation. This procedure is very useful when the absorption spectrum of the samples under investigation is different before and after the interaction with opportune radiation: this was, indeed, our case. The products of irradiation have, than, been studied by thin layer chromatography (TLC) and partially characterized by mass spectrometry.

The organization of the paper is as follows. In section II we briefly describe the preparation and specific characteristics of the samples used throughout the experiments. In section III the exper-imental set-up and the irradiation conditions are exposed. Section IV resumes the experimental results and includes a brief discussion.

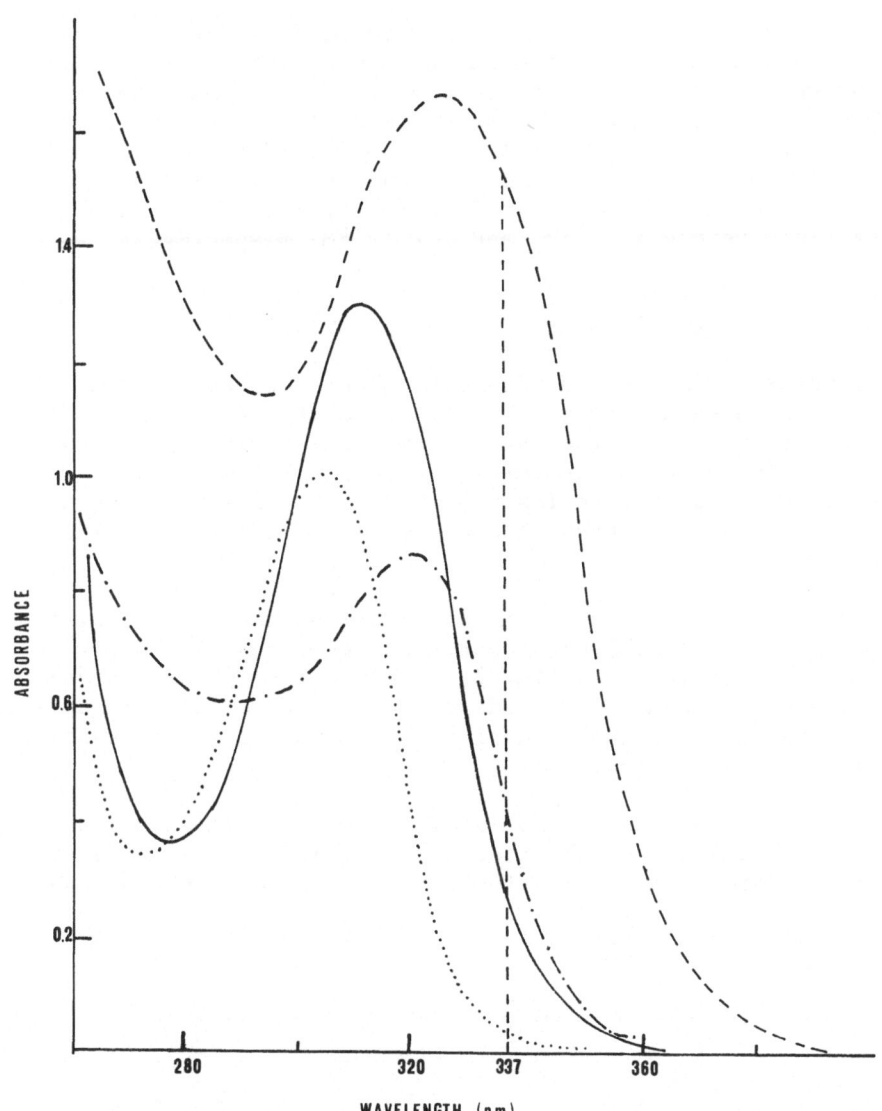

Fig. 1.   MIT (···), DIT (——), T3 (-·-), and T4 (---) absorption
spectra in NH$_4$OH solution. The vertical broken line
indicates the laser wavelength.

SAMPLES CHARACTERISTICS AND PREPARATION

The irradiation of the four iodoamino acids has been carried out
in different solvents. In this regard, NH$_4$OH (0,22 M), NaOH (0,1 M)
and HCl (1 M) have been used, to study the effects of the different
phenolic oxhydryl ionization on the absorption properties of the

molecules. The iodoamino acids were weighed in solid state with a
precision balance and dissolved in the appropriate solvent. All
solutions were clarified by centrifugation for 5 minutes at 12.000
rpm. The concentrations were then checked with a spectrophotometer,
using the appropriate molar extinction coefficients[3]. The final
concentration was $3,28 \times 10^{-4}$M in all but the HCl solutions, where,
due to the lower solubility in this solvent, the T3 and T4
concentrations were 4×10^{-5}M, and 1×10^{-5}M, respectively.

EXPERIMENTAL SET-UP AND LASER IRRADIATION CONDITIONS

The experimental set-up used throughout the experiments is shown
in Fig. 2. We used a commercial N_2 laser system (L.I.S.S., Mod.
780 U), emitting laser pulses at 337 nm. The typical pulse energy
was 1 mJ, while the pulse temporal length was 7 nsec. Thus, we
irradiated the samples with laser pulses of typically 150 kW peak
power, at a pulse repitition rate of 10 Hz. The corresponding mean
power was 10 mW.

In typical conditions 1,5 ml of solutions under investigation
were poured in 1 cm pathlength quartz cuvettes and vertically exposed
to the laser radiation (see Fig. 2). In order to obtain an uniform
irradiation of the samples we made use of a quartz lens (10 cm focal
length). The need for quartz optics is due to the strong absorption
of glass in the UV region.

To check the percentage of laser radiation transmitted through
the cuvette (that is, not absorbed by the samples), we measured the
pulse energy before and behind the cuvette: almost 100% of laser

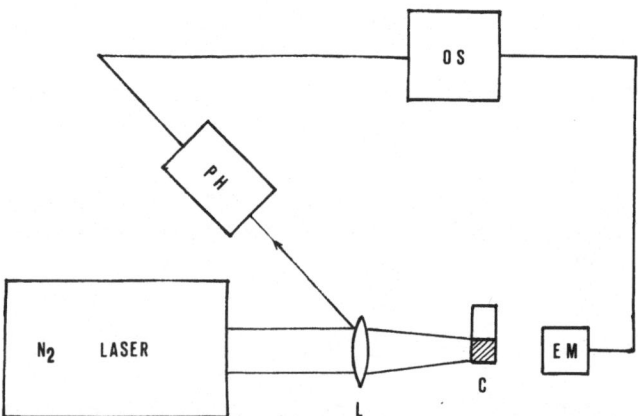

Fig. 2. Experimental set-up (L = quartz lens; C = quartz cuvette;
        PH = photodiode; E.M. = energy meter; OS = oscilloscope).

light was absorbed by the samples. The preservation of this condition during the irradiation time was checked by leaving the energy
meter in the position of Fig. 2 during the experiments. We also
checked the stability of laser energy, by using a photodiode, which
measured the small percentage of laser light reflected off the focusing lens. The pulse energy could be adjusted by slightly changing
the flowing laser gas mixture ($N_2:SF_6$).

No use was made of a sample cooling system, since the substances
under investigation showed very little sensitivity to temperature
variations (between 4°C and 25°C), over a three-day period of time.

The relationship between the photodeiodination of the iodoamino
acids and the total absorbed energy was monitored by exposing the
samples to the laser source for different intervals of time, namely
1, 2, and 3 hours. The average energy absorbed by samples amounted
to about 36 J/h.

After laser irradiation the absorption spectra of photoproducts
were recordered using a Beckman spectrophotometer (Model UV 5230) in
the range 280–400 nm.

Differential spectra of all molecules were obtained by using a
equimolar non-irradiated sample, or blank.

The products of irradiation were, then, furtherly analyzed by
TLC and mass spectrometry. TLC was carried out on silica gel plates
60, coated with fluorescent indicator. Mass spectra were obtained by
using a KRATOS spectrometer (Model MS50): a Fast Atomic Bombardment
(FAB) source produced negative ionization of the samples, exploiting
the high electronic affinity of I atoms and the relative weakness of
C–I bonds with respect to other bonds of the molecules under investigation.

RESULTS AND DISCUSSION

In Figure 3 the absorption spectra of T4 in 0.22 M NH$_4$OH
irradiated from 0 to 3 h are shown (other experimental details are
in the Figure caption). The spectra clearly show: a) a significant
increase in the absorption (at any wavelength) related to the amount
of absorbed energy; b) a modest blue-shift (6÷8 nm) of the absorption
peak. The difference spectra (Fig. 4) moreover, indicate the formation of free iodine as judged by the shoulder around 350 nm. The
iodination has been confirmed by TLC. However, rather surprisingly,
photodegradation of T4 did not yield to any of the following related
compounds MIT, DIT, T3 or tyrosine (see Fig. 5). The absorption
spectrum of the irradiated material does not change with time, being
constant up to 2-3 weeks, thus indicating the formation of a stable
product.

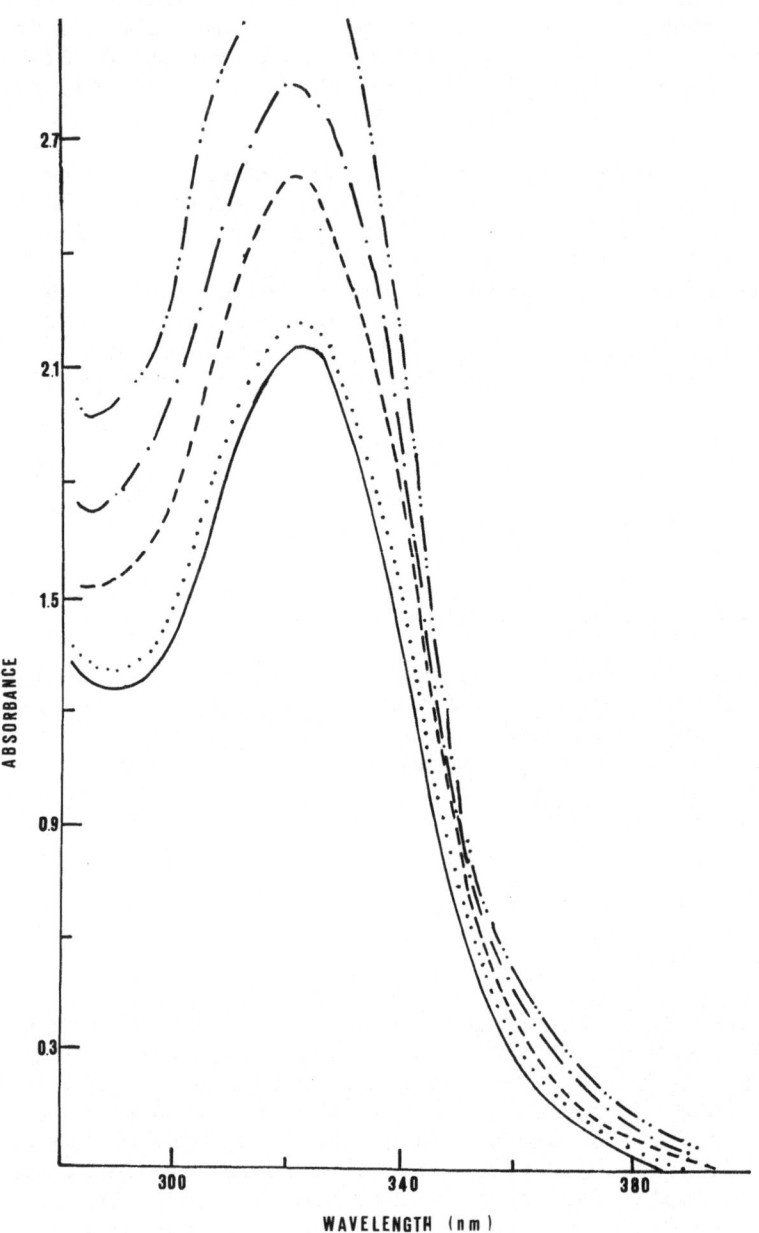

Fig. 3. Absorption spectra of T4 (continuous line) and T4 irradiated
         for 10 min (···), 1 hour (---), 2 hours (-·-), and 3 hours
         (-··-) in NH₄OH solution.

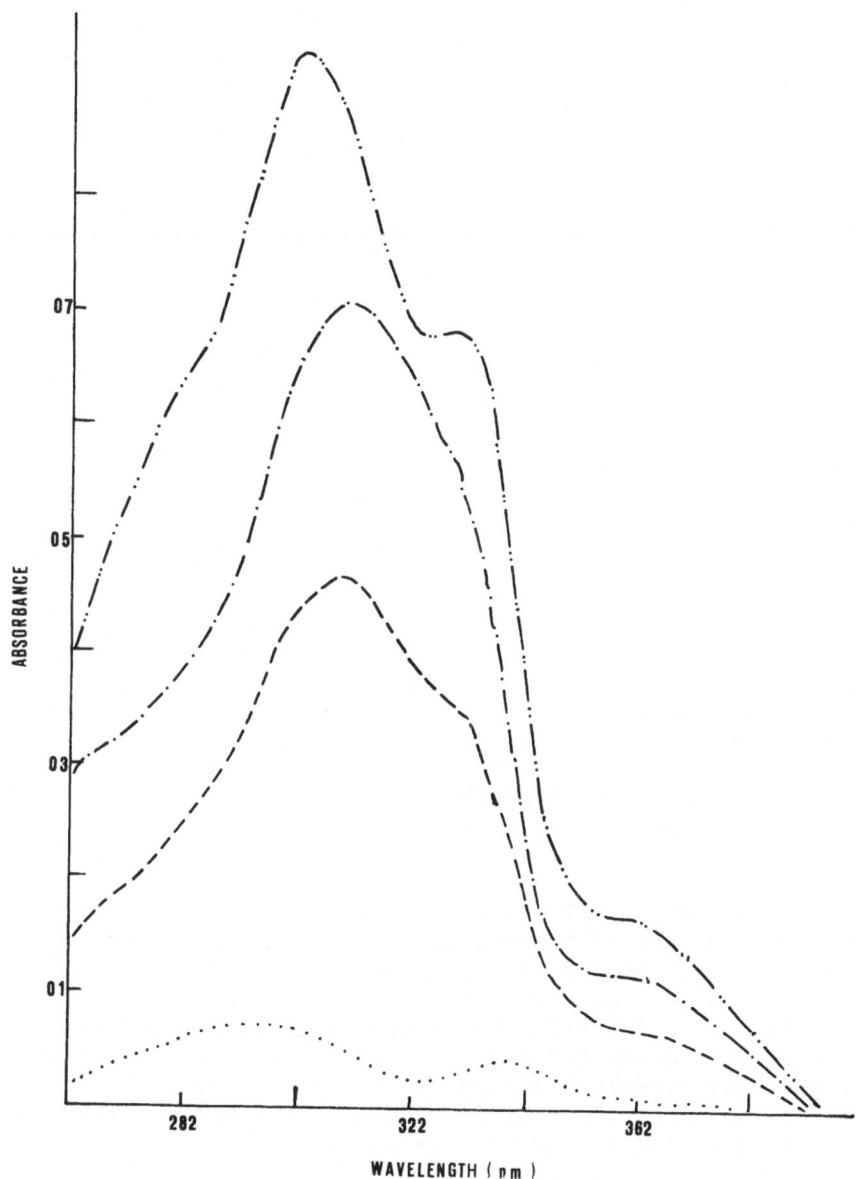

Fig. 4.   Difference spectrum (T4, T4$_{ir}$) after irradiation times of
         10 min (···), 1 hour (---), 2 hours (-·-), and 3 hours
         (-··-).

        Equimolar amounts of MIT, DIT and T3 were irradiated in the same
experimental condition as from T4. While for the T3 molecule we
observed an effect similar to that of the T4, i.e. an increase of the
molar extinction coefficient and a 3-5 nm "blue-shift" of the absorp-

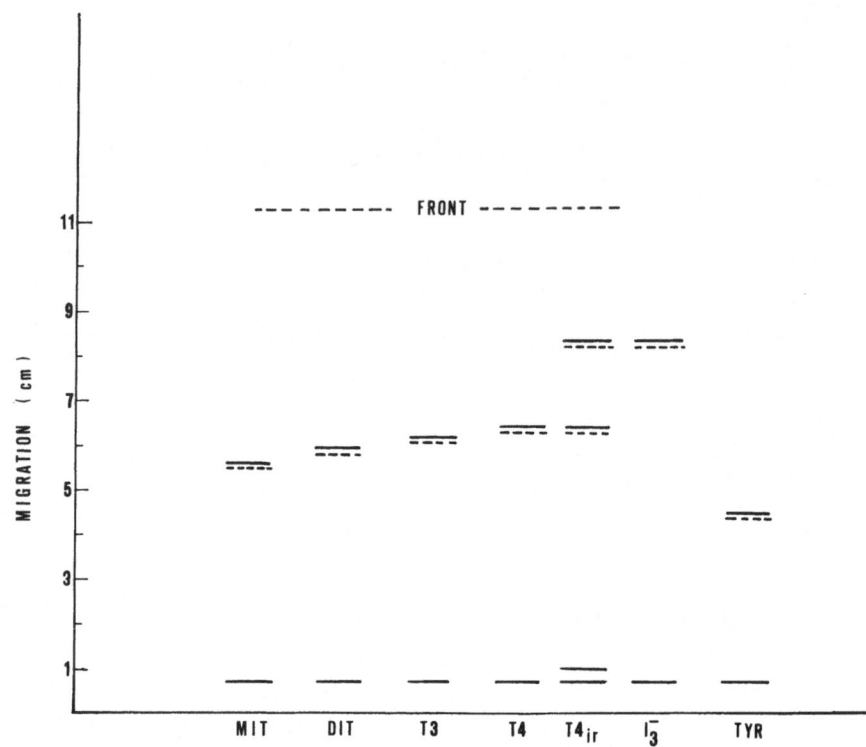

Fig. 5. Thin layer chromatography of the indicated species ($\overline{\cdots}$ spot size).

tion peak, the MIT and DIT showed a different behavior. In fact, in both molecules we still observed an increase in absorption in the region of 350 nm (free iodine), but, in contrast with what observed with both T3 and T4, the absorption in the 300 nm region shows an opposite trend as a function of the absorbed laser energy: it decreases at increasing irradiation times. This is clearly shown in Fig. 6, where the absorption spectra of MIT are depicted for different irradiation times. A similar behavior is found studying the effect of irradiation of the DIT spectrum.

The similarity between the MIT and DIT spectra and between T3 and T4 presumably derives from the common molecular scheleton of the respective couples of compounds: tyrosine for MIT and DIT, and thyronine for T3 and T4.

Also in this case, we carried out chromatographic analysis. Fig. 7 shows the obtained chromatograms. The presence of free iodine is evident both in the DIT and in the T3 irradiated (DIT$_{ir}$, T3$_{ir}$) molecules. Moreover, it is possible to notice that in the DIT and T3, there are three not identified products, which are very likely due to the laser action on the molecules.

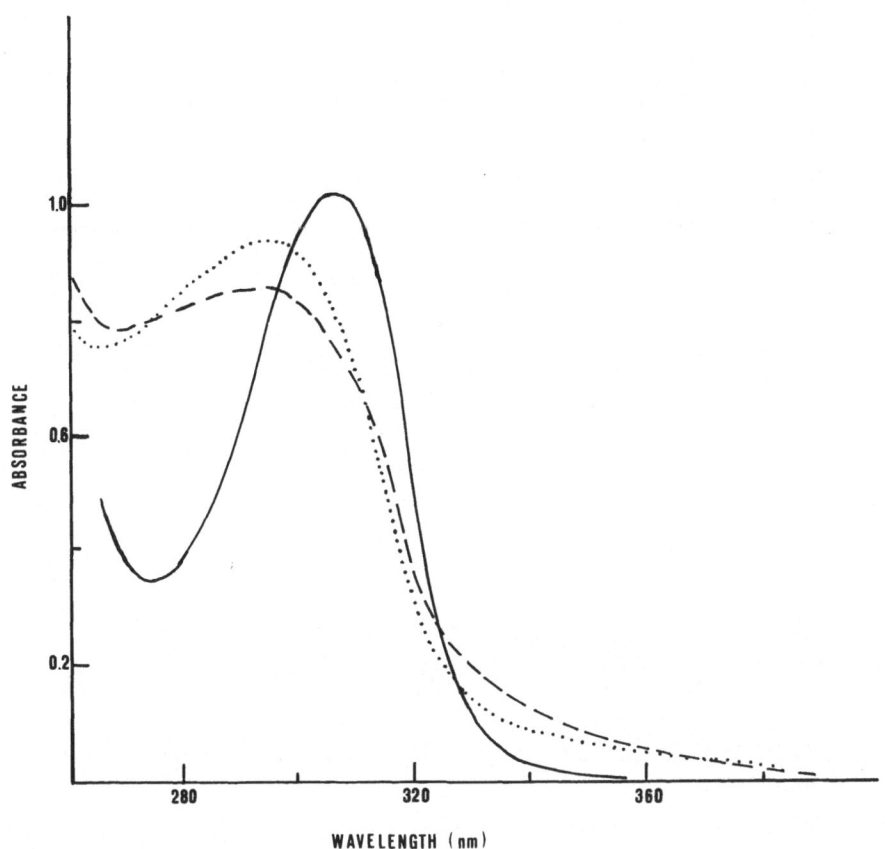

Fig. 6.   MIT (continuous line) and MIT$_{ir}$ absorption spectra after
          1 hour (...), 2 hours (---) of laser irradiation in NH$_4$OH
          solution.

        As control experiments we also irradiated samples of tyrosine
and free iodine in NH$_4$OH solution and pure solvent.

        Thus, it was possible to check if the laser radiation was inter-
acting with iodine itself, and if there was any possible radiation-
solvent interaction. The absorption spectra did not show any change,
and the "difference spectra" were identically flat in both cases.

        The present results suggest that the N$_2$ laser radiation has a
specific action on iodoamino acids, since no other artifactual
effects take part on the photochemical process.

        Since the ionization of all aminoacids is strongly dependent of
pH, we also carried out the same set of experiments and analysis in
more alkaline conditions, i.e. 0.1 M NaOH.

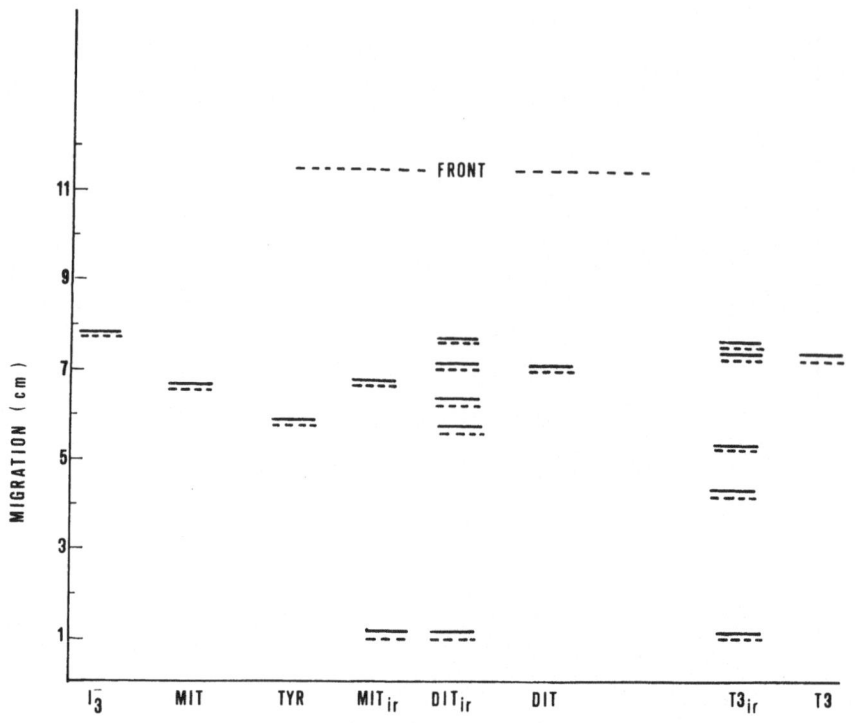

Fig. 7. Thin layer chromatography of the indicated species ($\overline{\cdots}$
         spot size).

        Again, T4 absorption spectrum behaved similarly (Fig. 8): as it
can be easily seen, the main features of Fig. 3 (NH$_4$OH solvent) are
retained. The same holds true for the T3 spectrum. It is of some
interest that, in contrast with what observed in ammonia solutions,
MIT and DIT spectra in NaOH 0.1 M did not show signs of increase
in absorption at 350 nm (Fig. 9). This is probably due to the dismutation reaction that iodine undergoes in strong alkaline solutions
to non-absorbing species as I$^-$ and IO$^-_3$. Perhaps this reaction is
not quantitative in the case of T4 and especially T3 (data not shown)
since a small shoulder (nearby 350 nm) is noticed in this case.

        Finally, we repeated the above experiments in acidic solution
i.e. 1 M HCl as solvent. In these conditions it appears that they
are almost completely transparent to the 337 nm laser radiation. In
fact the absorption spectra for MIT, DIT, T3 and T4 also after 1 hour
of laser irradiation show practically no difference with controls.
However, this was expected since, in the wavelength range of
interest, the molar extinction coefficients are very small for all
four molecules under investigation.

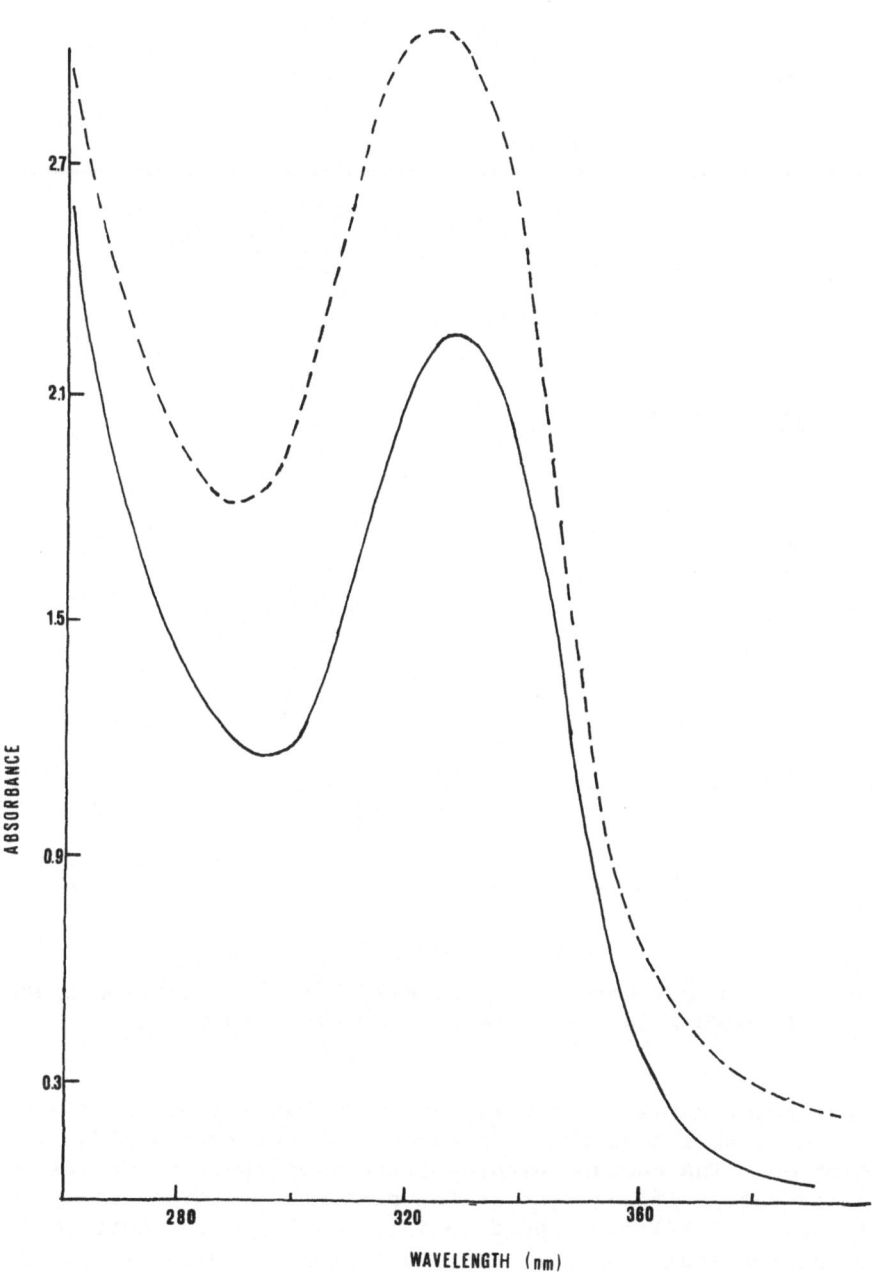

Fig. 8.   T4 (continuous line) and T4$_{ir}$ (---) absorption spectra in NaOH solution, after 1 hour of irradiation.

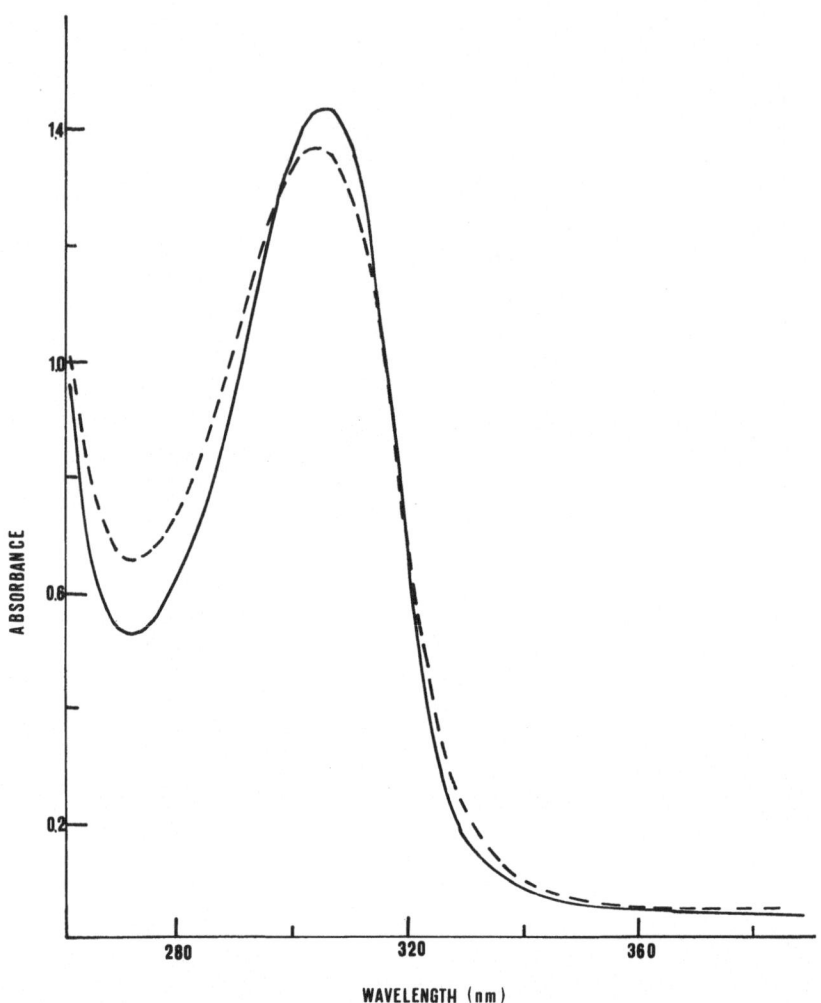

Fig. 9. MIT (continuous line) and MIT$_{ir}$ (---) absorption spectra, in NaOH solution, after 1 hour of irradiation.

A further series of experiments were carried out by mass spectrometry analyzing the iodoamino acids before and after laser irradiation. The common feature of the mass spectra for all four molecules before irradiation is the presence of three main peaks: the first one (127) correspond to iodine, the second corresponds to the intact molecule, and, finally, a third one which is due to the presence of chemical complexes produced by the molecule-solvent interaction (in particular with Na, whose atomic weight is 22).

The upper parts of Fig. 10, 11, 12, 13 show the mass spectra of the four iodoamino acids in NaOH solution, before laser irradiation.

In the lower parts of the same Figures, the spectra after 1 hour of
laser irradiation are shown.

The common feature after exposure to laser radiation, for all
the molecules, is the strong reduction of the iodine peak very likely
due to evaporation of molecular iodine after laser photodissociation.
It appears that different photoproducts with different M.W. are
found. These observed differences in molecular weights are summarized in Table 1.

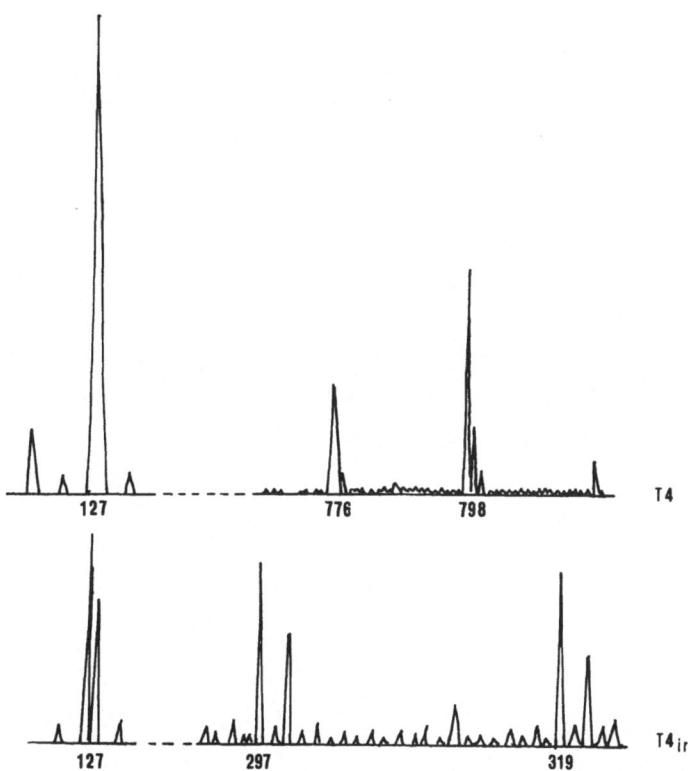

Fig. 10.   Mass spectrum of T4 (upper part) and T4$_{ir}$ (lower part) in
           NaOH solution after 1 hour of irradiation.

The observation of this table and the mass spectra as well leads
to some interesting considerations. It appears that, as a consequence of the irradiation all the iodine originally present in the
iodoamino acids is totally displaced. This fact is clearly in contrast with our former hypothesis of a selective effect of laser
radiation on T4 and T3 compared to the mono and diiodotyrosine molecules. One possibility is that the energy irradiated (36 J) in our
experimental conditions is too high to detect specific effects. In

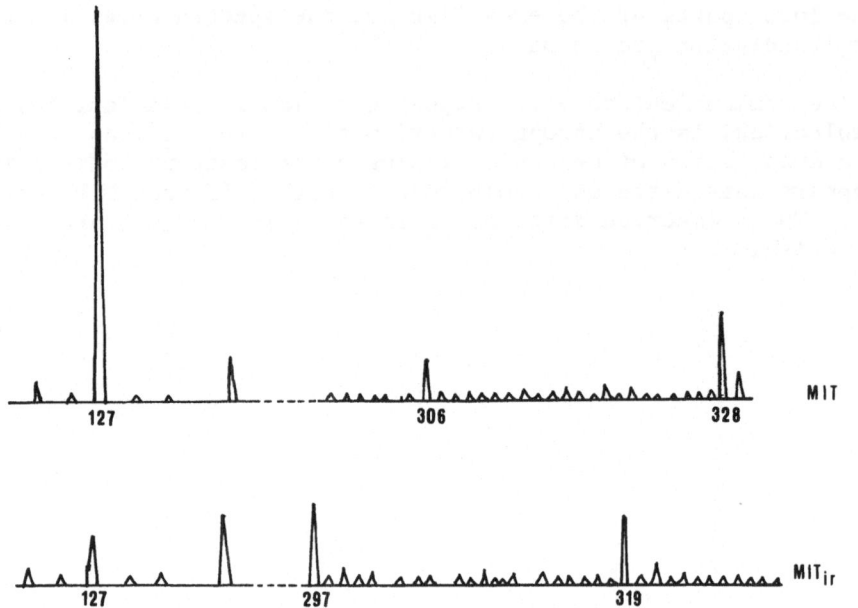

Fig. 11.   Mass spectrum of MIT (upper part) and MIT_{ir} (lower part)
in NaOH solution after 1 hour of irradiation.

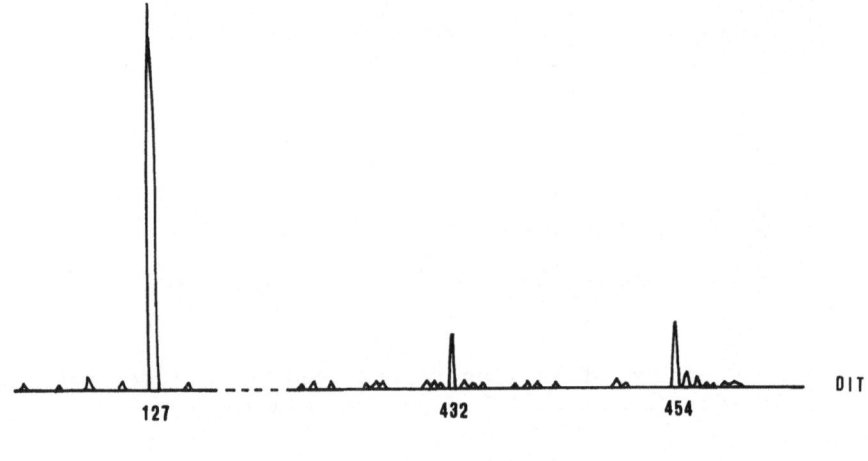

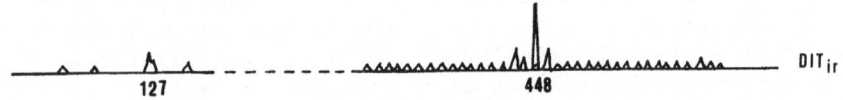

Fig. 12.   Mass spectrum of DIT (upper part) and DIT_{ir} (lower part)
in NaOH solution after 1 hour of irradiation.

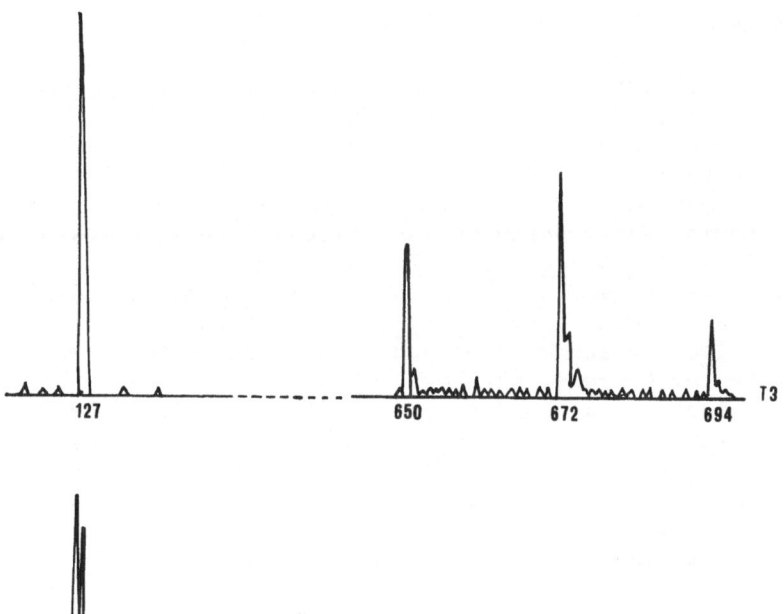

Fig. 13. Mass spectrum of T3 (upper part) and $T3_{ir}$ (lower part) in NaOH solution after 1 hour of irradiation.

Table 1. Molecular Weights of Analyzed Iodoamino Acids.

	MIT	T4	DIT	T3
Control	307	777	433	651
Irradiated	297	297	448	448

Molecular weights of analyzed iodoamino acids.

this instance a lower radiating energy should be strongly recommended. A second observation is relative to the formation of two unknown photoadducts, which, strangely enough, have the same molecular mass when originated from MIT and T4 (297 u.m.a.) or DIT and T3 (448 u.m.a.). The latter finding breaks the parallel behavior of $T4_{ir}$ – $T3_{ir}$ and MIT_{ir} – DIT_{ir}, as far as absorption spectra are concerned. The reason for this behavior is at the moment still unknown.

As for the last point it obviously needs further work. The problem is, at moment, under investigation.

REFERENCES

1. M. L. Wolbarsht, ed., "Laser Applications in Medicine and
    Biology," Academic Press, New York, Vol.L+4 (1971).
2. R. Pratesi and C. A. Sacchi, eds., "Laser in Photomedicine and
    Photobiology," Proc.Conf.Florence, Springer Verlag (1980).
3. H. J. Cahnmann, "The Thyroid and Biogenic Amines," Rall and
    Kopin, eds., North-Holland Elsevier Publ. Co. (1972).
4. A. Ehrl, Atomkernenergie, 22:272 (1974).
5. N. Autissier and A. M. Buthican, Ann.Pharm.Fr., 34:265 (1976).
6. E. Loh, J.Raman Spectroscopy, 3:327 (1975).
7. G. Salvatore and H. Edelhoch, "Hormonal Proteins and Peptides,"
    Academic Press, New York (1973).
8. G. Salvatore, J. B. Stanbury, and J. E. Rall, "The Thyroid
    Gland," M. Vischer, ed., Raven Press, New York, pp. 443-487
    (1980).
9. G. Palumbo and G. Ambrosio, Arch.Biochem.Biophys.Acta, 212:37-42
    (1981).
10. H. J. Cahnmann, J. Pommier, and J. Nunez, Proc.Nat.Acad.Sci.USA,
    74:5333-5335 (1977).

THE INFLUENCE OF OXYGEN-RADICALS ON THE MECHANISMS
OF CONFORMATIONAL CHANGES AND DENATURATION OF
HAEMOGLOBIN INDUCED BY UV IRRADIATION

H. Malak

Department of Physics
Academy of Agriculture
60-637 Poznan, Poland

INTRODUCTION

Photooxidizing properties of haemoglobin are stimulated by the presence of a haem group in this compound. Haem is a very good photosensitizer to the processes generating such oxygen species as $O_2^{\cdot}$, $^1O_2$, H_2O_2, $OH^{\cdot}$ (1-3).

The attempts to elucidate the mechanism of photooxidization of haemoglobin were grounded on two models of configuration of the bond in an oxyheme complex. In the first model of Pauling-type[4.5] the bond between oxygen and iron (II) is assumed to be covalent. On the base of this model Posnani et al.,[6] have suggested the possibility of singlet oxygen $^1O_2$ generated in Type II processes 1,2 to take part in photooxidization of haemoglobin (Fig. 1). The alternative model of Weis-type[7] taking the configuration of the oxygenhaem complex bond as $Fe^{3+}O^-$ (superroxoferrihaem) 8 has been assumed by Demma and Salhany[9,10]. They reported to have observed a photolysis of HbO_2 to MetHb with the formation of superoxideanions O_2^- due to the influence of light. The self-oxidation processes of haemoglobin also support the latter model[11-14].

In this paper I would like to present the results of the studies on the influence of oxygen and species on the mechanism of haemoglobin photooxidation in the Type I processes.

RESULTS AND DISCUSSION

The observations performed with the methods of optical molecular spectroscopy revealed the changes in conformation and denaturation of haemoglobin occurring in the solution of HbO_2 with a concentration

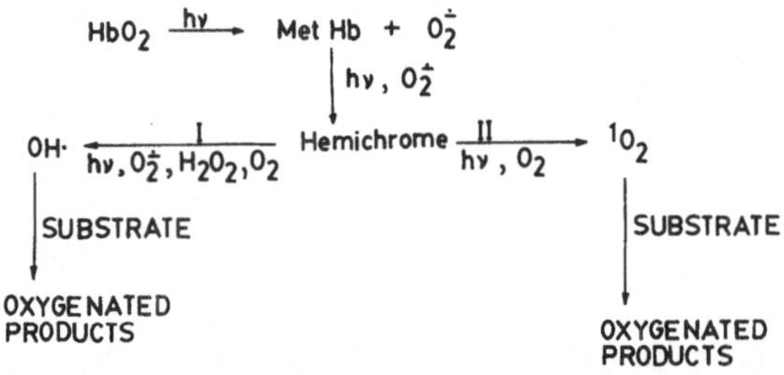

Fig. 1. Photooxidation mechanisms of haemoglobin.

10^{-5}M due to the UV irradiation with an intensity of 55 J. Figure 1 illustrates a scheme of these changes. The observed formation of MetHb conformation from HbO_2 conformation under the influence of light is analogous to the results reported by Posnani et al.,[6] and Demma and Salhany[9,10].

The intermediate haemichrome state between MetHb configuration and denaturation of haemoglobin was identified which is a very important result as this conformation was so far observed only in α and β subunits of haemoglobin [15,16]. The changes in number of $O_2^{\cdot}$ radicals corresponding to the conformational changes and denaturation of the irradiated haemoglobin was estimated by reduction of ferricytochrome c (Figure 2)[17].

The observed increase in number of $O_2^{\cdot}$ radicals in the conformation MetHb is connected with a photolysis of $Fe^{3+}O_2^{\cdot}$ bond in haemoglobin. Generation of $O_2^{\cdot}$ radicals in the haemichrome conformation in photooxidation processes of Type I results from a transfer of an electron from haem to the oxygen dissolved in the solution. On the other hand a decrease in number of $O_2^{\cdot}$ radicals observed for the irradiated solution (see Figure 2) may indicate the appearance of hydroxyl radicals OH: which are very strong oxidizers of haemoglobin. The process of formation of OH· radicals under the presence of $O_2^{\cdot}$ radicals and H_2O_2 was proposed by Heber and Weiss [18] to be:

$$O_2^{\cdot} + H_2O_2 \quad\text{------------}\quad OH\cdot + OH^- + O^2$$

Recent studies have proved the reaction to be highly efficient under the presence of small amounts of iron[19-21]. The presence of iron in the irradiated haemoglobin solutions is possible taking into regard the destruction of haem groups caused by reattack of $O_2^{\cdot}$ or H_2O_2,[15,22].

Hydrogen peroxide required for the reaction (1) to occur can be produced in dismutation of O_2 radicals[23]:

$$2H_2^+ + O_2^{\cdot} + O_2^{\cdot} \quad\text{------------}\quad 2H_2O_2$$

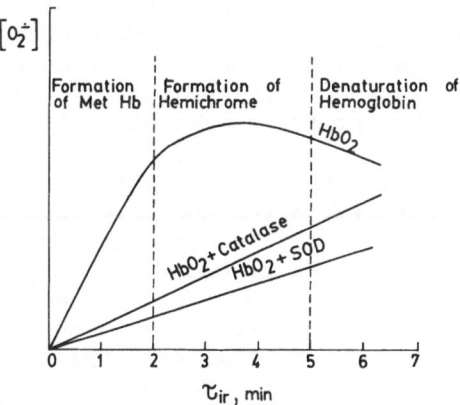

Fig. 2. Formation of superoxide anion $O_2^{\pm}$ in irradiated solutions of hemoglobin in the presence of superoxide dismutase SOD – 30 µg/ml and catalase 1200 units/ml.

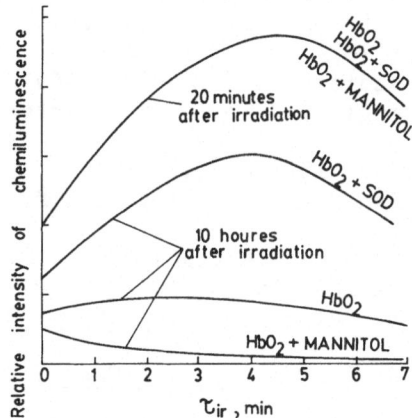

Fig. 3. Chemiluminescence activity of irradiated haemoglobin in the presence of SOD 30 µg/ml or Mannitol 10^{-3}M. Experiments were carried out 20 minutes or 10 hours after the irradiation.

With increasing pH of the solution the efficiency of this reaction increases spreading up the destruction of the irradiated haemoglobin.

Under the presence of superoxide dismutase – SOD (an inhibitor of $O_2^{\pm}$) or catalase (an inhibitor of H_2O_2). in the irradiated solutions of haemoglobin, a considerable inhibition of these photoprocesses was observed (Figure 2), which points to a dominant role of $O_2^{\pm}$ and H_2O_2 in the photooxidation of HbO_2.

Haemoglobin irradiated with UV radiation also changes its chemiluminescence properties in the reaction with H_2O_2. The MetHb and
haemichrome conformations formed under the influence of light, induced much higher intensity of chemiluminescence (I_{Cl}) in the reaction with H_2O_2 than that of the native HBO_2 (Figure 3).

No influence of SOD and Manniton inhibitors of OH· on I_{Cl} of
haemoglobin was found directly after the irradiation. This influence
was observed however after several hours since the stop of the irradiation. Then we could find a decrease in I_{Cl} of the irradiated
haemoglobin under the presence of Mannitol and no such changes under
the presence of SOD (Fig. 3). These facts point to the essential
role of $O_2^{\overline{2}}$ in dark processes following the irradiation in self-
oxidation of haemoglobin and destruction of a haem groups.

REFERENCES

1. C. S. Foote, Oxygen and Oxy-Radicals in Chemistry and Biology,
    pp. 425-433, Academic Press, New York (1981).
2. C. S. Foote, in: "Pathology of Oxygen," pp. 21-37, Academic
    Press, New York (1982).
3. J. D. Spikes, Adv. Radiat. Biol., 3-29 (1969).
4. L. Pauling, in: "Hemoglobin" pp. 57-65, Butterworth Co.
    Publishers, Ltd., London (1949).
5. L. Pauling, Nature 203, 182-183 (1964).
6. L. D. Possnani, R. Banerjee, C. Balny, and P. Dauzon, Nature
    226, 861-862 (1970).
7. J. J. Weiss, Nature 202, 83-84 (1964).
8. A. J. Thomson, Nature 265, 15-16 (1977).
9. L. S. Demma and J. M. Salhany, J.Biol.Chem., 252, 1226-1230
    (1977).
10. L. S. Demma and J. M. Salhany, J.Biol.Chem., 254, 4532-4535
    (1979).
11. R. Wever, B. Oudega and B. F. Van Gelder, Biochem.Biophys.Acta
    302, 475-478 (1973).
12. H. P. Misra and I. Fridovich, J.Biol.Chem., 247, 6960-6962
    (1972).
13. W. J. Wallace, J. C. Maxwell, and W. S. Cauhey, Biochem.Biophys.
    Res.Commun., 57, 1104-1110 (1974).
14. M. Brunori, G. Falcioni, E. Fioretti, B. Giardina, and G.
    Rotilio, Eur.J.Biochem., 53, 99-104 (1975).
15. A. Tomoda, K. Sugimotco, M. Suhara, M. Tekeshita, Y. Yoneyama,
    Bioch.J., 171, 329-335 (1978).
16. E. Rachmlewitz and J. White, Nature New.Biol., 241, 115-117
    (1973).
17. W. Koppenol, K. Van Buuren, J. Butler and R. Braams, Biochem.
    Biophys Acta 449, 157-168 (1976).
18. F. Haber and J. Weiss, Proc.Roy.Soc. London, A., 147, 332-351
    (1934).

19. D. Rowley and B. Halliwell, Febs Lett 138, 33-36 (1982).
20. R. Richmond, B. Halliwell and J. Chauhan and A. Darbre,
        Anal.Biochem., 118:328-335 (1981).
21. J. McCord and E. Day, Febs Lett 86: 139-142 (1978).
22. H. Sutton, P. Roberts and C. Winwerbourn, Biochem.J.,
        155:503-510 (1976).
23. J. McCord and I. Fridovich, Photochem.Photobiol., 17, 115-123
        (1973).

ASPECTS OF DERMATOLOGICAL PHOTOPATHOLOGY

I. A. Magnus

Institute of Dermatology
London, UK

INTRODUCTION

Dermatological photobiology encompasses manifold disparate areas of knowledge; these were surveyed some 8 years ago by the present author[1] and summarized in a diagram. See Fig. 1 for such a survey in summary.

Much progress has been made in some departments since than and progress can perhaps be foreseen in some others. For instance recently physicists have tackled the optical properties of the skin with interesting and plausible results, physical chemists have been studying the basic photochemical properties of photosensitizers such as porphyrins and furocoumarins, and it looks possible that more career immunologists may be enticed to enter the photobiology realm.

THE RADIATION

In the case of skin, this will of course be both solar and artificial. The general features of solar power and its spectral distribution are well known but very much more routine day-to-day data are needed. This is outside detailed consideration here and is or should be within the provinces of national meteorology services. It is a pity that weather bureaux do not make available forecasts of at least records of solar ultraviolet performance; it has to be left to academics, mostly those working with the botanists and dermatologists, to gather.

The development of artificial radiation, especially from the laser is of course the concern of lighting engineering an technology;

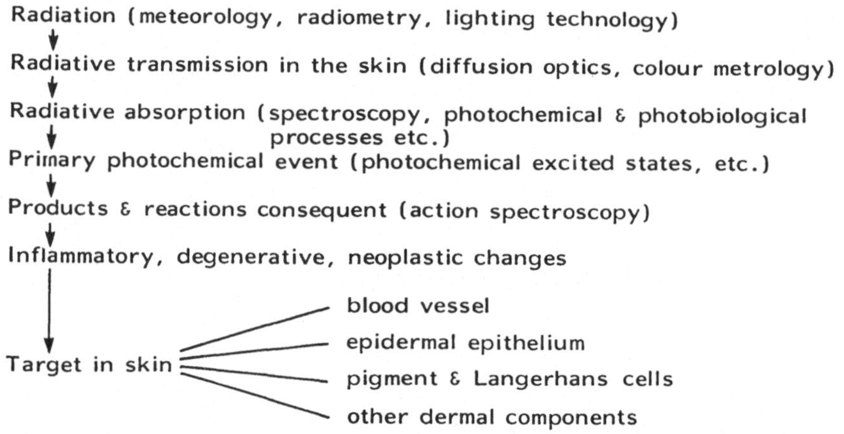

Radiation (meteorology, radiometry, lighting technology)
↓
Radiative transmission in the skin (diffusion optics, colour metrology)
↓
Radiative absorption (spectroscopy, photochemical & photobiological
↓   processes etc.)
Primary photochemical event (photochemical excited states, etc.)
↓
Products & reactions consequent (action spectroscopy)
↓
Inflammatory, degenerative, neoplastic changes

Target in skin
— blood vessel
— epidermal epithelium
— pigment & Langerhans cells
— other dermal components

Fig. 1. A flow-chart illustrating some of the various aspects of
         dermatological photobiology and photopathology (after
         Magnus, 1976).

this has been continuously expanding and developing in the last years
especially in the laser domain. There are break-through-promising
devices like the free-electron laser for instance. But alas, by far
the most work in the field of clinical and investigative dermatology
over the last decades has been done with conventional gas discharge
lamps of wide spectral output spectrally modified with filters or
monochromators. Thus our optical technology continues to be not far
from that of the era of Galrileo, Huygens and Newton, viz. that of
three centuries or so ago. Only very recently have lasers been used
in dermatological research and clinical investigative work; their
application in dermatological therapy is confined essentially to
cosmetic surgery, see for instance Rosen[2]. Phototherapy with
porphyrin for cancer also may employ argon ion or dye lasers[3,4].
The greatest disincentive to using the laser of course has been its
capital cost, especially in cosmetic surgery where knife, X-radiation
and various chemical approaches are much less expensive and have been
long used. To most of us the unique property of the laser that has
application in dermatology is its unexamined monochromaticity. Hence
the argon ion laser with its green emission, which would be strongly
absorbed by hemoglobin seemingly, suits vascular lesion ablation.
Thus good results in so-called portwine lesions or stains are
claimed[2]. The very recent claims for the argon-fluoride excimer
laser (193 nm) for ablative photodecomposition will be watched with
interest.

OPTICAL PROPERTIES OF THE SKIN

     The next item to consider is radiation transport through the
skin to the site where a photochemical reaction takes place. Over

the years measurements have been made from time to time of skin
transmission, reflection and remission by a few workers. Much more
work of this kind is needed. Until the last 3-4 years no attempts at
theoretical models of the optical properties of the skin have been
advanced. But things have changed; a first model was suggested by
Anderson[5]. The most recent treatment is due to Diffey[6], which we
will consider briefly here.

Diffey uses a simplification based on the general transport
equation of radiation through matter and the Kubelka-Munk approximation treatment for turbid media in which radiation is supposed to
travel as 2 diffuse fluxes moving in opposing directions. But an
important feature of his ideas is that he considers as well a relatively stronger forward directed component of the flux with low angel
scattering. This was based on the notion that melanin granules will
act as independent scattering centers and, being approximately 1μm in
diameter, would lead to Mie type scattering predominating. This
would result in a forward passing beam, "semicollimated" and with low
angle scatter as being much the predominant effect to take account of
in "skin optics". Previous models of skin optics transmission have
accordingly led to low calculated values; Diffey concludes that

1. Scattering is more important than absorption in the stratum
    corneum optical properties.
2. Mie rather than Rayleigh type scattering results in scattering
    being largely independent of wavelength in the range of 250-
    400 nm.
3. Absorption coefficients in the stratum corneum peak at about
    280 nm, then falls rapidly with longer wavelengths to about
    310 nm, thereafter falling much more gradually.

Diffey has pointed out that the experimental values reported in
the literature by Everett et al.[7] and by de Gruijl[8] for total as
against direct UVR transmission through stratum corneum or epidermis
agree well with his calculated values and support his theories. By
contrast the models of Anderson et al.[9] and of Wan et al.[10]
relying solely on Kubelka-Munk theory, underestimate scattering and
find absorption coefficients greater than scattering coefficients.
See Figs. 2 and 3 which summarize some aspects of Diffey's results.

PHOTOCHEMISTRY

This of course separates the photobiological from the non-
photobiological. The origins of the response in the skin are in
primary photochemical events. Laser induced photochemical changes
may presumably entail multiphoton absorption rather than classical
Stark-Einstein rules; this needs to be sorted out. The photochemical
excited states of some of the photosensitizers important in human
photodermatoses are now being examined, e.g. porphyrins, psoralens,

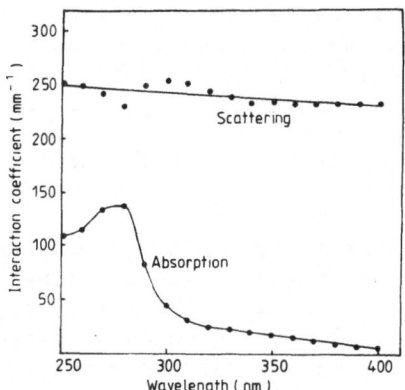

Fig. 2. Diffey's calculated values for scattering and absorption
         coefficients in human stratum corneum with respect to
         wavelength change (from Diffey, 1983).

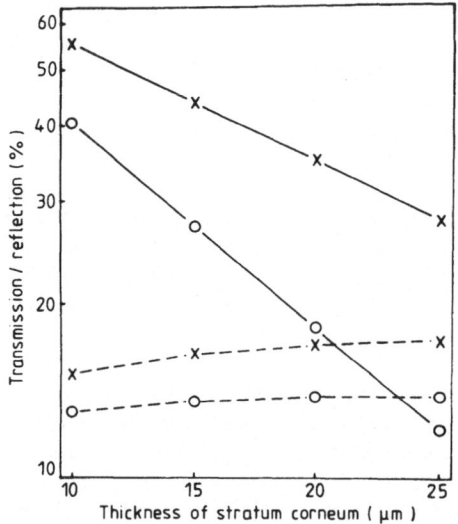

Fig. 3. Diffey's calculated relative values for % transmission/
         reflection for 300 nm (o) and 360 nm (x) UVR with respect
         to change of stratum corneum thickness (from Diffey, 1983).

certain drugs and materials used in sun-barrier preparations e.g.
PABA and cinnamates. The site of these happenings may be in the
cytosol of cells, on cell membranes, or intracellular organelles such
as mitochondria or lysomes[11,12].

INFLAMMATORY REACTION DUE TO LIGHT OR UVR

The photochemical reaction is followed in the case of acute
vascular inflammatory responses, so it is believed, by the release of
chemical mediators which lead to widening of the inner bore of the
vessels, i.e. vasodilation, and to an increase in permeability of the
walls of these vessels. These seem to be 2 different pathophysio-
logical changes effected by separate mechanisms i.e. different
mediators. The idea of chemical mediators of inflammation is old and
probably first put on a form basis in a famous lecture by Dale[13].
Research in this domain, a very active industry, had led to the
discovery of many new chemical mediators, e.g. prostacyclins and
leucotrienes. The result is at present perplexing and obscure. A
decade or so ago when prostaglandins first came into prominence they
seem to be candidates for resolving the basic problems of the inflammatory responses. It was simplistic in expecting quick neat solutions, the mechanism of the most commonplace photobiological reaction, normal sunburn, has not yielded itself. So although much
effort has been expanded and much new data collected, results may be
said to be disappointing in that no overall pattern of mechanism has
emerged.

What we can say for the moment is that prostaglandins may appear
during the earlier stages of development of UV-B and UV-A erythema
but the amount observed and timing of their release, rise and disappearance do not fit in with the degree or time course of the
erythema reaction itself; furthermore inhibitors of cyclo-oxyhenase,
such as indomethacin, merely reduce erythema but do not abolish it.
The role of bradykinin and other so-called kinins is not clear; their
effect may be potentiated by prostaglandins. Again the part played
by histamine in the normal sunburn, UV-B and UV-A erythema is
obscure. However, it is still a possible candidate for a role in
some immediate type reactions, i.e. those appearing within about half
an hour of exposure. A role for histamine in solar urticaria seems
more feasible in some instances of this disorder, but not in all.

THE FINAL PRODUCT-LESION AND MEASURING IT

On the morphology and type of lesion depends how it can be
measured or fitted into some kind of scale. The final product-lesion
can vary from erythema to tumor, from histamine to Vitamin D. We
will look at erythema first as it is much the most familiar photobiological reaction. Unfortunately reliable methods for assessing or
measuring it that can be used generally be all or most workers as a
routine have not been established. Some individual workers have
elaborated methods that appear to be useful in their own hands but
not in others[14,15]. Many have used instruments, varying in degree
of elaborateness, for skin reflectance spectroscopy[16,17]. Where
the apparatus has appeared to be relatively simple to use it has

often in my experience given unreproducible results or is too insensitive in measuring. I have found other instruments appear to be
sensitive enough but too difficult to use generally. In the spectral
reflectance equipment of Dawson an "index of vasodilation" and "index
for pigmentation" have been proposed[18]. His apparatus would seem
to be easy to use and his suggestions seem an improvement in making
routine measurements. But "corrections" have to be introduced to
avoid interference between measuring the 2 indexes. However, this
approach seems to be a first step in the right direction of practical
quantitative assessment of erythema and pigmentation.

In the meantime the matter of measuring erythema and/or pigmentation as practiced by most other workers is not entirely satisfactory. Subjective grading is by far the commonest procedure, e.g.
using an ordinal scale of 0, ±, +, ++, +++. This seems to be of some
use given a trained observer. Basic type reflectance spectrophotometric assessments with cut-off filters and simple photocell have
been used by a number of workers and appear to have some use in
certain kinds of field work, but, the indications are that such
equipment is not sensitive enough for following small charges. Thus
strong and weak erythema reactions as judged by the human eye may
show obvious differences in depth of color, whereas the values
of the reflectance measurements are slight and not significantly
different.

Other non invasive methods of assessing erythema would be by
thermal measurement; a number of instruments are available, but none
seem suitable for general routine use. These include blood flow
measurement using laser Doppler-shift velocimetry or the technique
of photoplethysmography. These last 2 methods may have some application[19,20].

THE ACUTE VASCULAR SKIN REACTION THAT IS EASILY MEASURABLE

Only the skin weal, or hives, or solar urticaria can be measured
easily and reproducibly. This is of rather limited practical use but
of considerable theoretical interest. One can probably obtain more
quantitative information from wealing lesions than any other dermatological response.

Useful dose-response curves, time course data and action
spectrum studies can be produced. The dose-response tests have
especially proved useful in doing double-blind therapeutic trials.
In such trials it can be shown that, as may be expected, antihistamines have a suppressive effect of the weal in most patients.
But the degree of suppression with standard H_1 antihistamine seems
disappointing, viz. amounting to no more than a reduction of mean
lesion area by a factor of about 2 (Fig.4). This is of little or no
clinical use. H_2 histamines given alone have no definite effect and

H_1 and H_2 combined have no more definite effect on the weal size than the H_1 drug taken alone. H_1 and H_2 drugs combined however did seem to have some effect in reducing flare, the transient axon reflex erythema commonly associated with weal or hives responses.

Time course experiments in 8 patients have shown the phenomenon of a bi-phasic response; the first phase peaks at about 20 minutes, then weal size subsides a little to increase again at about 40 minutes before the weal subsides completely at about 60 minutes. This occurred in all 8 patients studied, whether given antihistamines or placebo (Fig. 5).

The appearance of histamine in the blood has some resemblance in time course to that if the wealing reaction[21].

Action spectrum studies are also easily measured in solar urticaria. The results of such work , however, are usually not easy to interpret with respect to discovering the nature of the photo-sensitizer; in other words one should have strong suspicions as to what photosensitizer is and knowledge of its absorption spectrum before expecting action spectrum studies to be elucidating. Unfor-tunately few biological absorption spectra seem to be sufficiently characteristic for an individual substance to be suspected so in the case of solar urticaria the photosensitizers cannot be identified. The one exception is in the rare instances of patients with solar urticaria due to porphyria. In these cases the action spectrum for

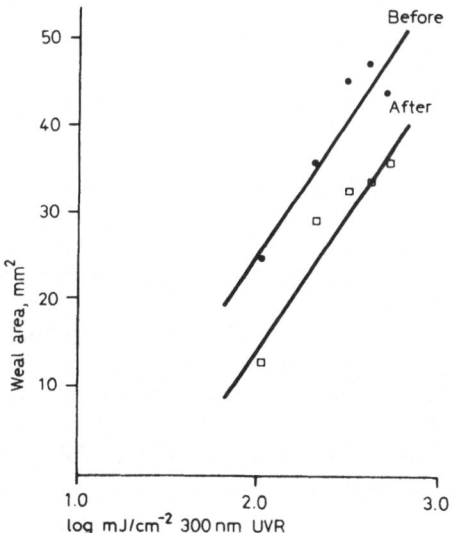

Fig. 4. Mean dose response data in 8 patients with solar urticaria provoked by 300 nm UVR; before and after H_1 antihistamine (from Michell et al., 1980).

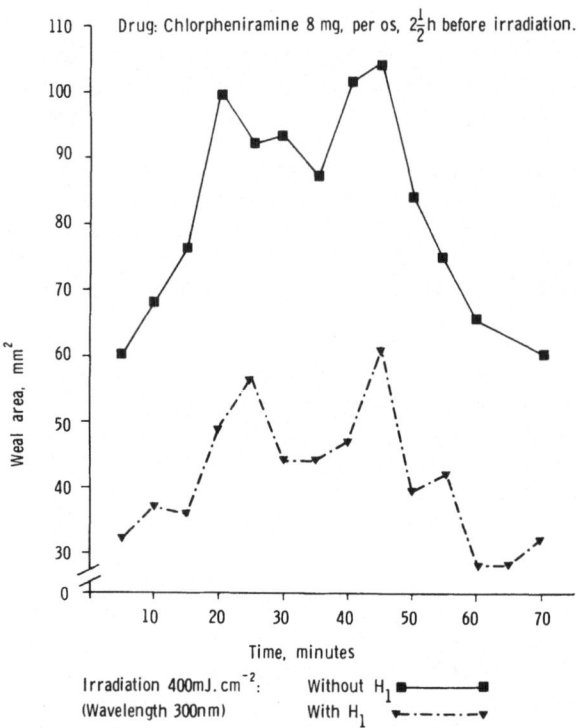

Fig. 5. Time course of solar urticaria weal size showing biphasic
         response (data from Michell et al., 1980).

solar urticaria has resembled the absorption spectrum for porphyrin,
namely with a characteristic maximum of activity and absorption in
the 400 nm region (Fig.6).

QUANTITATIVE HISTOLOGICAL DATA

     A change that seems to develop parallel with the normal sunburn
reaction and with responses to UV-A plus psoralens is the so-called
"sunburn cell" (SBC). This is a well known cellular change, usually
known as "apoptosis", provoked in many tissues other than skin by
many processes or stimuli[22]. In the skin apoptotic cells or (SBCs)
are easily counted and these can be used to construct the usual
quantitative data viz. action spectra, time courses, dose-response
curves. Young and Magnus[23,24] have used the SBC to compare the
photosensitizing effect of certain psoralens.

QUANTITATIVE DATA IN EXPERIMENTAL SKIN PHOTOTUMORIGENESIS

     Data from experimental phototumorigenesis (or photocarcino-
genesis) is quite simply collected by merely counting the number of

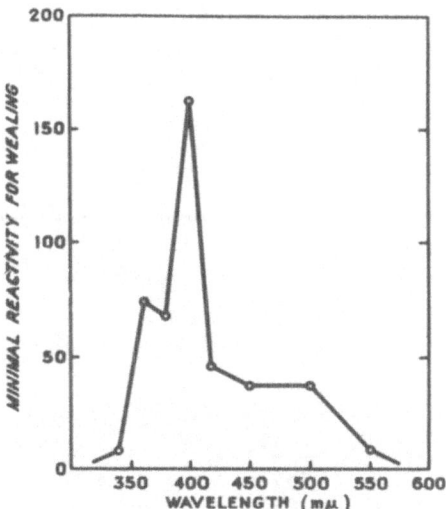

Fig. 6. Erythropoietic protoporphyria action spectrum for wealing; note peak response at 400 nm as for absorption spectrum of protoporphyrin in most polar solvents (from Magnus et al., Lancet:448 (1961)).

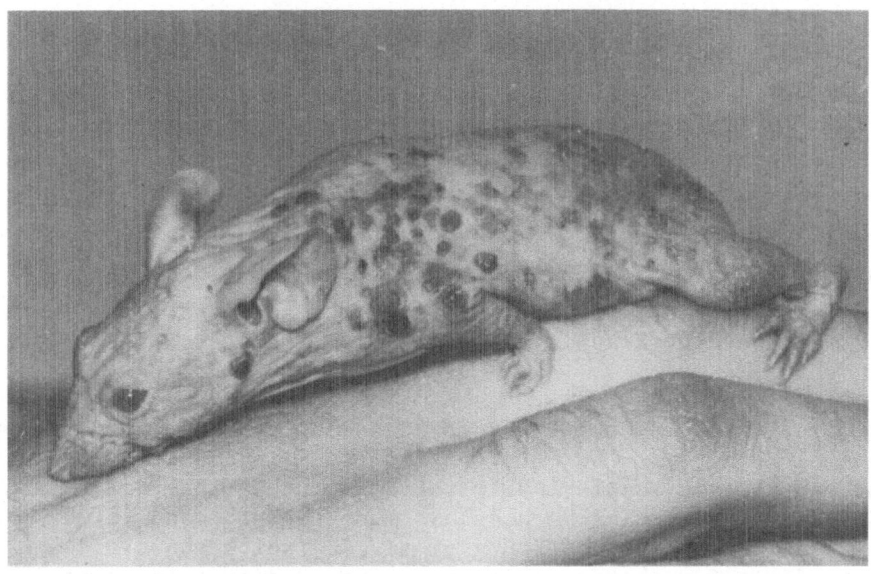

Fig. 7. Example of tumors provoked after 34 weeks repeated exposures to solar simulated UVR. Hairless mice, skin painted with 0.01% 8-methoxypsoralen in arachis oil base (data from Dr. Anthony Young, 1983).

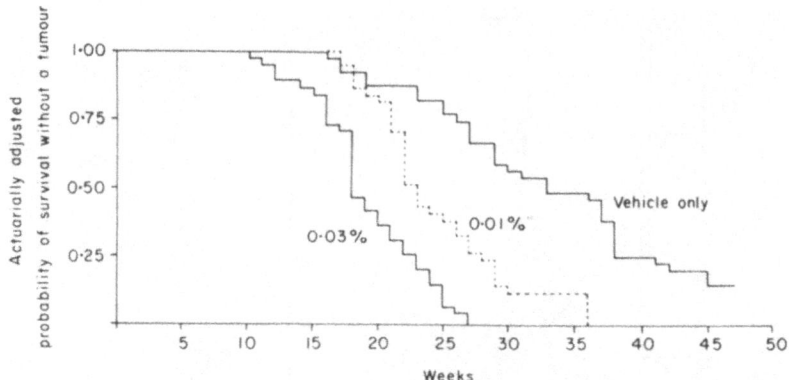

Fig. 8. Actuarial data on phototumorigenic potential of 8-methoxy-
         psoralen in hairless mice exposed to solar simulated UVR
         (from Young et al., 1983). Note dose response effect
         between the 0.01 and 0.03% psoralen, and the vehicle without
         the psoralen.

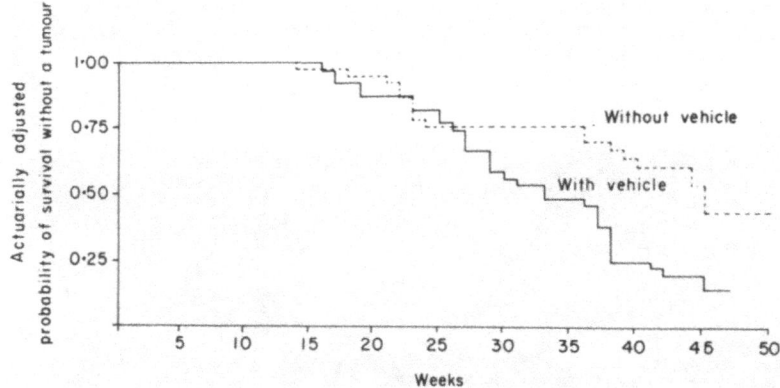

Fig. 9. Actuarial data on phototumorigenesis potential of solar
         simulated UVR in hairless mice. To show phototumorigenesis
         enhanced by arachis oil base (from Young et al., 1983).

macroscopic lesions. Usually it is necessary to map the tumors on
each animal (Fig. 7). The point of issue and interest is more how to
present and analyze statistically the data rather than how to collect
it. The use of actuarial methods has much to recommend it[25]. Fig.
8 shows results obtained by Young et al.[26] which shows a dose-
response effect for phototumorigenesis from 2 different concentrations of 8-methoxypsoralen applied to the skin of hairless mice,
which were then subjected to exposure with a filtered xenon arc
source to give solar simulated radiation. Fig. 9 shows that even the
vehicle (arachis oil plus isopropyl myristate) has a phototumorigenic
effect greater than solar simulated radiation given alone. This is

of considerable interest to the cosmetic industry which has very wide
uses for arachis oil.

ENVOI

In this paper some of the differing disciplines encountered in
studying dermatological photobiology are described. It is the very
variety of these disciplines which is so challenging and engaging a
feature of this work.

Acknowledgements

We thank authors and holders of copyright material for per-
mission to reproduce certain figures in this article. They are Dr.
Brian Diffey and the Institute of Physics for Figs. 2 and 3, S.
Karger, A.G. Basel and the Editors of "Dermatologica" for Fig. 4,
the Editors of the "Lancet" for Fig. 6 and of the "British Journal
of Dermatology" for Figs. 8 and 9. Dr. Anthony Young kindly supplied
Fig. 7.

REFERENCES

1. I. A. Magnus, "Dermatological Photobiology," Blackwell, Oxford
    (1976).
2. K. A. Arndt, J. M. Noe, and S. Rosen, eds., "Cutaneous Laser
    Therapy," J. Wiley & Sons, Chichester (1983).
3. H. Kato, C. Konaka, J. Ono, Y. Matsushima, K. Nishimiya, J. Lay,
    H. Sawa, H. Shinohara, T. Saito, K. Kinoshita, T. Tomono, M.
    Aida, and Y. Hayata, Effectiveness of HPD and radiation
    therapy in lung cancer, in: "Porphyrin Photosensitization,"
    D. Kessel and T.J. Dougherty, eds., Plenum, New York (1972).
4, R. G. Vincent, T. J. Dougherty, U. Rao, and D. R. Doiron, Hemato-
    porphyrin derivative in the diagnosis and treatment of lung
    cancer, in: "Porphyrin Photosensitization," D. Kessel and
    T.J. Dougherty, eds., Plenum, New York (1972).
5. R. R. Anderson and J. A. Parrish, Optical properties of human
    skin, in: "The Science of Photomedicine," J.D. Regan and J.A.
    Parrish, eds., Plenum, New York (1982).
6. B. L. Diffey, Phys.Med.Biol., 25:647 (1983).
7. M. A. Everett, E. Yeargars, R. M. Sayre, and R. L. Olson, Photo
    chem.Photobiol., 5:533 (1966).
8. F. R. de Gruijl, "The Dose-Response Relationship for UV-Tumori-
    genesis," PhD Thesis, State University of Utrecht (1982).
9. R. R. Anderson, J. Hu, and J. A. Parrish, Proceedings of the
    Symposium on Bioengineering of the Skin, London MTB Press
    (1980).

10. S. Wan, R. R. Anderson, and J. A. Parrish, Photochem.Photobiol., 34:493 (1981).

11. A. C. Allison, I. A. Magnus, and M. R. Young, Nature, 209:874 (1966).

12. S. Sandberg, J. Glette, G. Hopen, C. Ola Solberg, and I. Romslo, Photochem.Photobiol., 34:471 (1981).

13. H. H. Dale, Bull.Johns Hopkins Hospital, LIII:297 (1934).

14. K. W. Hausser and W. Vahle, Wiss.Veröff.Siemens-Konzern, 6:101 (1927).

15. D. Berger, F. Urbach, and R. E. Davies, The action spectrum of erythema induced by ultraviolet radiation: Preliminary Report, in: "XIII Congressus Internationalis Dermatologiae," E. Jadassohn and C.G. Schirren, eds., 2:1112, Springer-Verlag, Berlin (1968).

16. D. Langen, Strahlentherapie, 63:142 (1938).

17. F. Daniels and J. D. Imbrie, J.Invest.Derm., 30:295 (1958).

18. J. W. Feather, J. B. Dawson, D. J. Barker, A theoretical and experiemntal study of the optical properties of the skin in vivo, Proceedings of the Symposium on "Bioengineering of the Skin," MTB Press, London (1980).

19. E. Tur, R. H. Guy, M. Tur, and H. I. Maibach, J.Invest.Derm., 80:499 (1983).

20. A. V. J. Challoner, Photoelectric plethysmography for measuring cutaneous blood flow, in: "Non-invasive Physiological Measurements," P. Rolfe, ed., Academic Press, London (1979).

21. P. Michell, J. L. M. Hawk, A. Shafrir, M. F. Corbett, and I. A. Magnus, Dermatologica, 160:1C8 (1980).

22. E. Duvall and A. H. Wyllie, Hospital Update, 9:297 (1983).

23. A. R. Young and I. A. Magnus, Br.J.Dermatol., 104:541 (1981).

24. A. R. Young and I. A. Magnus, J.Invest.Derm., 79:218 (1982).

25. R. Peto, M. C. Pike, N. E. Day, R. G. Gray, P. N. Lee, S. Parrish, J. Peto, S. Richards, and J. Wahrendorf, Guidelines for simple, sensitive significance tests for carcinogenic effects in long-term animal experiments, in: "Annexe to International Agency for Research on Cancer Supplement 2, IARC Monographs on the evaluation of the carcinogenic risk of chemicals to humans," IARC, Lyon (1980).

26. A. R. Young, I. A. Magnus, A. C. Davies, and N. P. Smith, Br.J. Dermatol., 108:507 (1983).

PHOTOTHERAPY OF NEONATAL HYPERBILIRUBINEMIA: PHYSICAL ASPECTS

R. Pratesi

Istituto di Elettronica Quantistica del CNR,
and Istituto di Fisica Superiore dell'Università
Firenze, Italy

1. FOREWORD

The availability of coherent optical sources (lasers) has permitted a rapid development of photosurgical and photocoagulative techniques during the last decade [1]. Photomedicine [2] appeared another very interesting field for laser applications in view of the potential progresses in the already existing phototherapeutical procedures, and for the importance of laser techniques in the photobiological field [3]. The laser photochemotherapy of tumors [4] represents a noticeable example of this prevision.

Phototherapy of hyperbilirubinemia in the newborn is a widely accepted procedure to lower bilirubin concentration in plasma. However, despite intense clinical and experimental efforts, the clinical procedure is still largely empirical and the mechanism responsible for the clearance of bilirubin in light-exposed babies is not yet clearly defined [5].

Visible lasers are very convenient sources of narrow-band, intense radiation to be used for more detailed studies of bilirubin photochemistry "in vitro" and "in vivo". For instance, the small spectral differences among the various bilirubin photoproducts can be more easily shown using nearly-spaced laser lines or frequency tunable lasers.

In 1980 an investigation was promoted by the CNR Special Project on Laser Applications in Medicine to explore the utilization of longer wavelength, narrow-spectral regions in the phototherapy of hyperbilirubinemia in order to reduce the potential side-effects associated with the near ultraviolet, violet-blue radiation emitted

by fluorescent lamps used for phototherapy [6]. Preliminary observations showed that laser photolysis of bilirubin "in vitro" could
be efficiently achieved also with the 514 nm green laser line [7].
Confirmation of this result with narrow-spectrum fluorescent green
lamps "in vitro" led to clinical comparison of green and white
lights. A greater efficiency of green lamps than white fluorescent
lamps was reported [8,9]. Further research was then devoted to
compare green lamps with the "special" blue lamps [10], and to understand the possible mechanisms responsible for the observed and the
unexpected good efficiency of green light.

In this paper we present a general introduction to the physical
aspects of phototherapy (PT) of hyperbilirubinemia (HBR), and report
some recent results obtained by the research group in Florence
[7-11].

2.   INTRODUCTION

Hyperbilirubinemia in neonates is more commonly known as jaundice
of the newborn. It is estimated that 50% of all infants develop at
least mild jaundice during the first week of life, and that about
10% of neonates have HBR of sufficient severity to require therapy
[12].

Prior to the 1940s, when exchange transfusion was introduced
as treatment, bilirubin (BR) encephalopathy was a leading cause of
neonatal death and cerebral palsy.

BR is formed in the organism mainly by the degradation of hemoglobin (Hb) released into the blood plasma from decomposing erythro-
cytes. Normal BR concentration in newborns rarely exceeds 1.5 to
2.0 mg/dl but when concentrations rise to 5 – 6 mg/dl, a quantity of
the pigment is transferred to the skin to produce the yellowish skin
color known as jaundice. This visible indication is the symptom of
HBR.

The removal of BR from the organism is to a large extent complicated by the insolubility of BR in aqueous media, so that BR is
transported in the organism being bound in water-soluble complexes.
In particular, BR circulates in the blood channels in a complex with
serum albumin (SA), and with participation of liver cells is removed
to the common bile ducts mainly in the form of esters with one or
two glucuronic acid molecules. If BR plasma concentration reaches
10 mg/dl or above in low-birth-weight infants, or 15 mg/dl or more
in full-term infants there is the danger of BR concentration then
exceeding albumin-binding capacity, and free, unconjugated BR circulating in the plasma, passing through the blood-brain barrier, and
causing neurological damage.

In 1956 an observant ward sister accidently discovered that irradiation of jaundiced infants with sunlight or artificial light decreased the concentration of BR in blood. Because of the high mortality rate associated with exchange transfusion, and of the perception that light at moderate intensity levels cannot be harmful to infants, the new form of light therapy rapidly gained acceptance. Today phototherapy (PT) is widely used for treatment of severe neonatal jaundice, despite a lack of understanding concerning its mechanism of action or potential toxicity. While PT of HBR has been used for 25 years, the treatment is still largely empirical, and there are multiple unknown factors.

PT of neonatal jaundice is administered as a whole body irradiation usually by banks of 8 to 18 fluorescent lamps (20 W) adjacent to each other with some portion of the radiation from lamp emission within the visible spectral region between 400 and 500 nm. Exposure to an irradiance of $\sim$ 1 mW/cm^2 in this range will cause a visual fading of the yellowish skin color within a few hours accompanied by a reduction of plasma BR.

2.1 Structure of Bilirubin

BR constitutes a set of structural isomers of a tetrapyrrole compound having the general formula shown in Fig. 1.

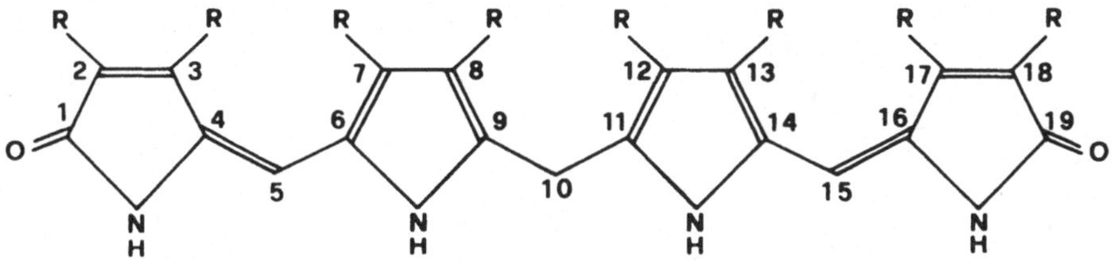

Fig. 1

where R denotes one of the following groups: Methyl (Me); Vinyl (Vin), or carboxyethyl. Mostly one of the structural isomers of BR is formed in the organism, i.e., BR IXα.

Cis-trans isomerism in the BR molecule is possible because the positions of the substituents relative to the two enamide C = C bonds can be different (Figure 2).

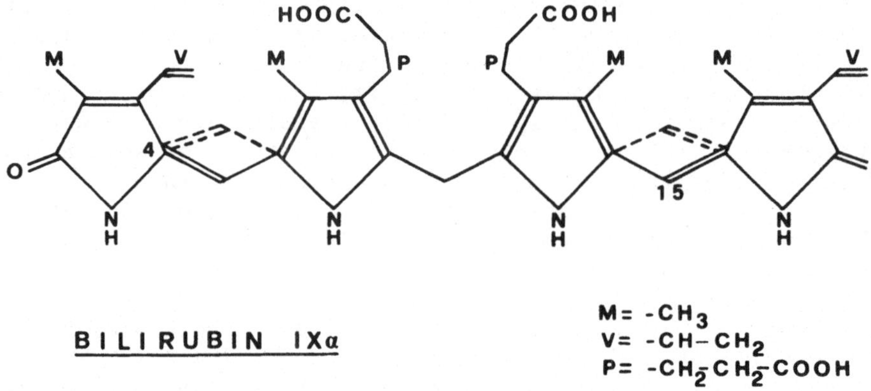

BILIRUBIN IXα

M= -CH$_3$
V= -CH–CH$_2$
P= -CH$_2$CH$_2$-COOH

Fig. 2

The Z-configuration of the two meso C-4 - C-5, and C-15 - C-16 double bonds allows the molecule to adopt the preferred highly hydrogen-bonded conformation (5Z, 15Z-BR, or simply Z,Z-BR) (Fig. 3)

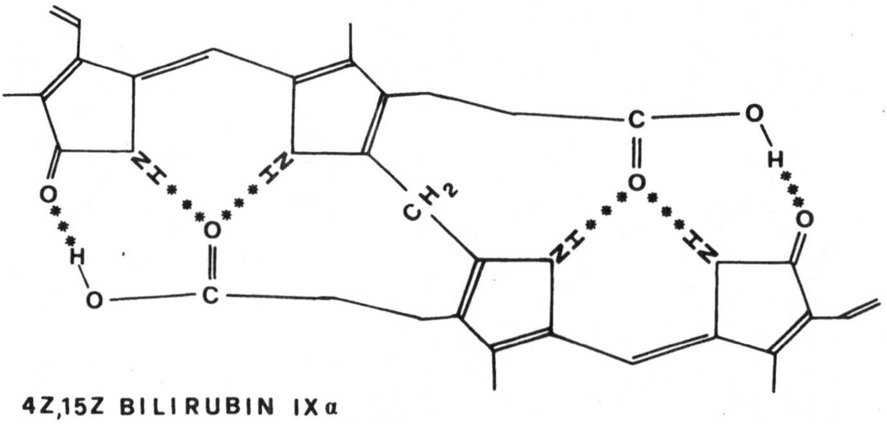

4Z,15Z BILIRUBIN IXα

Fig. 3

involving the propionic acid residues, the idrogen atoms of the pyrrole nitrogen atoms and the endocyclic carbonyl groups. X-ray diffraction data have shown that the Z,Z conformation of the BR molecule is planar. Furthermore the Z,Z-BR molecule is chiral, and its rigidity imparts to BR an optical activity. The Z,Z form explains the majority of the chemical and physico-chemical properties of BR, including the non-polar, water-insoluble character of BR.

3. HISTORICAL DEVELOPMENT OF PHOTOTHERAPY

The primary process in the PT doubtless involves light absorption by BR, and the primary question for the past 25 years has been how does light act to lower serum BR levels. Since 1938 it has been known [13] that BR is susceptible to photodestruction. Upon exposure to light BR transforms into several kinds of photoproducts, whose composition and properties have been studied throughout the past years.

3.1 Photooxidation of BR

Until 1978 the autooxidative process has been believed to be the main mechanism responsible for PT, although several experimental observations were not compatible with it. The proposed mechanism of the photochemical oxidation of BR includes the following stages [14].

The first stage is the photoexcitation of BR to the excited singlet state, and, as a result of intercombination transition, to the triplet state:

$$BR + h\nu \rightarrow {}^{1}BR^{*}; \quad {}^{1}BR^{*} \rightarrow BR.$$

The half-life of the state (9 µs) is entirely sufficient for the transfer of the triplet energy of the molecular oxygen. The singlet oxygen formed under these conditions can react in its turn with a BR molecule in the ground-state, oxidying it (self-sensitized reaction):

$$^{3}BR^{*} + O_{2}(^{3}\Sigma_{g}^{-}) \rightarrow BR + O_{2}(^{1}\Delta_{g})$$

$$O_{2}(^{1}\Delta_{g}) + BR \rightarrow BR \text{ oxidation products.}$$

The photooxidation of BR is in general detected spectrophotometrically by monitoring the decrease of the absorbance at the BR absorption maximum (the photoproducts are, in fact, transparent in the blue-green spectral interval (see Figure 4).

3.2 Configurational Photoisomerization of BR

The decrease of serum BR associated with PT was thought to be due to oxidative destruction of the pigment. However, substantial quantities of unconjugated BR are found in the bile when PT is administered. This fact, and the failure to find much of the expected photooxidation products in excreta of infants or experimental animals after PT, led to the abandonment of the idea that photodegradation is the major pathway. In the absence of light, excretion of BR by the liver is insignificant.

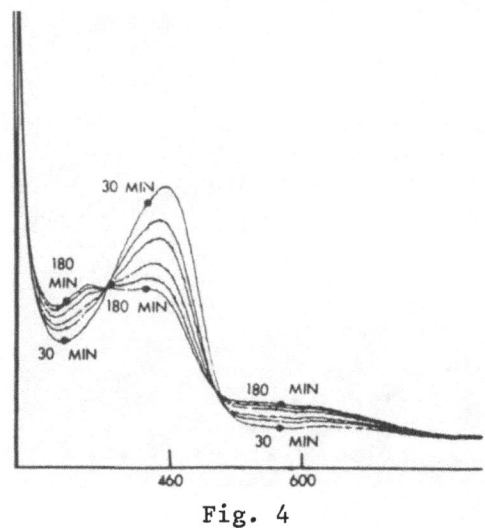

Fig. 4

McDonagh [15] first suggested that the apparent hepatic excretion of BR during PT proceeds via the formation of an isomer of BR that would be excreted with good efficiency but might revert to the natural BR in the bile. McDonagh [16] supplied evidence for the "in vivo" photoisomerization of BR based upon kinetics of biliary excretion of the pigment in irradiated Gunn rats.

3.2.1 A "twist" to phototherapy [17]. In addition to cis-trans isomerism, other geometric or configurational isomers are possible, involving rotation of one or both of the external pyrrole rings about the C = C axis. The double-bond character of the C-4 – C-5, C-15 – C-16 bonds prohibits free rotation about the C = C axis. However, when Z,Z configurational isomer absorbs light, the double bond may become momentarily so excited that it behaves like a single bond, and rotation about the C = C axis can occur. Figure 5 shows the photoreaction that transforms Z,Z-BR into the three possible configurational isomers, namely: 5E,15Z; 5Z,15E; and 5E,15E isomers. These Z,E-,E,Z-,E,E-BR isomers (called collectively photobilirubin = PBR) have been found to revert to native Z,Z-BR, and would be expected to hydrogen-bond to molecules of solvent water, and thus to become more soluble in that solvent. Photoisomerization of BR is easily detected by absorbance spectroscopy or in more detail by absorbance difference spectroscopy, as shown in Figure 6.

3.2.2 Photophysics of BR-PBR photoconversion. The results [18-20] of several different studies clearly indicate that upon excitation of BR/HSA with light, the most rapid relaxation mode of the excited BR involves twisting about one or the other of the C = C double bonds, and that this process occurs in the lowest excited singlet-state of the molecule. In fact: i) as the temperature is

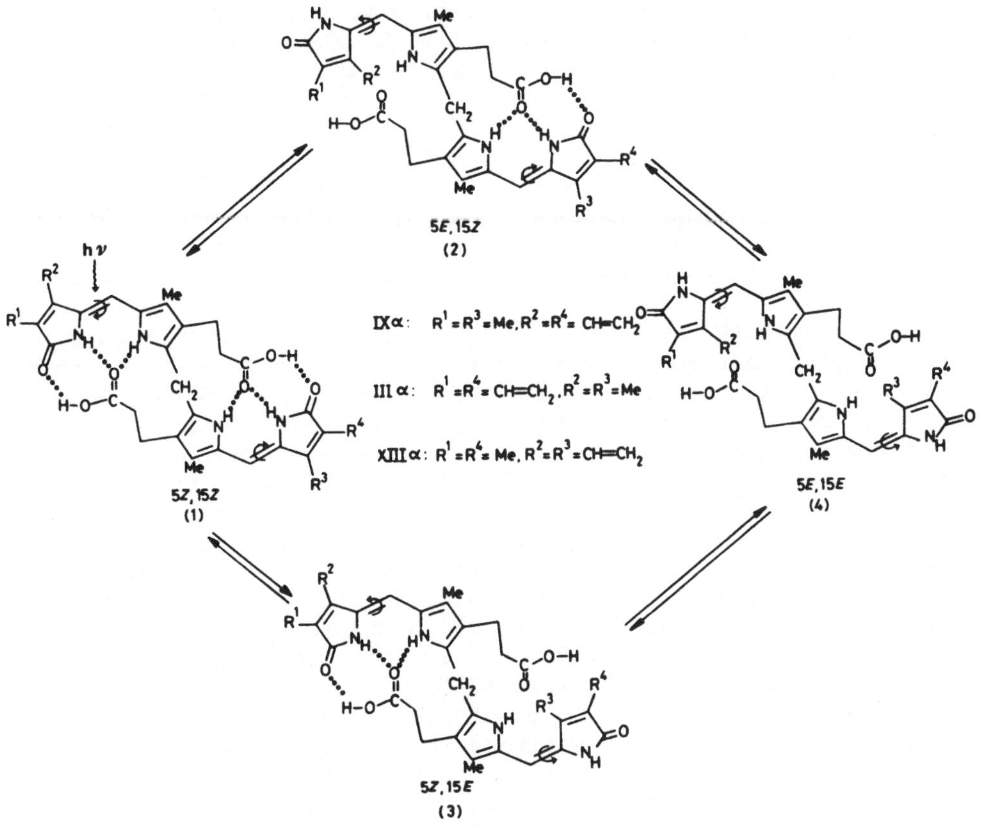

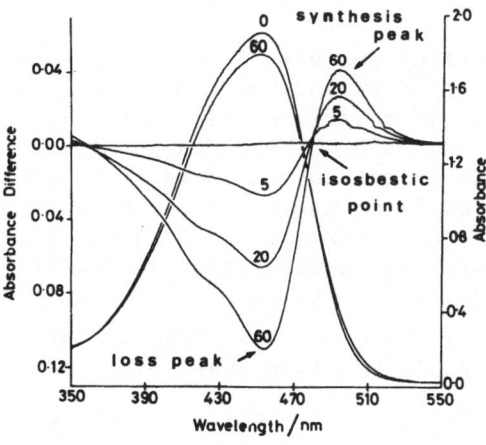

5E, 15Z
(2)
5Z, 15Z
(1)
5E, 15E
(4)
5Z, 15E
(3)
hν
IXα: R1 = R3 = Me, R2 = R4 = CH=CH2
IIIα: R1 = R4 = CH=CH2, R2 = R3 = Me
XIIIα: R1 = R4 = Me, R2 = R3 = CH=CH2

Fig. 5

synthesis peak
isosbestic point
loss peak
Absorbance Difference
Absorbance
Wavelength / nm
0
60
20
5
0·04
0·00
0·04
0·08
0·12
2·0
1·6
1·2
0·8
0·4
0·0
350
390
430
470
510
550

Fig. 6

varied the quantum yield (QY) for fluorescence from BR (/HSA) exhibits a reciprocal relation with the QY of photoisomerization. The fluorescence and isomerization yields are, respectively:

Temperature	Quantum Yield	
	$300°K$	$77°K$
Fluorescence	0.002	0.44
Isomerization	0.2	0.01

ii) photoexcited BR exhibits only a very small (< 0.01) yield of triplet states near room temperature (RT) in a variety of solvents. Fluorescent yields under the same conditions are < 0.001. Thus, electronic relaxation of excited singlet BR near RT proceeds predominantly via a radiationless pathway that does not include intersystem crossing; iii) at RT the BR/HSA → PBR/HSA isomerization QY when excited at 465 nm (isosbestic point) is within the experimental errors equal to the fraction of PBR/HSA in the photoequilibrium mixture. This observation is consistent with a nearly totally efficient formation of a twisted singlet-state intermediate and an isomerization yield essentially equal to the fraction of twisted molecules that decay to photoisomer; iv) difference spectra between photoexcited and non-photoexcited BR/HSA in aqueous buffer recorded at various times from 2 to 100 ps after excitation with 0.4 ps laser pulses revealed a transient assigned to the lowest excited singlet-state. The life time of the state was found to be 19 ps at $22°C$ and exhibited a temperature dependence that paralleled that of the fluorescence yield. No other transient was observed. These results are consistent with a rapid decay of excited singlet molecules to a mixture of ground-state isomers via some twisted state that is even shorter lived. This simple mechanistic scheme of the BR photo-isomerization process is schematically depicted in Figure 7.

3.2.3 Rate-equation analysis of BR → PBR photoisomerization. The temporal evolution of BR and PBR concentrations during light excitation can be easily described using a rate-equation approach. The relevant energy levels of BR and PBR involved in the present scheme of BR photoisomerization are shown in Figure 8. Light induces transitions of BR molecules from the S_o ground state to S_n excited states, from which very fast relaxation processes leave the molecules in the lowest S_1 excited state (^1BR). A twist about one of the C = C bonds gives rise to an excited-state intermediate with twisted geometry, designated by tBR*: $^1BR* \xrightarrow{k_t} tBR*$ ($k_t \equiv$ rate of twisting of $^1BR*$). Radiation-less transitions from this state yield ground-state BR: $tBR* \xrightarrow{k_Z} BR$ ($k_Z \equiv$ rate of decay of tBR* to BR); or its ground-state twisted conformation PBR: $tBR* \xrightarrow{k_E} PBR$ ($k_E \equiv$ rate of decay of tBR* to PBR). Once formed, PBR molecules follow the same pathway of BR upon light excitation. Excited-state PBRs decay

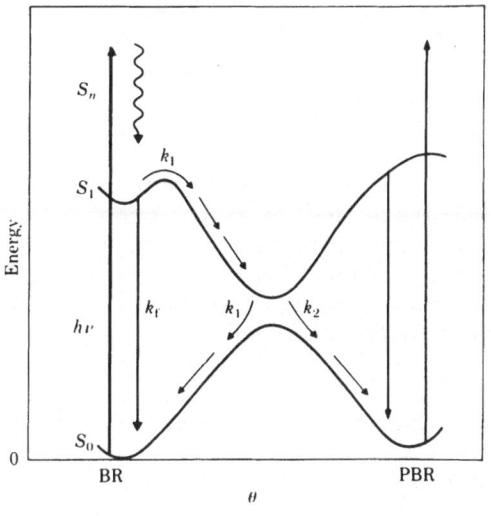

Fig. 7

to the common excited-state intermediate, designated by tBR*: ^{1}PBR* $\xrightarrow{k_t'}$ tBR* ($k_t' \equiv$ rate of twisting of ^{1}PBR*) from which they revert to native ground-state BR (tPBR* $\xrightarrow{k_Z}$ BR), or decay to ground-state PBR (tPBR* $\xrightarrow{k_E}$ PBR).

The rate-equations which regulate the time evolution of the molecular populations of the levels can be written as:

$$\dot{N}_1 \simeq -A_1 N_1 + k_Z N_t^*$$
$$\dot{N}_1^* \simeq A_1 N_1 - k_t N_1^* \tag{1}$$

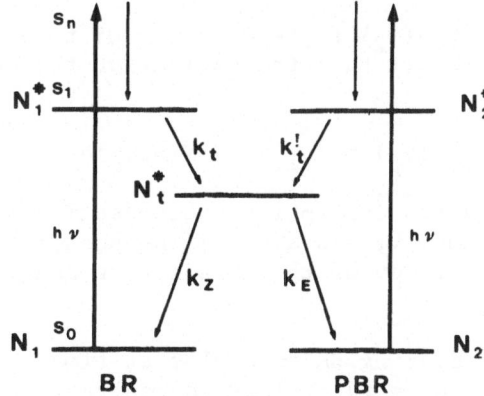

Fig. 8

$$\dot{N}_2 \simeq -A_2 N_2 + k_E N_t^*$$

$$\dot{N}_2^* \simeq A_2 N_2 - k_t' N_2^*$$

$$\dot{N}_t^* = k_t N_1^* + k_t' N_2^* - (k_Z + k_E) N_t^*$$

where $N_{1,2} \equiv N_{BR,PBR}$; $N_t^* \equiv N_{tBR*}$ denote the molecular densities of BR, PBR in the ground-state and in the common twisted excited-state intermediate, respectively; dots and asterisks indicate time derivative and excited-states, respectively. $A_{1,2}$ are the light excitation rates of BR and PBR, respectively. In Eqs. (1) fluorescence and intersystem-crossing transitions have been neglected due to their extremely low QY.

If it is assumed that $S_n \rightarrow S_1$ relaxation processes and k_t, k_t' twisting rates are fast compared to other radiationless relaxation modes of the excited-singlet states, then the quantum yield of tBR* starting with either BR or PBR is very close to 1. The QY of the isomerized product would then depend solely on the values of k_E and k_Z. Because the rates k_Z, k_E are also very fast, the $N_{1,2}^*$ populations have always nearly their steady-state values. Thus, $\dot{N}_{1,2}^* \sim \dot{N}_t^* \sim 0$, and Eq. (1) gives:

$$N_t^* = \frac{A_1 N_1 + A_2 N_2}{k_Z + k_E} \tag{2}$$

and

$$\dot{N}_1 = -\phi_1 A_1 N_1 + \phi_2 A_2 N_2$$

$$\dot{N}_2 = \phi_1 A_1 N_1 - \phi_2 A_2 N_2 \tag{3}$$

where

$$\phi_1 \equiv \phi_{BR} \equiv \frac{k_E}{k_Z + k_E}$$

$$\phi_2 \equiv \phi_{PBR} \equiv \frac{k_Z}{k_Z + k_E} . \tag{4}$$

In our approximation ϕ_1 (ϕ_2) represents the QY to go from BR (PBR) to PBR(BR). We now recall that the excitation rate $A_{1,2}$ has the expression [21]:

$$A_{1,2}(\nu_p) = \int \sigma_{1,2}(\nu) \, N(\nu, \nu_p) d\nu \tag{5}$$

where $\sigma_{1,2}(\nu)$ denotes the absorption cross-section of BR, PBR at the excitation frequency ν, and $N(\nu, \nu_p)$ the photon fluence rate in the frequency range $\nu, \nu + d\nu$ of the irradiating beam with peak frequency ν_p.

If we start irradiating BR molecules at time $t = 0$, solutions $n_{1,2}(t) = N_{1,2}(t)/N_o$ (N_o = initial BR concentration) of Eq. (3) which satisfy the initial conditions $N_1(0) = N_o$; $N_2(0) = 0$, are:

$$n_1(t) = n_1(\infty)[1 + R \exp(-t/\tau)]$$

$$n_2(t) = n_2(\infty)[1 - \exp(-t/\tau)] \tag{6}$$

where

$$R(\nu_p) \equiv \frac{\phi_1 A_1}{\phi_2 A_2} = \frac{\phi_1 \int \sigma_1(\nu) N(\nu, \nu_p) d\nu}{\phi_2 \int \phi_2(\nu) N(\nu, \nu_p) d\nu} \tag{7}$$

$$\tau^{-1}(\nu_p) = \phi_1 A_1 + \phi_2 A_2 = \int [\phi_1 \sigma_1(\nu) + \phi_2 \sigma_2(\nu)] N(\nu \nu_p) d\nu$$

and

$$n_1(\infty) = \frac{1}{1 + R} \; ; \; n_2(\infty) = \frac{R}{1 + R} \tag{8}$$

denote the steady-state, or photoequilibrium (PE), concentration of BR and PBR, respectively.

Difference Absorbance Spectrum (DAS). From Eq. (6) we obtain the expression of the difference absorbance $\Delta\alpha(t,\nu)$ of the PBR/BR mixture at time t:

$$\Delta\alpha(t,\nu) \equiv \alpha(t,\nu) - \alpha(0,\nu) = \sigma_1(\nu) N_1(t) + \sigma_2 N_2(t) - \sigma_1(\nu) N_o$$

or

$$\Delta\alpha(t,\nu) = -N_o n_2(\infty)[\sigma_1(\nu) - \sigma_2(\nu)](1 - e^{-t/\tau}) =$$

$$= -N_o[\sigma_1(\nu) - \sigma_2(\nu)] n_2(t). \tag{9}$$

Equation (9) then shows that the DA spectrum varies in time without changing its shape, which depends only on the difference between the BR and PBR cross-sections. The maximum value of $\Delta\alpha$ occurs for the frequency that maximizes this difference. Isosbestic points occur at frequencies for which $\sigma_1 = \sigma_2$.

Monochromatic excitation. For monochromatic excitation at frequency ν_p Eq. (8) reduces to:

$$n_1(\infty) = \frac{\phi_2 \sigma_2(\nu_p)}{\phi_1 \sigma_1(\nu_p) + \phi_2 \sigma_2(\nu_p)} \; ; \; n_2(\infty) = \frac{\phi_1 \sigma_1(\nu_p)}{\phi_1 \sigma_1(\nu_p) + \phi_2 \sigma_2(\nu_p)} \tag{10}$$

and the expression of the PBR concentration rise-time becomes:

$$\tau^{-1}(\nu_p) = N_T[\phi_1 \sigma_1(\nu_p) + \phi_2 \sigma_2(\nu_p)] \tag{11}$$

where N_T represents the total photon fluence rate of the source.

If the QY ratio ϕ_2/ϕ_1 is substantially independent of frequency, the spectral dependence of PBR relative abundance at PE is expected to depend only on the ratio of the BR, PBR absorption cross-sections. Moreoever, the spectral distribution of PBR rise-time is a weighted superposition of the BR, PBR cross-sections.

Quantum yields. At the isosbestic points ($\sigma_1 = \sigma_2$) the ratio $(n_2/n_1)_{PE}$ in the photostationary mixture given by Eq. (10) is equal to the ratio ϕ_1/ϕ_2 (= k_E/k_Z) between the corresponding QYs. Moreover, Eq. (10) gives:

$$(n_2)_{PE,ISO} = \frac{\phi_1}{\phi_1 + \phi_2} \simeq \phi_1$$

and similarly for ϕ_2. That is, the QY of conversion of BR to PBR is equal to the fraction of PBR at PE at the isosbestic points, as already pointed out by Lamola et al [18,19]. When the absorption cross-section of PBR, $\sigma_2(\nu)$, is known, measurements of the differential spectrum and PE rise-time give the values of ϕ_1 and ϕ_2.

Finally, when $t \ll \tau$, Eqs. (6) and (9) give:

$$n_2(t) \simeq \phi_1\sigma_1(\nu_p)N_T \cdot t;$$

$$\Delta\alpha(t,\nu) \simeq -N_o[\sigma_1(\nu) - \sigma_2(\nu)]\phi_1\sigma_1(\nu_p)N_T \cdot t, \qquad (13)$$

i.e., the spectral efficiency for BR production depends initially only on the absorption cross-section of BR and on its QY. As PBR molecules are formed, they start to revert in part to BR molecules according to the value of $\phi_2\sigma_2$: the formation process of PBR is, thus, slowed down, and PE is reached with the time-constant τ, which depends on the $\phi\sigma$ product of both species.

3.2.4 Laser investigation of BR → PBR photoreaction. Owing to the large QY for BR → PBR photoisomerization, a photoequilibrium (PE) state is reached for photon fluences much smaller than those necessary to produce substantial degradation of BR through photooxidation or other irreversible processes. Until 1978 action spectra of BR photolysis have been determined using experimental conditions suitable for the latter processes (high irradiance, long exposure times, missing out the initial transient of the reaction), and the greatest effects have been found with wavelengths matching the absorption spectrum of BR.

The spectral behavior of configurational isomerization of BR has been investigated only very recently using laser and narrowband filtered lights [22].

Spectral dependence of PBR formation. The formation of PBR and the achievement of the photoequilibrium (PE) state of the BR $\rightleftharpoons h \rightleftharpoons$ PBR reaction have been investigated using the 457, 465, 488, 501 and 514 nm lines of an argon laser at excitation irradiances of 25, 50, 500 $\mu W/cm^2$, and 15 mW/cm^2. Solutions of BR (10 μM) and HSA (115 μM) containing 0.15 M NaCl and 0.01 M phosphate buffer have been used. Figure 9A shows the differential absorbance spectra (DAS) in the 430 – 520 nm range using the above lines at 0.5 mW/cm^2 irradiance.

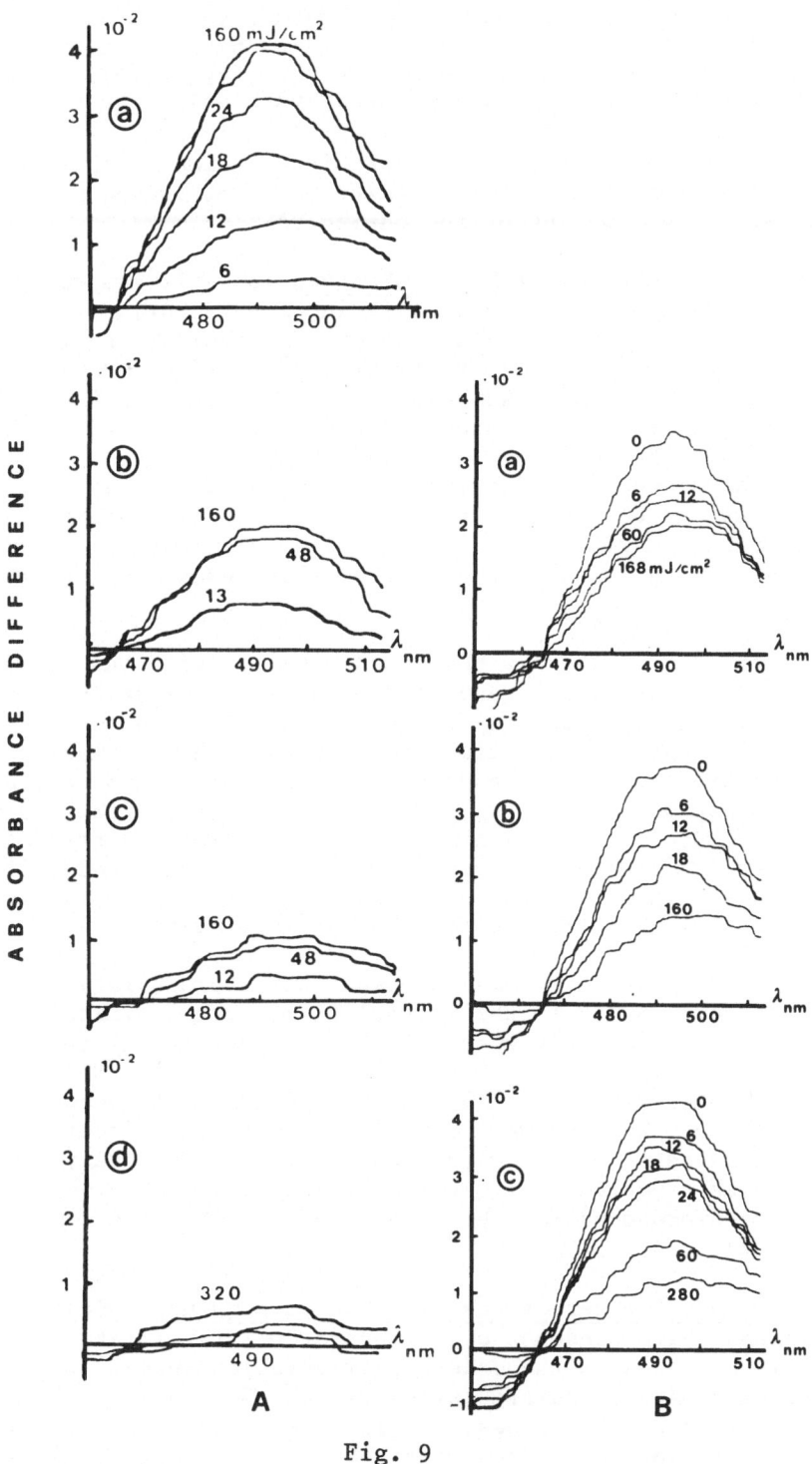

Fig. 9

An isosbestic point is present at $\lambda \simeq 465$ nm, and the DAS maximum occurs at $\lambda \simeq 490$ nm. Similar patterns are obtained at the lower and higher irradiances used. As can be seen, the 457-line produces the highest value of PE concentration of PBR, while the 514-line is quite ineffective for PBR formation. At high irradiance the PE is reached in quite a short time: further irradiation of the solution produces slow degradation of the PBR/BR mixture, as demonstrated by the corresponding lowering of the entire DAS.

Spectral dependence of PBR → BR reversion. It was soon apparent that the PE value produced by irradiating a BR solution with (narrow band) light at a certain wavelength (450 nm) could be shifted to a new PE value by exposing the PBR/BR mixture to light of different wavelength (510 nm) [23]. We have studied the inverse PBR → BR reaction in greater detail using the laser lines λ_i = 488, 501, and 514 nm. Figure 9B shows the formation of new PE values of PBR/BR mixture obtained upon irradiation of BR at 457 nm, using the λ_i sequence. As can be seen: i) the complete reversion of PBR to BR cannot be achieved with these lines; ii) greater reversion efficiency is associated with the 514-line, followed in descending order by the 501 and 488 lines.

A comparison of the new PE values obtained with the irradiation sequence $457/\lambda_i$ with those generated by direct excitation with the λ_i lines (Figure 9A) show that the DAS values for the two-step 457/ /488 and one-step 488 excitation are equal within the experimental errors. On the contrary, the green line excitation produces DAS values that depend on the irradiation sequence, the effect being more pronounced with the 514-line. Moreoever, the reverted DAS of the two-step irradiation are now asymmetrical, with a red-shifted maximum and a larger absorption in the long-wave wing. The effect is much less evident with the 488-line (Figure 9B-a).

As a further step, we have checked the reversibility of the new mixtures back again to the initial PE mixture by irradiating them with the 457-line. Figure 10A shows schematically the behavior of the DAS value at λ = 490 nm as a function of energy fluence for the sequence $457/\lambda_i$; for λ_i = 501 nm the full DAS are shown below each time cycle. The initial PE values formed with direct excitation were in all cases recovered with a good degree of accuracy. However, the DAS shapes revealed a larger absorption at longer wavelengths, that is more pronounced for the green than for the blue lines (see Figure 10B-2).

Effect of the order of the irradiation sequence. Finally, we have examined the dependence of the one-step intermediate and two-step final PE values on the order of irradiation sequence. BR solutions have been irradiated with the 488, 501, and 514 lines to give three different PE values, and then irradiated with the 457-line up to the final PE. The DAS at 490 nm for these $\lambda_i/457$ sequences

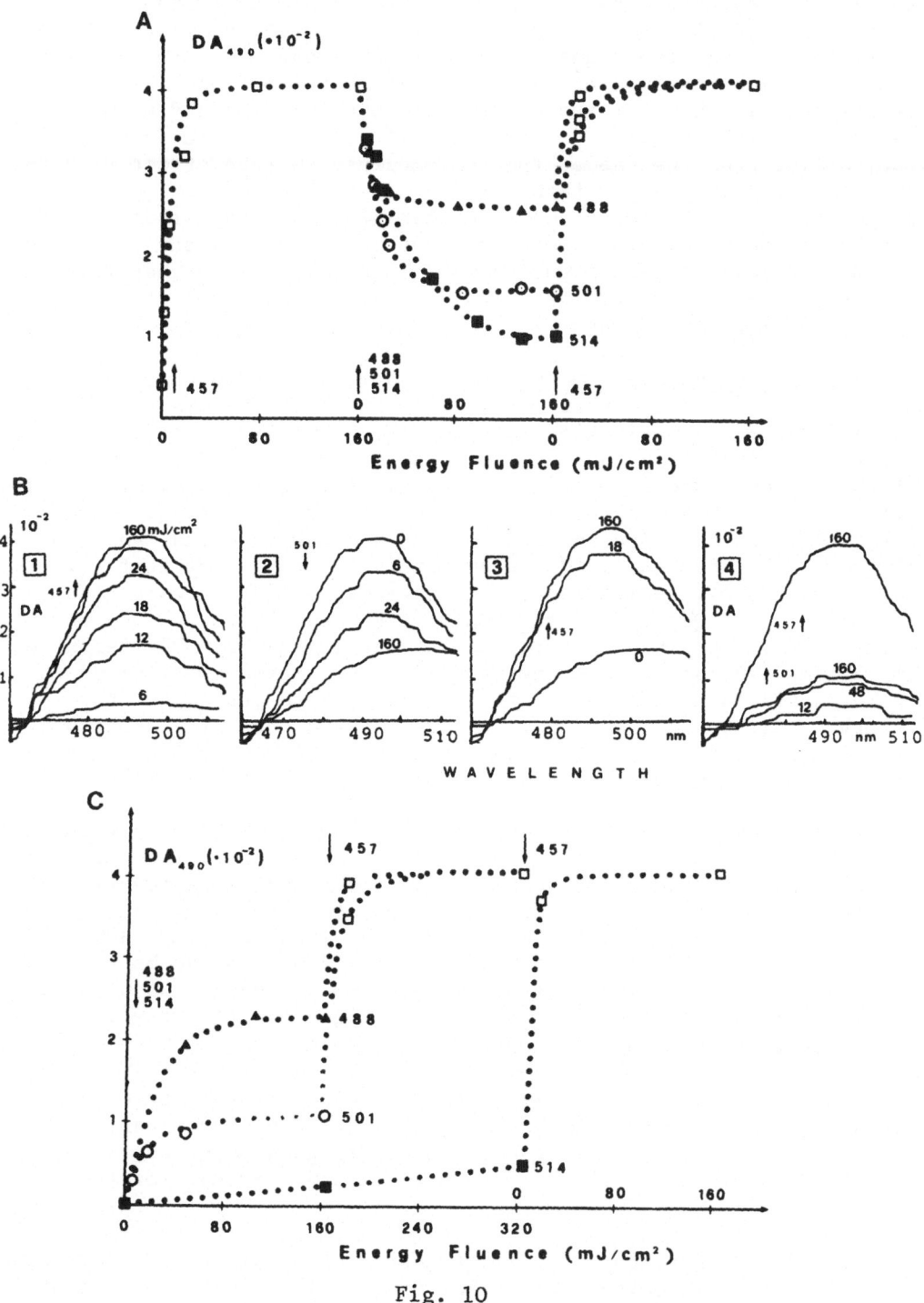

Fig. 10

are shown in Figure 10B. As can be seen, the final PE values of
the $\lambda_i/457$ sequence equal with good degree of accuracy the corresponding $457/\lambda_i/457$ PE values. Again the $\lambda_i = 488$ nm intermediate
PE values for the two sequences are nearly equal. On the contrary,
the green-line intermediate PE values are different and the difference
is greater for the 514-line. Moreover, the DAS obtained with direct
excitation with any line are symmetrical and similar in shape, while
they differ from the DAS of the intermediate PE (the effect with the
488-line is much smaller than with the green lines) (in Figure 10C
only the 501-DAS is reported). Furthermore, the comparison of DAS
shapes of the three- and two-step cases shows that i) the absorption
in the long-wave region after the $457/\lambda_i$ sequence is higher than the
corresponding value after the $\lambda_i/457$ sequence; ii) this effect
increases with increasing λ_i.

3.2.5 Action spectrum of PBR formation. The above results
are in general agreement with the first-order reversible photoreaction
assumed for BR isomerization, at least for blue-line excitation, and
described by the rate-equation approach of Section 3.2.3. In case
of multi-step irradiation with different excitation wavelengths λ_p,
integration of Eqs. (3) with the proper initial conditions shows
that the final PE value depends only on the last λ_p used, and thus
it equals the PE value produced by direct excitation at that λ_p.
For a quantitative comparison, the knowledge of PBR cross-section,
σ_2, and BR, PBR QYs is needed. Only very recently the PE concentration of PBR has been determined in a direct way by HPLC technique
at various excitation wavelengths [22]. Absorption spectrum of
isolated PBR is not yet sufficiently accurate to reproduce the
observed DAS [24]. Therefore, we have derived the $\sigma_2(\lambda)$ spectrum
from the differential absorption spectrum published by Ennever et al
[22]. The σ_1 and σ_2 spectra are shown in Figure 11.

Photoequilibrium PBR concentration. In Figure 12 we present
the PBR concentrations (stars) at PE measured by Ennever et al [22]
by HPLC using narrow-band (10 nm) excitation light as a function of
the peak wavelength. By fitting these points with Eq. (10) of
Section 3.2.3 for the monochromatic case in the range $390 - 490$ nm,
we found for the QY ratio the value $\phi_2/\phi_1 \simeq 2.3$. As can be seen,
Eq. (10) describes accurately the wavelength dependence of n_{PBR} at
PE over this 100 nm range; outside this range, the accuracy of σ_2
is not sufficient to permit a better agreement.

Equation (9) at PE allows us to derive $n_2(\infty)$ from the experimental values of $\Delta\alpha$ at some wavelength. Our PE n_2 values are
reported in Figure 12 (triangles) and are in good agreement with
Ennever's values and with the theoretical curve (except for the 514-
line).

Figure 12 represents the first action spectrum for PBR formation
since 1978. It is worth noting that the maximum of PBR concentration

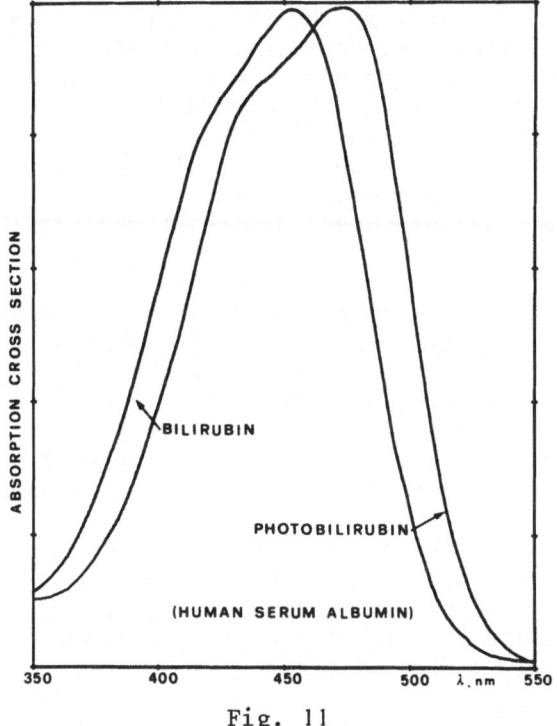

Fig. 11

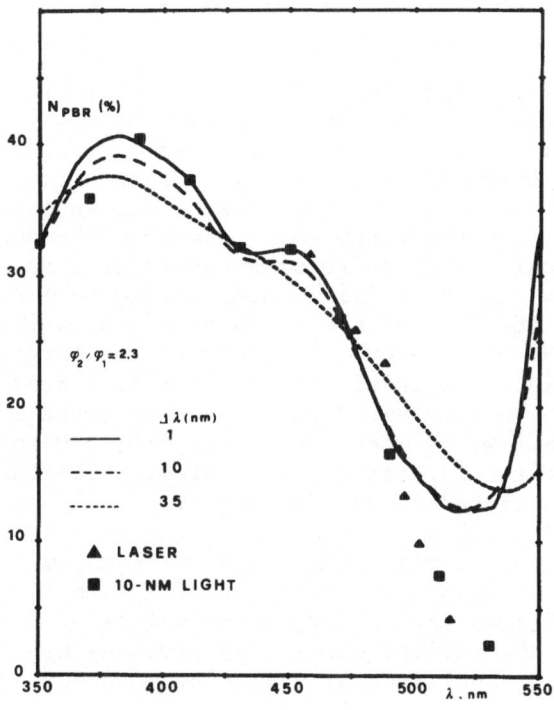

Fig. 12

occurs at $\lambda \simeq 382$ nm, i.e., in the UV, a region which is avoided
in PT for its well-known cytotoxicity. Moreover, a deep minimum
of n_2 is present in the green part of the spectrum at $\lambda \simeq 516$ nm.
Here, the larger absorption and QY of PBR than BR combine to keep
the PBR concentration low. As λ increases, $\sigma_2 \rightarrow \sigma_1$ (and both to
zero) and n_2 increases again up to the value corresponding to the
isosbestic point. However, owing to the very small values of σ_1,
σ_2 the photoisomerization process occurs at very small rate in this
long-wave wing.

The effect of the monochromaticity on the n_2 spectral distribution has been investigated using Eqs. (8) and (7) for a spectrally
Gaussian distribution of the excitation light. In Figure 12 the
dotted curves correspond to spectral band widths $\Delta\lambda = 10, 35$ nm.
The 10 nm light behaves here as a monochromatic light; in the case
$\Delta\lambda = 35$ nm the average of σ_1 and σ_2 over a larger distribution gives
rise, as expected, to a broader minimum and a red-shift.

The PE values of PBR concentration produced by standard commercial fluorescent lamps used for PT are shown in Table 4. There
is a decrease from violet, blue to green light, with a slight
increase with green-yellow light (as already noted in Figure 12).
These values are shown also in Figure 12.

Tungsten-halogen lamps, also used for PT, produce "in vitro"
PBR concentration equal to 14.5%, that is only slightly smaller
than that reported by Ennever for 10 nm bandwidth radiation at $\lambda =$
490 nm.

Photoequilibrium rise-time. In Figure 13 we present the time
evolution of PBR concentration in solutions (BR:HSA $\equiv$ 1:2) irradiated
with several laser lines. A similar pattern is reported in Ref. 5
for narrow-band (10 nm) filtered light. Equation (10) has been used
to derive $n_2(\infty)$ from the measured values of the absorbance difference
$\Delta\alpha(\nu,t)$. In this spectral range (457 – 514 nm) PE concentrations
decrease monotonically with increasing wavelength, as already noted
in Figure 12. The largest initial rates of the BR $\rightarrow$ PBR reaction
are associated to the 465 and 476 nm laser lines, while the 457-line
has a lower initial efficiency, but produces the largest PE photoisomerization. Since the initial rate ($t \ll \tau$) depends only on
$\sigma_1(\lambda)$, the 457-line was expected to produce also the largest initial
photoisomerization. With 10 nm light, the 410, 430 and not the
450-lights produce the largest initial rates, while the largest PE
PBR density is associated to the 390-light.

In Figure 14 the PE rise-time τ (in units mJ/cm^2) is plotted
vs the excitation wavelength. The value of τ with 10 nm light
(squares) is 2 – 3 times the corresponding values with laser lines
(triangles). The theoretical curve $\tau(\lambda)$ given by Eq. (11), with
$\phi_2/\phi_1 = 2.3$, (solid line) is nearly flat between 400 and 500 nm,

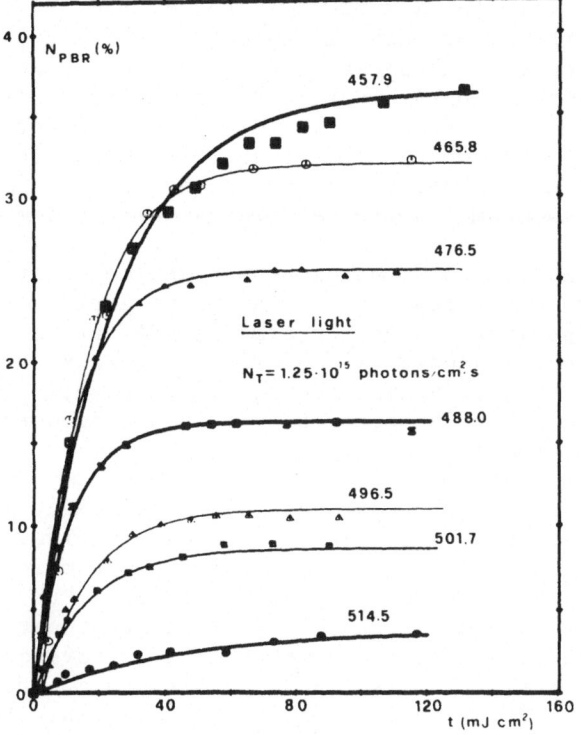

Fig. 13

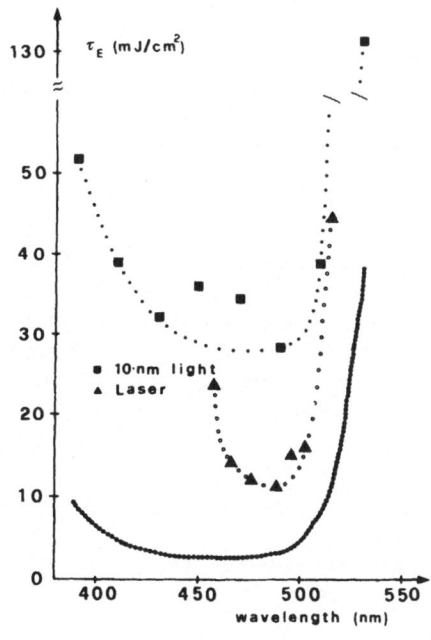

Fig. 14

and increases sharply for $\lambda > 500$ nm. Apart from the two points at 450 and 470 nm, the 10 nm curve is in qualitative agreement with the theoretical one. With laser light also the 457 and 465-lines seem to deviate from a possible flat distribution. Unfortunately, the lack of argon laser lines with $\lambda < 457$ nm does not allow us to know whether the short-wave shape of the $\tau(\lambda)$ curve is similar to that of 10 nm light. Finally, in both cases, the minimum value of τ occurs for $\lambda \simeq 490$ nm.

The discrepancies between incoherent and coherent lights, and experimental and theoretical values cannot be explained at this moment. Both experiments have the same BR/HSA ratio, and not two different light power densities (80 – 120, 500 $\mu W/cm^2$ for filtered and laser light, respectively). Further investigation with continuously tunable laser is needed to obtain a more detailed action spectrum of PBR formation.

3.2.6 Clinical model. For concluding this Section we recall that the mechanism of PT accepted until today involves various steps: i) formation of PBR, which probably occurs mainly in the extravascular tissue near the skin surface. Some PBR also may be formed directly in plasma as the blood circulates near the skin surface; ii) movement of PBR from peripheral extravascular tissues to blood. Most likely this is a passive diffuse-controlled process (it seems to be bidirectional and quite rapid because the circulating isomeric pigment in irradiated obstructed rats reached PE concentration within 3.5 hours); iii) rapid hepatic removal of PBR from circulation; iv) secretion of PBR from the hepatocyte into biliary caniculus.

Thus, PT causes unconjugated BR to be transported from skin to bile. The BR departed is replaced in the skin by BR from the blood resulting in a decrease in the concentration of circulating BR (Figure 15).

3.3 Structural Photoisomerization of BR

An unexpected result has been recently found by Ennever et al. when the dosage of PBR in serum of infants exposed to PT has been determined by PHLC technique. In premature infants the percentage

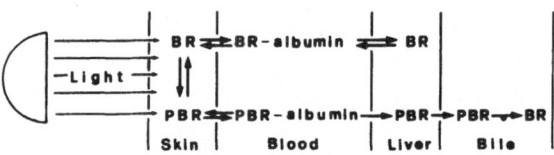

Fig. 15

Table 1.   Time Evolution of Infant Serum PBR during PT [5]

Fluence Spectral Density ($\mu W/cm^2 \cdot$ nm)	PBR concentration (%); hours of PT:			
	0	2	4	8
6	6.5(0.8)*	9.5(0.8)	10.8(0.9)	11.6(0.8)
12	6.3(1.0)	10.5(0.9)	12.1(0.7)	12.5(0.7)

* Standard error of the mean

of the total BR converted to PBR reaches quickly large PE values
(∿ 12%), comparable to the "in vitro" PE values (∿ 14.5%). Table 1
reports these data. Moreover, the rate of decline in PBR concentration after cessation of PT has been found to be very slow. The
serum PBR half-life, calculated from the initial rate of disappearance, ranges from 12 to 21 hours, with a mean of 15 hours.

These data establish that the clearance of configurational
photoisomers in preterm infants is inefficient. Since far greater
therapeutic responses to PT have been documented, these data indicate
that alternative photochemical reactions must contribute to the
clinical effectiveness of PT [5].

3.3.1 Lumirubin. Already in 1976 McDonagh [26] found that
irradiation of BR in aqueous solution with visible light in the
presence of serum albumin and the absence of oxygen leads to formation of a yellow photoderivative, slightly more polar and less
lipophilic than BR. Today, we know that in prolonged photolysis of
BR (i.e., irradiation times several time τ) the configurational
isomerization of BR is invariantly accompanied by a slower reaction
which appears to involve structural isomers of BR. Very recently
HPLC analysis has revealed the presence of two additional BR isomers,
called by McDonagh lumirubin (LR) and isolumirubin (ILR) [27]. NMR
spectroscopy indicates that BR undergoes an intramolecular cyclo-
addition reaction, which involves the endovynil group at C-3 [28-30].

According to McDonagh formation of Z-LR from BR is not prevented by
albumin binding. At present it is uncertain whether Z-LR is formed
from BR in a concerted one-photon process (4Z, 15Z-BR → Z-LR) or
in a sequential two-photon process with 4E, 15Z-BR as intermediate

(4Z, 15Z-BR → 4E, 15Z-BR → Z-LR). However, as noted earlier,
photolysis of BR on HSA results in rapid formation of the 4Z, 15E
configurational isomer. This isomer cannot be a direct precursor
for Z-LR. Nevertheless, on continued irradiation there is a gradual
formation of Z-LR and E-LR without the appearance of detectable
amounts of 4E, 15Z-BR. Furthermore, LR formation is not markedly
enhanced in systems where the configurational isomerization leads
to both of the E,Z and Z,E-BR isomers. These observations indicate
that a direct one-photon mechanism may be predominantly responsible
for Z-LR formation [28].

Like other compounds with a dipyrrylmethenone chromophore LR
undergoes facile reversible configurational isomerization to more
polar and thermally unstable E isomers on exposure to light, reaching
Z → E equilibrium without detectable reversion to the parent BR.
The reaction is readily detectable by difference spectroscopy and
HPLC, and it accounts for the minor HPLC peak appearing on prolonged
photolysis of BR.

Continued prolonged photolysis of the Z,E mixture or irradiation
of Z-LR with very intense light, leads to slow formation of configurational isomers of BR and overall destruction of pigment.
Unlike the E,Z isomers of BR, Z-LR does not undergo rapid stoichiometric photoreversion to Z,Z-BR.

Therefore, the "fast" anaerobic photochemistry of BR may be
summarized by:

$$(E)\text{-}LR \rightleftharpoons (Z)\text{-}LR \qquad\qquad (4Z,15E)\text{-}BR$$
$$(Z,Z)\text{-}BR \qquad\qquad (E,E)\text{-}BR$$
$$(E)\text{-}ILR \rightleftharpoons (Z)\text{-}ILR \qquad\qquad (4E,15Z)\text{-}BR$$

On the time-scale of the Z → E conversion, LR formation is slow and
irreversible [28].

3.3.2 Clinical observations. On the basis of the experimental
observation the presence of LR in serum of infants exposed to PT
has been investigated by HPLC technique. Table 2 contains the data
on LR concentration in the serum of infants treated with high and
low dose PT for eight hours [5].

Steady state levels of LR are reached in approximately two
(four) hours in the low (high) dose PT, and are in the range ∿ 0.5 −
− 2.0% of total BR, much lower than PE PBR concentration (18 − 20%).
Unlike PBR, LR concentration maximal level depends on light intensity, being higher for higher irradiance. Furthermore, the clearance
of LR from the serum has been found to be more rapid than for PBR.

Table 2.   Time Evolution of Infant Serum LR during PT [5].

| Fluence Spectral | LR concentration (%); hours of PT: | | | |
Density ($\mu W/cm^2 \cdot nm$)	0	2	4	8
6	0.2(0.1)*	0.6(0.1)	0.7(0.1)	0.7(0.1)
12	0.2(0.1)	0.8(0.1)	1.3(0.3)	1.5(0.4)

* Standard error of the mean

LR was detected only in the serum samples obtained immediately following cessation of PT, and none was found in serum specimens obtained two or three hours after PT [5].

The conclusion of these observations are: i) in premature infants excretion of PBR is so slow that this pathway cannot account for the therapeutic effect of PT; ii) although LR is a minor photo-product in terms of serum concentration, the efficient excretion of this photoproduct may play a quantitatively important role in the therapeutic effectiveness of PT.

3.3.3 A twist "back" for phototherapy. If LR originates only from BR molecules, the quick formation of PBR is a counterproductive process since it considerably lowers the number of BR molecules from which LR is produced at a much lower rate. Although the rever-sibility of the PBR/BR reaction causes PBR molecules to revert quickly to BR to maintain PBR/BR equilibrium as the BR concentration is lowered during LR formation, diffusion of PBR to sites less accessible to light may prevent PBR reversion.

The optimum light for PT will then very likely i) keep BR con-centration as low as possible at the formation sites; and still ii) ensure the maximum absorption by BR molecules. As Figure 12 shows, the PBR concentration is minimum in the green (at 516 nm). Here, however, the absorption coefficient of BR is quite small and we need a green intensity $\sim$ 14 times ($\eta \equiv N_T(452)/N_T(516) = \sigma_1(452)/\sigma_1(516)$) the blue intensity to produce the same absorption by BR in the blue. On the other hand, the simultaneous presence of both blue and green lights with a suitable intensity ratio would permit high BR absorption and at the same time a large reduction of the PBR concentration produced by blue light. Moreover, Figure 12 shows that the minimum of PBR is not washed out by polychromatic light with band widths $\Delta\lambda$ = 10 and 35 nm. The proper choice of peak wavelength and spectral band width of a single polychromatic light beam could ensure the desired balance between the constraints i) and ii): the short-wave part of the spectrum excites BR to form LR; the long-wave part forces BR to twist "back" to native BR, thus keeping the overall concentration low.

In the following Section we present a discussion of the two-light irradiation scheme. The conclusions, however, are highly speculative, since the behavior of n_2 in the green has not yet been measured and our theoretical expression does not fit the only three experimental points at 510, 514 and 530 nm.

3.4 Control of the BR → PBR Reaction by Two-Wavelength Irradiation

In this Section we discuss the feasibility of a two-frequency control of BR configurational photoisomerization, and present experimental evidence of efficient quenching of PBR formation using simultaneous irradiatioon of BR solutions with laser blue and green lights.

3.4.1 Theoretical analysis.

When a solution of BR is irradiated by two polychromatic light beams of peak frequencies ν_p', ν_p'', the PBR concentration $n_2(\nu_p', \nu_p'')$ at PE and rise-time $\tau^{-1}(\nu_p'\nu_p'')$ are expressed by Eqs. (7) and (8):

$$n_2(\nu_p'\nu_p''\eta) = [1 + R^{-1}(\nu_p'\nu_p''\eta)]^{-1} \tag{14}$$

where

$$R(\nu_p'\nu_p''\eta) = R(\nu_p') \cdot f(\nu_p'\nu_p''\eta); \quad A_1(\nu_p'\nu_p''\eta) = A_1(\nu_p'')\left(1 + \eta\,\frac{\bar{\sigma}_1(\nu_p')}{\bar{\sigma}_1(\nu_p'')}\right)$$

$$f(\nu_p'\nu_p''\eta) = \frac{\eta + \bar{\sigma}_1(\nu_p'')/\bar{\sigma}_1(\nu_p')}{\eta + \bar{\sigma}_2(\nu_p'')/\bar{\sigma}_2(\nu_p')}; \quad \bar{\sigma}(\nu_p) = \int \sigma(\nu)\chi(\nu\nu_p)d\nu \tag{15}$$

$$\chi(\nu,\nu_p) = \frac{g(\nu,\nu_p)/\nu}{\int g(\nu,\nu_p)d\nu/\nu}; \quad g(\nu = \nu_p) = 1; \quad \eta = N_T'/N_T''$$

$$\tau^{-1}(\nu_p'\nu_p''\eta) = \tau^{-1}(\nu_p') + \tau^{-1}(\nu_p''). \tag{16}$$

In Eq. (15) the function $g(\nu,\nu_p)$ represents the spectral profile of the light intensity, and $\bar{\sigma}$ denotes the convolution of the absorption coefficient with the light profile.

It is important to note that under two-wavelength excitation the PBR concentration depends in general on the ratio of the light beam intensities $\eta (= N_T'/N_T'')$. Depending on the choice of the two wavelengths λ' and λ'', and of the intensity ratio η, the f-factor in Eq. (15) may be $^>_< 1$, and hence the possibility arises of either enhancing or depressing the PBR concentration produced by single beam irradiation at λ_p. The two-frequency absorption rate $A_1(\nu_p',\nu_p'',\eta)$ is always $\geq A_1(\nu_p')$, or $A_1(\nu_p'')$. Obviously, when $\lambda' = \lambda''$ we have f = 1 (i.e., single frequency excitation); moreover, when $\lambda'' \gg \lambda'$, so that $\bar{\sigma}_1(\nu_p'') \simeq \bar{\sigma}_2(\nu_p'') \simeq 0$, the effect of the second beam is no longer present, and f = 1.

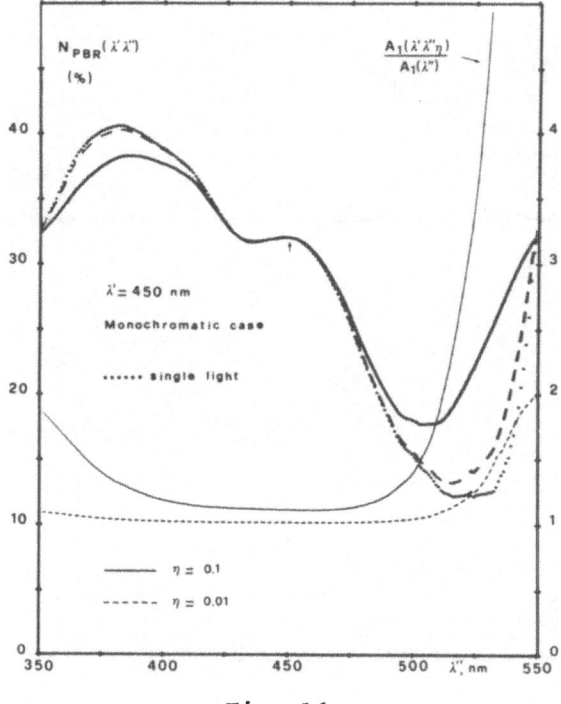

Fig. 16

Monochromatic source. In Figure 16 we present the plots of n_2 (λ', λ'', η) for monochromatic excitation at a fixed formation wavelength λ' = 450 nm, and for two values of the "reversion" beam intensity at λ'' (η = 0.1, 0.01). For comparison, the PBR concentration produced by a single beam is also shown (solid line). In this case, PBR concentration at PE has a well pronounced minimum in the green part of the spectrum, between 510 and 540 nm. For η = 0.1 (i.e., λ''-light ten times more intense than the blue λ'-light) the PBR concentration for λ'' > λ' has a minimum at $\lambda'' \simeq 505$ nm (dotted line); here the PBR concentration is substantially lower than that produced by direct excitation at λ' = 450 nm and only slightly larger than that generated by single-light irradiation at λ'' = 505 nm and $N_T'' = 10\ N_T'$. In this case, the overall absorption $A_1(\lambda'\lambda'')$ by BR is ~ 1.6 times $A_1(\lambda'')$. When η = 0.01 the $n_2(\lambda'\lambda'')$ curve approaches the single-light curve $n_2(\lambda)$, and $A_1(\lambda'\lambda'') \simeq A_1(\lambda'')$. Since for λ'' > 500 nm our σ_2 values are largely uncertain, the exact behavior of n_2 at long wavelength is not known. The two-wavelength scheme would work more efficiently with λ'' in the long-wave wing of σ_2 if the condition $\sigma_2(\lambda'') \gg \sigma_1(\lambda'') \simeq 0$ could be satisfied.

Narrow-band source. In clinical PT fluorescent lamps emitting a narrow-spectrum, typically 30 - 40 nm wide, are commonly used. PBR concentration at PE has been calculated using a spectrally Gaussian distribution of the light beams for several bandwidths ($\Delta\lambda$ = 1, 10, 35 nm) and intensity ratios η. Figure 17 shows the plots

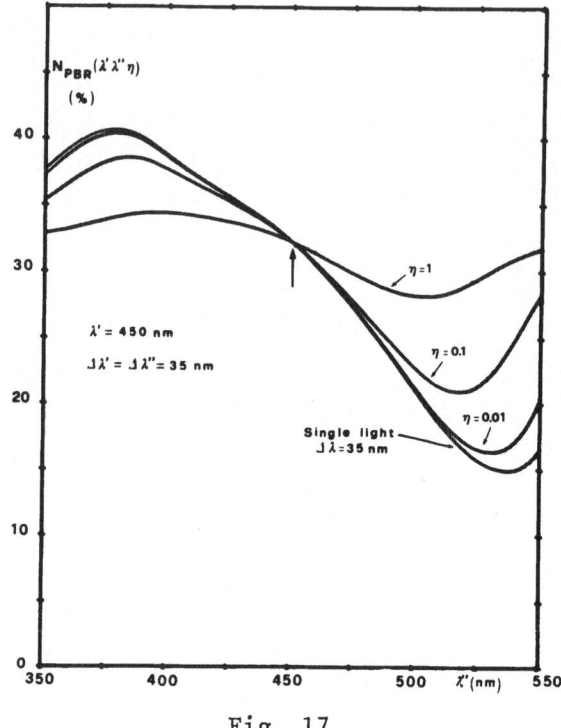

Fig. 17

of $n_2(\lambda'\lambda'')$ for λ' = 450 nm, $\Delta\lambda'$ = $\Delta\lambda''$ = 35 nm and η = 1, 0.1, 0.01.
The dotted curve represents the single-light case for 35 nm-band
width excitation. The presence of the second beam at $\lambda'' > \lambda'$ reduced
the PBR concentration produced by the λ'-beam and the PBR minimum
shifts towards longer wavelengths as the intensity of the "reversion"
beam is increased. For η = 0.1 the minimum of n_2 occurs at λ'' =
= 518 nm, where n_2 (450, 518, 0.1) = 20.8%. The 450 and 518 beams
would produce, separately, PBR densities $n_2(450)$ = 32.1% and $n_2(518)$=
= 16.7%, respectively. The total BR absorption A_1(450, 518, 0.1)
is now 1.7 times that produced by the 518-beam alone.

Figure 18 shows the effect of the band width $\Delta\lambda''$ of the green
"reversion" light on PBR density in the case λ' = 450 nm and $\Delta\lambda'$ =
= 35 nm, and η = 0.1. As $\Delta\lambda''$ is reduced from 35 to 10 and 1 nm the
n_2-minimum becomes sharper and shifts to shorter wavelengths, and
the corresponding PBR density decreases. The dotted lines represent
the single-beam excitations at the corresponding band widths. As
expected, narrower-band light is more efficient to quench PBR formation due to the smaller average over the spectral distribution
of σ_1 and σ_2. With η = 0.01 the effect is larger, and higher reversion can be achieved at still longer wavelengths. In the following
Table 3 we report the relevant numerical values for the case η = 0.1.

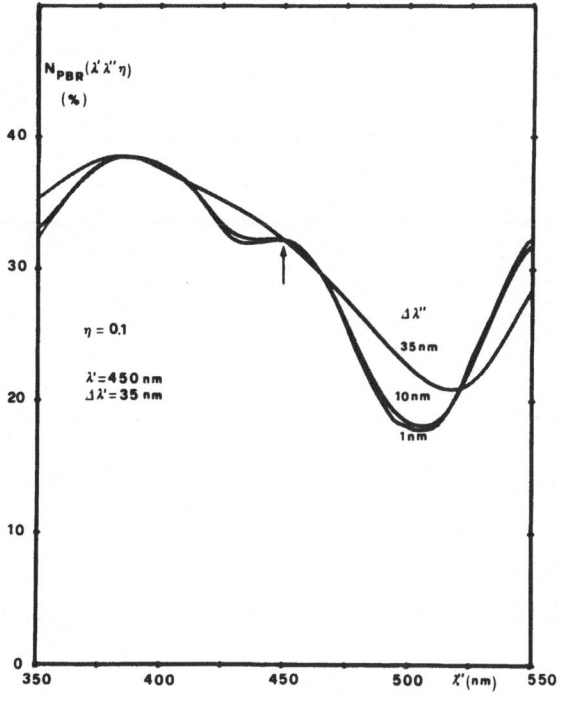

Fig. 18

In the case $\Delta\lambda'' = 1$ nm the two-light scheme would produce 1.35 reduction and 2.8 increase of blue light PBR concentration and BR absorption, respectively.

If the lowest PBR concentration is desired, a single, 1 nm band width green light at $\lambda_p = 518$ nm can ensure the same BR absorption as a 450, 35 nm blue light at a green/blue intensity ratio equal to 14. If only 35 nm band width green light is available minimum PBR concentration and equivalent blue light BR absorption are obtained with single green light at $\lambda_{min} = 536$ and $\eta = 1/23$.

In Section 5.1 we have computed the photon absorption rates $A_1(\lambda_p)$ and PE PBR concentrations n_2 using the spectral profiles of the fluorescent lamps currently used in PT. For the Westinghouse special blue and Sylvania green lamps the PBR densities are 32% and 15%, respectively, and the average absorption cross-sections ($\bar{\sigma}_1 = = A_1/N_T$) are 1.66 and 0.09, respectively. It follows that the same photon absorption rate produced by the blue lamp can be obtained using green lamps with a total photon fluence rate N_T 18 times the blue one, but at a much smaller PBR concentration. Simultaneous irradiation with the blue-green lamps with $\eta = 1/18$ doubles the absorption rate, but also increases PBR from 15% to 20.1%.

Table 3.

$\lambda' = 450$ nm		$\Delta\lambda''$(nm)	
$\Delta\lambda'' = 35$ nm	1	10	35
$\bar{\sigma}_1(450)\cdot 10^{16}(\text{cm}^2)$	1.818	1.800	1.652
$n_2(450)(\%)$	33.8	33.5	32.1
Single beam: λ_{min}(nm)	516	520	536
$n_2(\lambda_{min})(\%)$	13.1	13.4	14.9
$\bar{\sigma}_1(\lambda_{min})\ 10^{17},\text{cm}^2$	1.288	1.053	0.703
$\eta = 0.1:$ λ''_{min}(nm)	504	506	518
$n_2(450,\lambda''_{min})(\%)$	17.7	18.1	20.8
$n_2(\lambda''_{min})(\%)$	15.2	15.2	16.7
$\bar{\sigma}_1(\lambda_{min})\ 10^{17},\text{cm}^2$	3.34	3.055	2.365
$\dfrac{A_1(450,\lambda_{min},\eta)}{A_1(\lambda''_{min})}$	1.54	1.59	1.70
$\dfrac{A_1(450,\lambda''_{min},\eta)}{A_1(450)}$	2.84	2.70	2.43
$\bar{\sigma}_1(450)/\bar{\sigma}_1(\lambda_{min})$	14	17	23
$\bar{\sigma}_1(450)/\bar{\sigma}_1(\lambda''_{min})$	5	6	7

3.4.2 Experimental Results

Laser excitation. The control of configurational isomerization
of BR by two-light irradiation has been investigated by simultaneously
irradiating a solution of BR/HSA with the 457 nm blue and 514 nm
green laser lines. The 514-line, attenuated to equal the 457-line
intensity ($\eta = 1$), is switched-on when irradiation with the 457-line
has produced PE. After the new PE is reached, the 514-line intensity
is progressively increased at each new PE to give smaller values of
η. Figure 19 shows the differential absorbance spectrum at PE for
several values of η. As can be seen, efficient quenching of PBR
formation is achieved with $\eta = 1/12$ and 1/40. After the $\eta = 1/40$
PE was reached, the 514-line has been switched off: the DA_{490}
recovered only $\sim 2/3$ of the initial value. A progressive change of
the DA spectrum is apparent, with the disappearance of the isosbestic

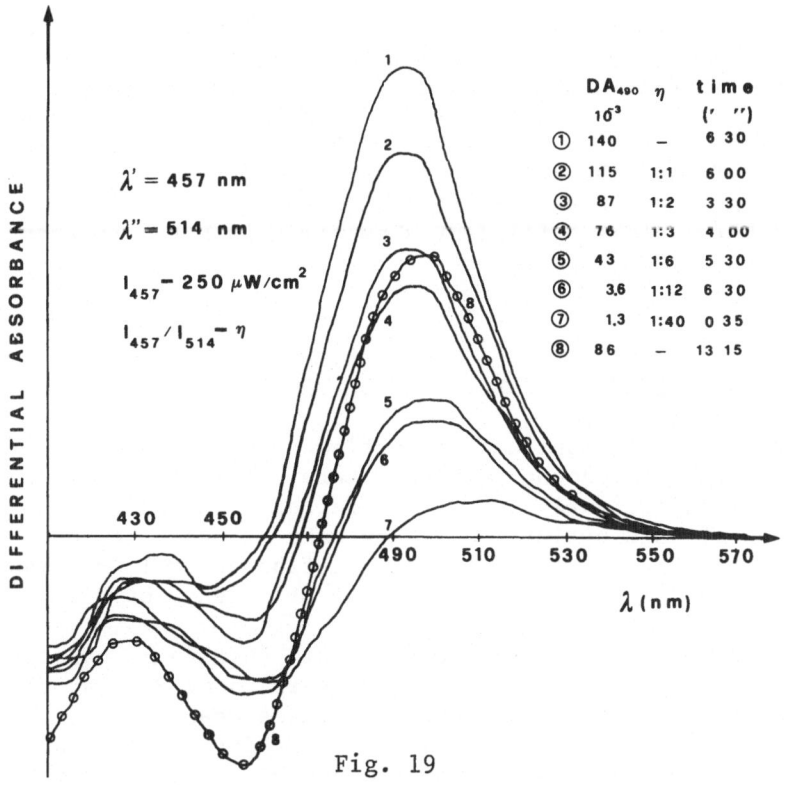

Fig. 19

point at 465 nm. This behavior could be mainly related to the formation of LR, and not to degradation of BR or/and PBR.

"In vivo" measurements. Two-light irradiation of Gunn rats has been carried out by Ballowitz [31]. No significant difference has been found in the decline of serum BR concentration with two different combinations of blue (Philips BAM blue 20 W/52) with green (Osram L20 W/63) and white lamps (these lamps correspond closely to Westinghouse special blue and Sylvania green lamps). However, unlike in "normal" icteric infants, configurational isomerization is a very efficient process in Gunn rats [28]. The presence of green light in this case would tend to decrease the PT efficiency. Moreover, the comparison of single blue light and two-light irradiation has been performed at the same "effective" irradiance, i.e., with the blue/white or blue/green light mixture adjusted to give the same (Gunn rat) phototherapeutical response as in the single light irradiation. The equivalence of the observed results appears simply as a good calibration of their "special" radiometer!

In conclusion, large reduction of the PBR concentration produced by blue light excitation of BR can be achieved by simultaneous irradiation with green light of suitable wavelength, band width and

intensity. In principle, the same desired absorption by BR molecules
as produced by a blue light matching the BR absorption maximum, but
at the minimum PBR concentration, can be achieved using a single
green light beam with an intensity $\sigma_1(\lambda')/\sigma_1(\lambda'')$ times larger than
the blue light intensity. This approach may result impractically
in clinical PT, and the relative absorption of BR and PBR in the
long-wave wing msut be experimentally investigated further in order
to see whether a still sufficient absorption by PBR with practically
no excitation of BR is possible.

4.   INFLUENCE OF SKIN OPTICAL LOSSES ON PHOTOTHERAPY EFFICIENCY

The effect of the optical properties of the skin on PT efficiency
has been taken into account only in a very qualitative way due to
the lack of a sufficiently accurate model of skin optics. Recent
improvements of skin optics theory now make it possible to obtain
more qualitative information about the influence of light transmission through the skin on the spectral efficiency of PT.

The extravascular tissue near the skin surface is considered
the main site for formation of PBR. However, some PBR may also be
formed directly in plasma as blood circulates near the skin surface,
in particular in the profuse system of upper dermal capillaries and
in the subpapillary venous plexus at a depth of 0.3 - 0.5 mm beneath
the skin surface [32].

Recently, Parrish's group [33] considered the human skin as a
turbid medium and applied the Kubelka-Munk phenomenological model
of radiation transfer to the dermis. The Kubelka-Munk model considers the absorption and scattering coefficients as properties of
the radiated object and simplifies the radiation into two oppositely
directed fluxes I and J (Figure 20). The fluxes are described by
a system of linear differential equations with constant parameters:

$$dI = 2(-KI - SI + SJ)dx$$
$$-dJ = 2(-KJ - SJ + SI)dx.$$

(17)

The sample's back-scattering (S) and absorption(K) coefficients for
diffuse radiation are defined as the fraction of diffuse radiation
either back-scattered or absorbed per unit differential pathlength
dx of the sample. Since the layer dx is subject to radiation at all
possible angles with respect to the x-direction as a result of
diffuse radiation, the mean path length of the radiation within dx
is evidently larger than dx. It has been proved that when the
incident radiation is diffuse, the mean path length is twice the
geometrical layer thickness [34].

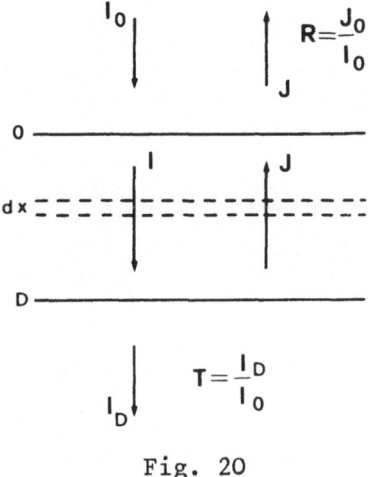

Fig. 20

With appropriate boundary conditions a closed form solution of Eq. (17) can be obtained, and the constants K and S can be expressed in terms of the measurable boundary conditions, as diffuse reflectance (R) and transmittance (T).

In the present analysis we consider the jaundiced skin as composed of multiple layers possessing distinct optical properties. Dermal BR is assumed to be confined within an optical thin layer underlying a 100 μm thick epidermal (ED) layer and a 200 μm thick dermal (D) layer. Values of ED and D absorption and scattering coefficients are derived from the "in vitro" measurements on human skin reported by Parrish and co-workers. The competing effect due to hemoglobin (Hb) absorption in the blood is taken into account by introducing a Hb-layer of suitable thickness between the D-layer and BR-layer.

The transmission of the ED-D-Hb multiple layer is given by:

$$T = \frac{T_{ED}T_D}{1 - R_{ED}R_D} \ \exp[-\alpha d_{Hb}] \tag{18}$$

where R_{ED}, R_D and T_{ED}, T_D denote diffuse reflectance and diffuse transmittance of the ED- and D-layers, respectively, with:

$$R = [(\tfrac{K}{S}+1) + \sqrt{(\tfrac{K}{S})^2 + 2\ \tfrac{K}{S}} \ \coth(\sqrt{K^2 + 2KS}\cdot 2d)]^{-1}$$

$$T = [1 + R^2 + 2R(1 + \tfrac{K}{S})]^{1/2} \tag{19}$$

where d denotes the thickness of the layer. In Eq. (18) the absorption coefficient of the Hb-layer $\alpha = \varepsilon c\ \ln(10)$ is expressed in terms of the molar extinction coefficient (ε) and concentration (c) of

HbO_2; d_{Hb} indicates the thickness of the HbO_2-layer. We have used values of the product $(cd)_{HbO_2}$ ranging from 0 to 10^{-4} cm M/ℓ, which correspond, at the HbO_2 concentration in normal blood ($\sim 2.2 \cdot 10^{-3}$ M/ℓ), to thicknesses up to 0.20 mm (the diameters of the dermal vessels vary between 0.1 and 0.3 mm).

Figure 21 shows the wavelength dependence from 350 to 600 nm of: i) K,S in the case of a 100 μm ED-layer and 200 μm D-layer; ii) the molar extinction coefficients of BR and HbO_2; iii) the multiple layer transmission T for several values of $(cd)_{HbO_2}$.

4.1 Effect of Skin Attenuation on BR Absorption Rate

Light absorption by BR molecules is the first step in the complex process which leads to BR elimination from the body. Therefore, the photon absorption rate, A_1, of BR is considered firstly as the suitable parameter to determine the influence of skin attenuation on the spectral efficiency of a light source for PT. From Eqs. (5) and (18) the photon absorption rate in the BR-layer is given by:

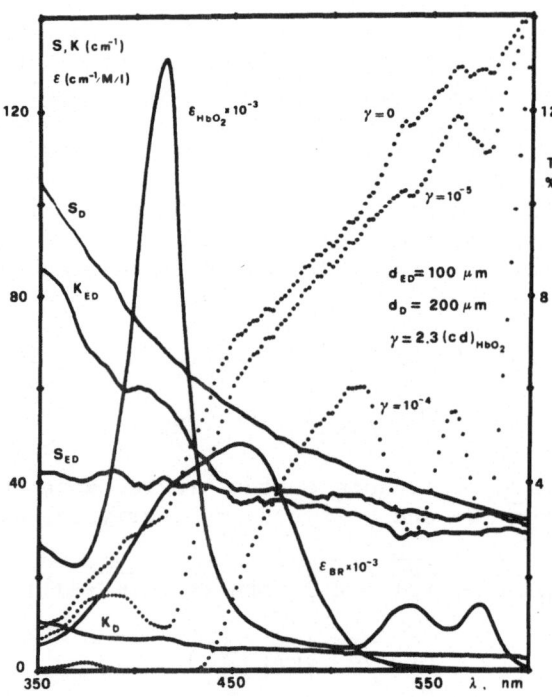

Fig. 21

$$A_1(\nu_p) = \int T(\nu)\sigma_1(\nu)N(\nu\nu_p)d\nu = N_T\overline{\overline{\sigma}}_1(\nu_p) \qquad (20)$$

where

$$\overline{\overline{\sigma}}_1(\nu_p) \equiv \int T(\nu)\sigma_1(\nu)\chi(\nu\nu_p)d\nu. \qquad (21)$$

Figure 22 shows the $A_1(\lambda_p)$ spectra computed assuming a spectrally Gaussian light source with band width $\Delta\lambda = 1, 35$ nm for $T(\lambda) = 1$ ("in vitro" case) and $\gamma \equiv 2.3$ (cd)$_{HbO_2} = 0, 10^{-5}, 10^{-4}$ cm M/ℓ.

The effect of the blood-free skin layers is a marked reduction of the absorption rate between 350 and 425 nm, and a red-shift $\delta\lambda = 10$ nm of A_1 for $\Delta\lambda = 1$ nm and $\delta\lambda = 30$ nm for $\Delta\lambda = 35$ nm. This is mainly due to the strong wavelength dependence in this region of: 1) S_D, which varies approximately as $\lambda^{-2.5}$; 2) K_{ED} which closely corresponds to the absorption spectrum of melanin: therefore, the short-wave part of the absorption curve of BR is sensibly less efficient than the long-wave part for BR photoexcitation. The effect of blood absorption is to further decrease $A_1(\lambda_p)$ and shift the A_1-maximum towards longer wavelengths ($\delta\lambda \simeq$ nm for $\Delta\lambda = 1, 35$ nm, respectively when $\gamma = 10^{-4}$ cm M/ℓ).

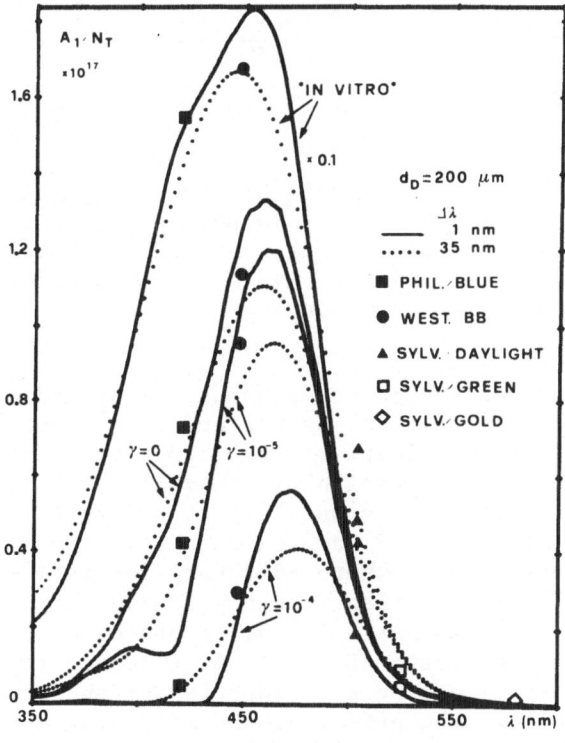

Fig. 22

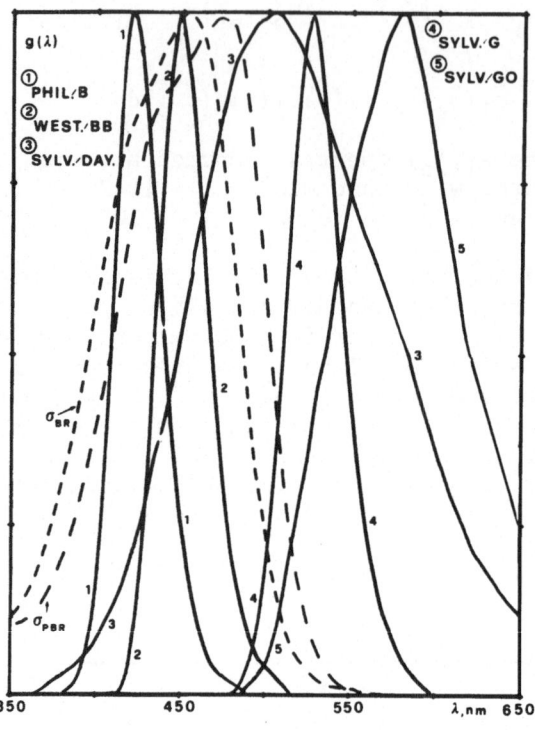

Fig. 23

Finally, the BR absorption rate has been computed in the case
of light emitted by the fluorescent lamps used for PT. The spectral
profiles of blue, special blue, white, green, and gold lamps are
shown in Figure 23. The Hg-lines have not been considered here.
Their effect is under investigation and it will be reported elsewhere.

Table 4 summarizes the computed values of A_1 and A_2; the numbers
in parenthesis indicate A-values as percent of the corresponding
values of the Westinghouse lamp. In the "in vitro" case (first
column) the Westinghouse special blue lamps ($\lambda_p \simeq 448$ nm, $\Delta\lambda = 34$ nm)
is, as expected, the most efficient lamp to excite BR. The Philips/
blue ($\lambda_p \simeq 421$ nm, $\Delta\lambda \simeq 33$ nm) lamp has the emission maximum shifted
towards smaller wavelengths, where BR absorption is lower and
results $\sim$ 8% less efficient than Westinghouse/BB lamps . The Sylv./
Daylight lamp ($\lambda_p \simeq 504$ nm, $\Delta\lambda \simeq 136$ nm) produces a relatively large
BR absorption, $\sim$ 40% of West./BB, while Sylv./Green ($\lambda_p \simeq 526$ nm,
$\Delta\lambda = 38$ nm) has only 5% of the BR absorption of special blue lamp.
The second column shows the effect of the filtering action of the
blood-free ED-D layers. The efficacy of West./BB is reduced by a
factor $\sim$ 15 owing to skin scattering and absorption; the effect is
more pronounced for the Phil./B lamp, whose efficiency is now only

Table 4.

LAMP	$\bar{\sigma}$, $(\tau N_T)^{-1}$ 10^{19} (cm²)	"In vitro"	$d_D = 200$ μm $\gamma = 0$	10^{-5}	10^{-4}	$d_D = 400$ μm $\gamma = 0$	10^{-5}	10^{-4}
PHILIPS/BLUE TL20W03/T	$\bar{\sigma}$ 1	1541.5	72.38	41.77	4.85	28.26	14.57	1.84
	(%)	(92.5)	(63.9)	(43.8)	(16.7)	(65.6)	(40.0)	(16.2)
	$\bar{\sigma}$ 2	1214.5	58.45	34.66	4.28	19.77	12.17	1.63
	(%)	(79.9)	(55.8)	(38.9)	(14.7)	(49.2)	(35.5)	(14.2)
	$(\tau N_T)^{-1}$	1312.6	62.6	36.80	4.45	21.11	12.89	1.69
	(%)	(83.9)	(58.3)	(40.5)	(15.3)	(51.5)	(36.9)	(14.8)
	n_2	.352	.347	.340	.327	.344	.339	.326
WESTINGHOUSE/BB F20T12/BB	$\bar{\sigma}$ 1	1666.9	113.3	95.22	28.94	43.06	36.39	11.32
	(%)	(100)	(100)	(100)	(100)	(100)	(100)	(100)
	$\bar{\sigma}$ 2	1519.9	104.8	88.98	29.01	40.13	34.26	11.45
	(%)							
	$(\tau N_T)^{-1}$	1564.0	107.39	90.85	28.99	41.01	34.90	11.41
	(%)							
	n_2	.320	.316	.314	.299	.315	.313	.297
SYLVANIA/ DAYLIGHT F20F12/DA	$\bar{\sigma}$ 1	667.4	48.10	41.78	18.35	19.10	16.80	7.58
	(%)	(40.0)	(42.4)	(43.9)	(63.4)	(44.3)	(46.2)	(67.0)
	$\bar{\sigma}$ 2	783.2	60.53	54.19	26.67	24.79	22.38	11.25
	(%)	(51.5)	(57.7)	(60.9)	(91.9)	(61.8)	(65.3)	(98.2)
	$(\tau N_T)^{-1}$	748.4	56.80	50.46	24.18	23.09	20.71	10.15
	(%)	(47.8)	(52.9)	(55.5)	(83.4)	(56.3)	(59.3)	(88.9)
	n_2	.267	.254	.248	.228	.248	.243	.224
SYLVANIA/GREEN F20F13/G	$\bar{\sigma}$ 1	92.15	9.13	8.55	4.84	4.18	3.91	2.20
	(%)	(5.5)	(8.0)	(9.0)	(16.7)	(9.7)	(10.7)	(19.4)
	$\bar{\sigma}$ 2	223.4	22.13	20.74	11.90	10.13	9.49	5.42
	(%)	(14.7)	(21.1)	(23.3)	(41.0)	(25.2)	(27.7)	(47.3)
	$(\tau N_T)^{-1}$	184.0	18.23	17.08	9.78	8.34	7.81	4.45
	(%)	(11.7)	(17.0)	(18.8)	(33.7)	(20.3)	(22.4)	(39.0)
	n_2	.150	.150	.150	.148	.150	.150	.148
SYLVANIA/GOLD F20F12/GO	$\bar{\sigma}$ 1	19.15	2.06	1.89	0.91	0.97	0.89	0.42
	(%)	(1.1)	(1.8)	(2.0)	(3.1)	(2.2)	(2.4)	(3.7)
	$\bar{\sigma}$ 2	40.28	4.20	3.88	1.99	1.96	1.81	0.92
	(%)	(2.6)	(4.0)	(4.3)	(6.8)	(4.9)	(5.3)	(8.0)
	$(\tau N_T)^{-1}$	33.94	3.56	3.29	1.67	1.66	1.53	0.77
	(%)	(2.1)	(3.3)	(3.6)	(5.7)	(4.0)	(4.4)	(6.7)
	n_2	.169	.174	.173	.164	.176	.174	.165

64% of that of West./BB. The Sylv./Daylight lamp maintains its
relative efficacy, $\sim$ 42%. with respect to West./BB. When the
presence of the skin vascular system is taken into account, the blue
radiation around the Soret maximum of HbO_2 (at 418 nm) is further
attenuated. The efficiency of Phil./B drops rapidly while those of
Sylv./Daylight and Sylv./Green remain nearly constant and are reduced
only for the highest value of γ: in this case the A_1-value of the
green lamp becomes equal to that of the violet-blue lamp; the A_1-
value of the white lamp is now 3.8 times that of the violet-blue
lamp. The efficiency with respect to West./BB decreases from 64%
to 17% for the Phil./B, increases from 42% to 63% for the Sylv./
Daylight, and from 8% to 17% for the Sylv./Green. It is worth noting
the large increase of the relative efficiency of white and green
lamps produced by the filtering action of the skin with respect to
the "in vitro" case. In the case of green lamp the relative BR
absorption is increased by a factor > 3.

The extrapolation of the K,S values measured in a 200 μm thick
dermal sample to a 400 μm thick dermal layer (last four columns)
leads to the same general conclusion: it is worth noting that at
large blood content the efficiency of white lamps approaches that
of the special blue lamps (relative efficiency $\sim$ 70%). If we consider that daylight lamp radiation with λ > 530 nm is not effective
for BR excitation, the nearly equivalence of special blue and daylight lamps indicates that a larger efficiency should be obtained
with a narrow spectrum fluorescent lamp with peak emission wavelength
between 450 and 500 nm, as found in the case of Gaussian source with
$\Delta\lambda$ = 35 nm. Note that now the green lamp efficiency is $\sim$ 1/5 of the
special blue one. The West./BB lamp remains the most effective lamp
over the entire range examined.

In conclusion, the above results show that the upper skin layers
and hematic vascular system affect the optimum value of the peak
wavelength of the light source to be used for PT. The balance
between light losses in the skin and good light absorption by BR
leads to an optimum peak wavelength for the photon absorption rate
around $\lambda \simeq$ 480 nm, i.e., to a $\sim$ 30 nm shift with respect to the BR
absorption maximum in solution.

4.2 Effect of Skin Attenuation on PBR Formation

According to Eqs. (8) and (7) with σ_j replaced by $\bar{\bar{\sigma}}_j$ (as defined
in Eq. (21)) the steady-state PBR concentration does not depend on
the total light intensity of the irradiating beam: therefore, $n_2(\infty)$
is expected to be influenced by the skin transmittance $T(\lambda)$ only
for large band widths and in those spectral intervals where a large
variation of $T(\lambda)$ is present. Figure 24 shows $n_2(\lambda)$ vs λ_p for $\Delta\lambda$ =
= 35 nm and $T(\lambda)$ = 1; γ = 0, 10^{-5}, 10^{-4} cm M/ℓ. The corresponding

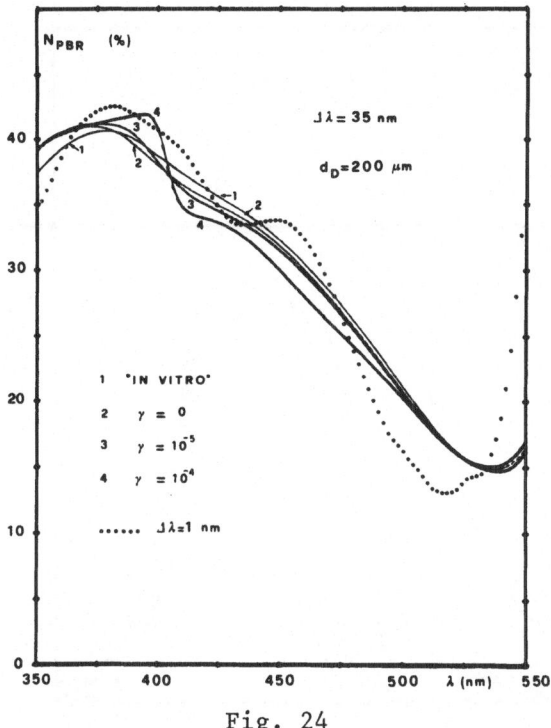

Fig. 24

curves idffer only around the Soret absorption peak of HbO$_2$ at 418 nm. The case $\Delta\lambda = 1$ nm is also shown for comparison (dotted line):. in this case, skin attenuation has no effect on the PE concentration of PBR.

On the contrary, the PE rise-time τ is sensibly affected by skin losses. Figure 25 shows the plots of $(\tau N_T)^{-1}$ vs λ_p for $\Delta\lambda =$ $= 1$, 35 nm (solid and dotted lines, respectively), for the same cases considered in Figure 24. Since the σ_2 peak value is only slightly smaller than σ_1, and $\phi_2 \simeq 2 - 4$ times larger than σ_1, the $\phi_2 A_2$ term in Eq. (7) predominates in the spectral range where $\sigma_2 \geqslant$ $\geqslant \sigma_1$, i.e., in the blue-green portion of the spectrum (470 - 530 nm). The wavelength of the τ^{-1} maximum for $\Delta\lambda = 1$ nm is shifted from the "in vitro" value 464 to 466, 468, 484 nm for $\gamma = 0$, 10^{-4}, 10^{-5} cm M/ℓ, respectively and $d_D = 200$ µm. When $\Delta\lambda = 35$ nm, the corresponding values are: 456, 468, 472, 482 nm.

In Table 4 we have computed the value of $(\tau N_T)^{-1}$ and $n_2(\infty)$ for various fluorescent lamps. The effect of skin losses on PE rise-time for increasing blood content is quite important and selective: the rise-time relative to West./BB lamp (i.e., τ/τ_{BB}): i) decreases for the Philips violet-blue lamp from 84% to 15%; ii) increases from 48% to 83% or 89%. depending on the dermal thickness, for the Sylv./

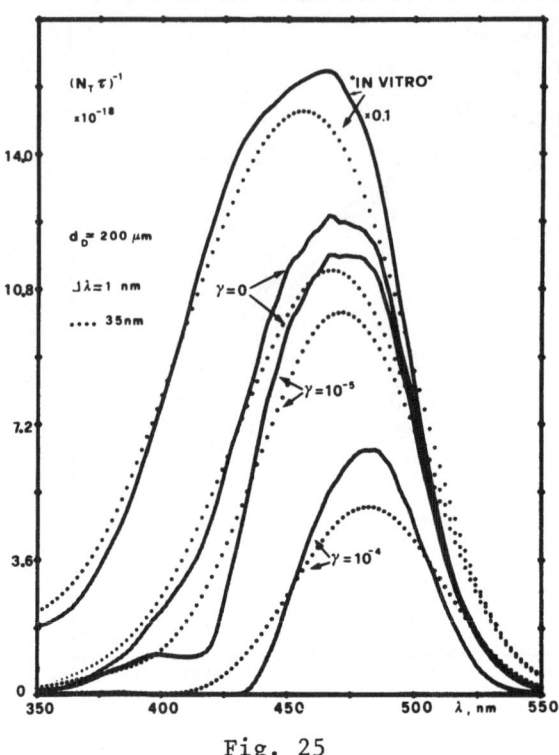

Fig. 25

Daylight, and from 12% to 34% or 39% for the Sylv./Green lamp. It
is important to note that green light produces PBR concentration
1/2 lower than special blue lamp, while the BR absorption and PE
rise-time are only 5 and 2.5 times smaller than in the corresponding
special blue case, respectively (in the case of highest dermal thickness and blood content here considered). Note also the good
efficiency of white light to produce BR absorption ($\sim$ 70%) and fast
isomerization ($\sim$ 90%) compared to special blue light. Both white
and green lights become much more efficient than the violet-blue
light when skin losses are taken into account. This agrees with
the experimental findings of Ebbesen and Tan [35,36].

In the latter case an irradiance with violet-blue (Philips TLAK
40 W/03) $\sim$ 2 times (1.66 - 1.84 in the 400 - 480 nm range) larger than
with white light (Philips TL20/54) produces only 1.1 increase in
24-h decline of infant serum BR.

A direct measurement with PHLC of PBR concentration at PE in
serum of newborns under PT with tungsten-halogen lamps gives $n_2(\infty)$
$\simeq$ 12%, that is very close to the "in vitro" value (14.5%) found
under the same irradiation conditions [5]. Moreover, $n_2(\infty)$ is not
affected by doubling the energy fluence rate of the source. On the

contrary, the steady-state concentration of LR doubles when the
light power is increased by 2 as expected for an irreversible
reaction.

5. THE OPTIMUM LIGHT FOR PHOTOTHERAPY

The optimum light for PT is yet undetermined, as well as the
entire irradiation procedure (i.e., continuous or intermittent illumination, optimum light power density at the body surface, etc.).
The determination of a complete and detailed action spectrum of
jaundiced infants has obvious moral implications, and comparisons
of non-toxic, high efficiency light sources are the only clinical
data available so far [35-37]. Therefore, "in vitro" [38-40] and
animal (Gunn rat) [41-43] action spectra have been used to guide
the clinical approach.

In addition to the difficulty to extrapolate the experimental
results to man, the optimum light for BR elimination from the
organism will clearly change each time a progress in the understanding
of BR photochemistry and metabolism is obtained. Moreover, the
influence of many factors, as, for instance, the filtering action
of skin, cannot be taken into account in a realistic way using "in
vitro" or animal models. On the other hand, PT has been introduced
since the beginning as a simple, fast, and relatively safe clinical
treatment, and a dramatic change of the present protocol is not to
be expected when the mechanism of action of light to clear off BR
from the organism will be definitely understood. Improvements of
PT efficiency and safety will, however, accompany a more rigorous
phototherapeutical approach.

5.1 Basic Mechanisms

During the "photooxidation period" (1958 – 1978) special lamps
with spectral emission matching the BR absorption maximum have been
developed and recommended for highest PT efficiency on grounds of
both "in vitro" and Gunn rat action spectra. In the subsequent
"photobilirubin period" (1978 – 1983) the use of special blue and
white lamps was continued, even if a different spectral light distribution would have to be expected for optimal PT. Detailed
investigations of BR phototransformation using laser light were
reported [7,43] but in experimental situations suitable for low QY
photoprocesses, thus missing BR configurational isomerization. Only
very recently Ennever et al.[22] have measured the "in vitro" action
spectrum of PBR and shown that optimum light for PBR formation falls
in the UV (390 nm): in any case, safety considerations would have
forced the use of longer wavelength light in clinical application,
i.e., blue light, once again. Today, at the beginning of the
"Lumirubin period" we need a better knowledge of LR properties "in

vitro" and "in vivo": if LR is formed only from excited BR, and PBR
formation is a counterproductive process, optimal light would very
likely produce the highest absorption by BR and at the same time
keep PBR concentration as low as possible at the formation sites.
However, if the toxicity of PBR (presently not known) was lower [44]
than that of BR, the formation of PBR should be maximized as it
transforms BR to safer photoproducts quickly and very efficiently
[45]. In the opposite case (i.e., same or greater toxicity than
BR) light matching the long-wave portion of BR absorption curve
should be preferred.

5.2   The "Green Light Story"

Vecchi and co-workers in Florence have recently demonstrated
that green light emitted by standard fluorescent lamps have clinical
efficiency higher than daylight lamps [8,9] and comparable [10] to
that of special blue lamps. This result came out unexpected according
to the established protocols, and followed from positive "in vitro"
measurements on the efficacy of laser and lamp green lights in
photolysis of BR [7].

I am not aware of any previous clinical use of green light.
Its efficacy "in vitro" and in Gunn rats was, however, already known.
Sisson and co-workers report i) good BR decomposition "in vitro"
[38] using green lamps with emission in the spectral range 540 – 600
nm (sic) ($\sim$ equal to that of daylight); and ii) 50 – 55% reduction
in serum BR in Gunn rats [41] under blue and green irradiation;
their conclusion is that green light (460 – 530 nm, sic) degrades BR
"in vitro" and "in vivo" and that the response is due to the energy
output with wavelengths below 500 nm. Since then little interest
has been devoted to the action of wavelengths above 500 nm, especially
for clinical applications.

Later, a detailed study of Ballowitz and co-workers [42] showed
that the largest reduction of serum BR concentration in Gunn rats
was associated to blue light (Philips 20W/52BAM-blue, nearly equivalent to Westinghouse/BB), and the lowest to UV and green (Osram
L20/63 green) lamps. Soon after McDonagh [16] found that the PBR
pathway is the fastest photochemical process of BR also in Gunn
rats, where very efficient formation, transport, and excretion of
BR occur, so that only traces of PBR are present in serum during PT.
Thus, green light with $\lambda > 500$ nm, scarcely efficient enough to form
PBR and very efficient to revert PBR, could not lead to efficient
PT of Gunn rats.

As usual, the Gunn rat results were transported to man. Fast
elimination of BR in babies with the Crigler-Najar (CN) syndrome

supported this assumption, even if the steady-state levels of PBR in babies (15%) during PT were much higher than those observed in the Gunn rat, and appeared to argue against its efficient excretion [46]. However, the congenital inability of Gunn rats to conjugate BR is present in man only in infants with CN-syndrome [47]. Dosage of PBR in blood serum of icteric infants with HBR of different pathological origin has, in fact, shown that PBR is formed in large amount (12%) during PT, but excreted with a time constant one order of magnitude larger than for the CN case [5].

The puzzling good clinical efficiency of green lamps with peak emission wavelength $\lambda_p > 500$ nm may rely on the following considerations: i) Sylvania green lamps ($\lambda_p \simeq 526$ nm, $\Delta\lambda \simeq 38$ nm) have sufficient blue light absorption ($\sim 6\%$ of West./BB "in vitro" and $\sim 20\%$ in skin) to form LR and at the same time much more green light (15% "in vitro" and 48% in skin) to keep PBR concentration low at the formation sites, and then increasing LR efficiency; ii) larger penetration of green light into skin tissues can remove BR from deeper sites less accessible to blue light; iii) the absorption spectrum of BR in skin, where it is bound to lipids and collagen, may differ very much from that of free or HSA-bound BR in plasma. Red-shifted absorption spectra with $\lambda_{max} \simeq 490$ nm have been reported for BR aggregates different from albumin complexes [43]. The larger effect associated more to 488-line than to 457-line found by Odell and co-workers in newborn Gunn rats could support this hypothesis, or simply depend on skin effect [43].

6. PHOTOTECHNOLOGY AND SIDE EFFECTS

6.1 Phototechnology

While important progresses have been registered in the understanding of PT mechanism, only a little progress has been made in the related phototechnology, and a suitable physical approach to evaluate the most effective irradiation procedure is still undetermined. The history of the development of the PT units presently in the clinical use is instructive [48]. These units usually consist of banks of 8 to 10 (or to 20) fluorescent lamps (20 W) adjacent to each other and covered with a Plexiglass-G sheet (to absorb UV radiation). The geometrical configuration was not based on any experimental dose-response study. It was simply decided that this was the largest number of lamps that could be easily placed above an incubator without interfering with nursing care. The only major improvement in the PT unit is the development, made 25 years ago, of special fluorescent lamps emitting mostly in blue regions and matching the absorption band of BR.

6.2   LED-Arrays for Phototherapy

The light sources currently used for PT are fluorescent lamps
emitting narrow-spectrum visible radiation in the violet-blue region,
or broad-spectrum radiation, such as white and daylight lamps. Light
power density at the body surface ranges typically from 1 to 3 mW/
cm^2. These lamps are efficient and cheap; however, the high intensity Hg-lines are always present superposed on the continuous fluorescent emission of the phosphor. The glass envelope of the lamp
cuts-off the residual UV-C and UV-B lines, while the plastic sheets
of the illuminator or/and incubator block all wavelengths below 380
nm. On the other hand, the 400 - 450 nm violet-blue radiation has
been found to cause toxic and mutagenic effects on cultured mammalian
cells. These lines cannot be easily filtered out without strongly
reducing the light intensity within the useful PT spectral region.
Even the recently proposed metal-vapor lamps [49] emit a large amount
of power in the 400 - 430 nm range, which should be ineffective for
PT according to our computation of Section 4.

Typical solid-state sources of narrow-band visible radiation
are light-emitting diodes (LED). High-efficiency visible LEDs are
being increasingly developed due to requirements for a variety of
optical applications. Usual peak radiation wavelengths are in the
green, yellow,and red. The development of high intensity LEDs at
other wavelengths is only a technological and economical problem,
and it may be triggered by some specific application.

Blue LEDs with λ_p = 480 nm and $\Delta\lambda$ = 90 nm are already available
at very low radiation power. Its spectral characteristics appear
suitable for efficient PT, as discussed in the other paper in this
volume. As soon as the technological developments improve the
power emission capability of this LED, a suitable shape panel with
tightly-packed arrays of LEDs could generate a sufficient power
density for PT applications, thus eliminating hazardous spectral
components. Good output beam collimation can then be acieved by
using the fabrication technique for the internal microlensing system
already available with IR-LEDs.

6.3   Side-Effects

In large part, the reason for the rapid acceptance of PT of
hyperbilirubinemia was the perception that light cannot be harmful
to infants. The argument was that the entire phylogenic development
of man occurred in visible light, and since the doses applied were
only one tenth the intensity of sunlight in temperate climates they
were clearly benign [48]. When PT was first introduced no side
effects were expected for the above reason and, possibly, because
none were expected, none were found [48]. As late as 1972 possible
"hazards" of PT appeared in the literature. Since then a number of

side effects have been demonstrated. Most of these clinical side effects have been reported following the use of broad-band spectrum fluorescent light, and are reported in several review papers [12].

Moreover, in recent years the growing awareness of blue-light-induced phenomena in plants, microorganisms, and animals has accelerated and expanded this research into an ever-increasing variety of blue-light effects in biological systems [50].

Still more recently, findings from several research areas have shown that even visible light emitted by fluorescent lamps is toxic and mutagenic. The effective wavelengths are in the 400 - 500 range. Visible light produces mutations; single-strand breaks; protein-DNA cross-links; chromatid damage and sister chromatid exchanges; and enhances neoplastic transformation of mammalian cells in culture [6].

In particular, from these experiments it turns out that chromosome damages are: present, when cool-white fluorescent light (the Westinghouse F15T8CW) is used (the effective wavelengths are around the 405 Hg peak) [51]; absent, when green fluorescent lamps (the Westinghouse F15T8G) are used with P-3 filter to cut-off the Hg lines of the lamp at 365, 405, and 436 nm [52]. In addition, Burki et al [53] have reported that the removal of wavelengths below 500 nm eliminates the toxic and mutagenic effects of white fluorescent light on mammalian cells.

The action spectrum of DNA strand breakage in normal human fibroblasts has been found to exhibit a second peak at 450 nm, and to be similar in the visible to the normalized riboflavin absorption spectrum [54]. A PT lamp that may achieve a reduction in the amount of DNA damage would emit wavelengths from 470 - 530 nm where BR absorption is still significant, but the ratio of the normalized strand breakage action spectrum to the normalized BR absorption spectrum is low.

On the other hand, alterations of human chromosome in lymphocytes have been found after exposure to relatively low dose of green laser light (514 nm).

Although "in vitro" phenomena should only extrapolate with caution to man, the potential side effects due to blue light PT [12], must be taken into account. It is debated whether, as a consequence of PT, peripheral blood lymphocytes exhibit increased sister chromatid exchange [55].

Moreover, recent measurements [56] have shown that there is essentially no VIS light transmitted through the chest or the abdominal wall of the body at wavelengths less than 500 nm, while in contrast, 10^{-4} of blue light less than 500 nm can reach the brain and testes. In the situation reported by Sisson [37] this means a dose of at least 25 mJ/cm^2 at the brain surface.

Finally, it is worth recalling that the threshold for non-thermal lesions of the retina is ten times greater for green light than for blue light [57]. Reduced risk of retinal damage to doctors and nurses is then expected with longer wavelength PT.

ACKNOWLEDGMENT

Stimulating discussions with Prof. G. Jori, F. Hillenkamp and J. A. Parrish are gratefully acknowledged.

This work has been supported in part by the CNR Special Project "Laser" and is part of a research program on phototherapy of jaundice in the newborn supported by the Italian CNR in Florence. Clinical investigations and "in vitro" laser photolysis of bilirubin are carried out by C. Vecchi and G. P. Donzelli at the Insitute of Pediatrics, and G. Sbrana and M. G. Migliorini at the Institute of Physical Chemistry. The section on the "Photoisomerization of bilirubin" has been written in collaboration with G. Agati and F. Fusi, Istituto di Farmacologia, Università, Firenze; and, the section on the "Influence of skin losses on phototherapy efficiency", in collaboration with G. Cecchi, Istituto di Elettronica Quantistica del CNR, Firenze.

REFERENCES

1. F. Hillenkamp, R. Pratesi and C. A. Sacchi, eds., "Lasers in Biology and Medicine", Plenum Press Ltd., New York (1980).
2. J. D. Regan and J. A. Parrish, eds., "The Science of Photomedicine", Plenum Press Ltd., New York (1982).
3. R. Pratesi and C. A. Sacchi, eds., "Lasers in Photomedicine and Photobiology", Springer, Heidelberg (1980).
4. R. Cubeddu and A. Andreoni, eds., "Porphyrins in Tumor Phototherapy", Plenum Press Ltd., New York, in press.
5. J. F. Ennever and W. T. Speck, Mechanism of Action of Phototherapy: New Aspects, in: "New Trends in Phototherapy", eds., F. Rubaltelli and G. Jori, Plenum Press Ltd. (1983).
6. R. Parshad, R. Gantt, K. K. Sanford, G. M. Jones and R. F. Camalier, Int. J. Cancer, 28:335 (1982).
7. G. Sbrana, M. G. Migliorini, C. Vecchi and G. P. Donzelli, Pediat. Res., 15:1517 (1981).
8. C. Vecchi, G. P. Donzelli, M. G. Migliorini, G. Sbrana and R. Pratesi, The Lancet, August 14, 390 (1982).
9. C. Vecchi, G. P. Donzelli, M. G. Migliorini and G. Sbrana, Pediat. Res., 17:461 (1983).

10. G. P. Donzelli, M. G. Migliorini, R. Pratesi, G. Sbrana and
     C. Vecchi, Laser-Oriented Search of Optimum Light for
     Phototherapy, in: "New Trends in Phototherapy", eds., F.
     Rubaltelli and G. Jori, Plenum Press (1983).

11. R. Pratesi, The Lancet, October 8, p. 859 (1983).

12. T. R. C. Sisson and T. P. Vogl, Phototherapy of Hyperbilirubin-
     emia, in: "The Science of Photomedicine", eds., J. D. Regan
     and J. A. Parrish, Plenum Press, New York (1982).

13. H. Fischer, K. Herrle and Z. Hoppe-Sayerl, Physiol. Chem.,
     251:85 (1938).

14. D. A. Lightner, "In Vitro" Photooxidation Products of Bilirubin,
     in: "Phototherapy of the Newborn: an Overview", eds., G. B.
     Odell, R. Schaffer and A. Simopoulos, pp. 34-55, Nat. Acad.
     Sci., Washington DC (1974).

15. D. A. Lightner, T. A. Wooldridge, and A. F. McDonagh, Proc.
     Nat. Acad. Sci. USA., 76:29 (1979).

16. A. F. McDonagh and L. M. Ramonas, Science, 201:829 (1978).

17. A. F. McDonagh, J. Pediat., 99:909 (1981).

18. B. I. Greene, A. A. Lamola and V. Shanck, Proc. Nat. Acad.
     Sci. USA., 78:2008 (1981).

19. A. A. Lamola, J. Flores and F. H. Doleiden, Photochem. Photo-
     biol., 35:649 (1982).

20. R. W. Sloper and T. G. Truscott, Photochem. Photobiol., 31:
     445 (1980).

21. J. I. Steinfeld, Molecules and Radiation, The MIT Press,
     Cambridge, p. 30 (1979).

22. J. F. Ennever, A. F. McDonagh and W. T. Speck, J. Pediat.,
     103:295 (1983).

23. A. F. McDonagh, L. A. Palma, F. R. Trull and D. A. Lightner,
     J. Amer. Chem. Soc., 104:6865 (1982).

24. A. A. Lamola, Personal communication.

25. T. R. C. Sisson, M. Ruiz, K. T. Wu et al., Pediat. Res., 12:
     535 (1978).

26. A. F. McDonagh, Photochemistry and Photometabolism of Bilirubin,
     in: "Bilirubin Metabolism in the Newborn (II)", eds., D.
     Bergsma and S. H. Blondheim, Excerpta Medica, Amsterdam,
     p. 30 (1976).

27. A. F. McDonagh and L. A. Palma, J. Am. Chem. Soc., 104:6867
     (1982).

28. A. F. McDonagh, Molecular Mechanism of Phototherapy of Neonatal
     Jaundice, in: "New Trends in Phototherapy", eds., F. Rubal-
     telli and G. Jori, Plenum Press (1983).

29. R. Bonnet, Recent Advances in the Chemistry of Bile Pigments,
     in: "New Trends in Phototherapy", eds., F. Rubaltelli and
     G. Jori, Plenum Press (1983).

30. H. Falk, N. Müller, M. Ratzenhofer and K. Winsaurer, Monatshefte
     für Chemie, 113:1421 (1982).

31. L. Ballowitz, F. Hanefeld and G. Wiese, The Gunn Rat: A Model
     for Phototherapy, in: "New Trends in Phototherapy", eds.,
     F. Rubaltelli and G. Jori, Plenum Press (1983).

32. A. F. McDonagh, L. A. Palma and D. A. Lightner, Science, 208:
    145 (1980).
33. R. R. Anderson and J. A. Parrish, J. Invest. Dermatol., 77:13
    (1981).
34. S. Wan, R. R. Anderson and J. A. Parrish, Photochem. Photobiol.,
    34:493 (1981).
35. F. Ebbesen, Dan. Med. Bull., 22:207 (1975).
36. K. L. Tan, J. Pediat., 90:448 (1977).
37. T. R. C. Sisson, N. Kendall, E. Show and L. Kechavar-Oliai,
    J. Pediat., 81:35 (1972).
38. T. R. C. Sisson, N. Kendall, R. E. Davies and D. Berger, Birth
    Defects, Orig. Article Series, 6:100 (1970).
39. H. A. Rachtel, J. Pediat., 87:110 (1975).
40. D. A. Lightner, T. A. Wooldridge, S. L. Rodgers and R. D.
    Norris, Experientia, 36:380 (1980).
41. S. M. Goldberg, S. Kendall and T. R. C. Sisson, Photodecompo-
    sition of Bilirubin "in vivo" (A), Clinical Res., 18:110
    (1975).
42. L. Ballowitz, G. Gentler, J. Krochmann, R. Pannitschka, G.
    Roemer and I. Roemer, Biol. Neonate, 31:229 (1977).
43. G. R. Gutcher, W. M. Yen and G. B. Odell, Pediat. Res., 17:120
    (1983).
44. R. Broderson, J. Pediat., 26:349 (1980).
45. P. Moggi, Personal communication.
46. A. A. Lamola, W. E. Blumberg, R. McClead and A. Fanaroff, Proc.
    Nat. Acad. Sci., 78:1882 (1981).
47. C. Vecchi, Personal communication.
48. T. P. Vogl, Introduction to the Mechanism of Phototherapy of
    Jaundice and Related Technology, in: "Lasers in Photomedicine
    and Photobiology", eds., R. Pratesi and C. A. Sacchi, p.
    136, Springer (1980).
49. S. Järig, D. Järig and P. Meisel, Metal Halide Vapor Lamps, in:
    "New Trends in Phototherapy", eds., F. Rubaltelli and G.
    Jori, Plenum Press (1983).
50. H. Senger, The Blue Light Syndrome, Springer Verlag (1980).
51. R. Parshad, K. K. Sanford, W. G. Taylor, R. E. Tarone, G. M.
    Jones and A. E. Baek, Photochem. Photobiol., 29:971 (1979).
52. R. Parshad, K. K. Sanford, G. M. Jones and R. E. Tarone, Proc.
    Nat. Acad. Sci. USA., 75:1830 (1978).
53. H. J. Burki and C. K. Lam, Mut. Res., 54:373 (1978).
54. B. S. Rosenstein and J. M. Ducore, Photochem. Photobiol., 38:
    51 (1983).
55. A. N. Cohen and J. D. Ostrow, New Concepts in Phototherapy:
    Photoisomerization of Bilirubin IXα and Potential Toxic
    Effects of Light, Pediatrics, 65:740 (1980).
56. S. Wan, J. A. Parrish, R. R. Anderson and M. Madden, Photochem.
    Photobiol., 34:679 (1981).
57. W. T. Ham Jr., H. A. Mueller and D. A. Sliney, Nature, 260:153
    (1976).

SURGICAL AND OPHTHALMOLOGICAL APPLICATIONS

FIVE YEARS OF LASER APPLICATION IN BRONCHOLOGY

(1637 ENDOSCOPIC RESECTIONS, 921 PATIENTS)

C. Personne, A. Colchen and L. Toty

Department of Thoracic Surgery, C.M.C. Foch
Suresnes, France

INTRODUCTION

Up to date the aim of our work has been to bring to light the capabilities of the laser in each speciality, to show their efficacy and to define their indications.

This information was necessary. It was the outcome of our own experience and the improvement of our technique. This period has now come to an end. For us, physicians and surgeons, a laser has become a device like any other, undoubtedly less unknown, still somewhat delicate and certainly more expensive.

The laser technique is no longer a technique that needs to be discovered, it has become widespread. We can accordingly no longer be satisfied with showing that laser coagulates, carbonizes or vaporizes. We can now determine the true place it will assume in each speciality. We have to say that it enables us to do things which were impossible before, to improve upon what we did badly and now do better, but at the same time we must not hide our disappointments.

We will try to apply this critical reflexion in our speciality: pneumology.

It was because of many unsolved problems in tracheal surgery that L. Toty undertook Laser YAG trials on animals[1]. C. Personne subsequently attempted the first treatment of humans in November 1978. Since then, C. Personnne and A. Colchen have carried out 1637 endoscopic resections on tracheal and bronchial lesions[2]. For us, laser resections are now routine procedures, about ten being performed weekly.

211

METHOD AND MATERIAL

The method we use has, needless to say, been improved upon over the years, but it has not varied in its main lines.

The laser is a Nd YAG.

The bronchoscope is a rigid bonchoscope specially adapted for the purpose. We will note how the choice of the endoscope is a main point.

The resections are carried out under general anesthesia.

Ventilation is achieved by the Sanders technique[3].

Each point of this protocol will be discussed item by item in this paper.

Indications: 1637 endoscopic resections have been performed since November 1978. 921 patients have been treated:

45 benign tumors
49 moderately malignant tumors
437 cancers   )
) iatrongenic
285 granulomas)
13 miscellaneous (8 amyloidosis)

The fact is that the number of sessions is nearly double the number of patients is due to the recurrences of stenoses and in particular of cancers. In practice, we treat the lesion in a single session, irrespective of its size.

Complications: 15 deaths. Three were directly connected with the procedure (massive hemorrhages in patients already submitted to several sessions). The other twelve were due to the respiratory distress of patients in a terminal condition. All but one had a cancer. Those twelve cases are failures, not real complications.

Twenty-five minor complications: pneumothorax or pneumomediastinum due to the ventilating technique. All were easily dealt with in a few hours.

Two hemorrhages had required an emergency lobectomy.

DISCUSSION

Each group of indications involves a different objective: cure or palliation. Each raises special technical problems. In spite of our attachment for a technique we coined, we must confess that, in

most cases, it is just an auxiliary technique with a palliative aim.
The lesions for which laser is indicated are very precise. They are
scattered in the pathology and each raises special technical problems.

True Benign Tumors

They are rare birds in bronchology. They fall into many histological groups (lipomas, hamartomas, histiocytofibromas, etc...), but
they all have some points in common. They are all more or less
spherical with poor vascularization, and are pediculated. They can
therefore always be removed in a single session. The only minor
difficulties concern the manipulation of the endoscope in order to
expose the pedicle. The retrieval of the growth is easily achieved
through a large bore bronchoscope.

Special attention must obviously be paid to the destruction of
the pedicle so as to avoid any recurrence. We have never observed
it. The laser can accordingly completely replace classical surgery
in the treatment of these benign isolated tumors.

The very special case of diffuse papillomatosis raises quite
different problems. When there are only a few dozen of vegetations,
it is possible to achieve a complete and radical cure in two or three
sessions. But, when the whole of the tracheal mucosa is invaded by
papillomas, no complete resection can be envisaged. Only an approximate cleaning can be carried out and the use of the laser is the best
procedure for this purpose.

Moderately Malignant Tumors

Carcinoids or adenocystic carcinomas are not, a priori, indications for endoscopic resections. The involvement of the bronchial
wall and surrounding tissue makes it necessary to perform a classical
excision in all operable cases. But there are always inoperable
cases, owing to a major organic dysfunction and particularly in old
age. If the laser can ensure a temporary improvement, the respite
will be worthwhile in the management of the slow growing tumors. But
another resection will be required within six months or a year's
time.

But, above all, there are the recurrences that follow surgical
excision. These are rare in the case of carcinoids and extremely
frequent in that of adenocystic carcinomas. The laser in such circumstances, as has been shown in some of our interventions, prevent
ultimate asphyxia for many years. It is therefore a palliative but
nonetheless irreplaceable solution.

From the technical point of view, the main problem raised by carcinoids is bleeding, which may be profuse and can only be controlled thanks to the facilities afforded by the rigid bronchoscope. We have never experienced any serious accident, but have sometimes been obliged to interupt the procedure and to resort to a second session.

As for adenocystic carcinomas, they seldom bleed, but often give other problems owing to their extension and the severity of the consequent airway obstruction.

Broncho Tracheal Cancers

They are the most frequent indication for endoscopic desobstructions, the most interesting also, since the laser provides a solution to an otherwise untreatable asphyxia condition.

It must be clearly understood that, with a thermic laser such as the Nd YAG, the aim is purely palliative, namely to restore the patency of the airway which will sooner or later be threatened by a growth obstructing either the trachea of the carina, or one of the main bronchi.

Two types of cancers must be considered:

1.   Undiagnosed broncho tracheal cancers, inducing over a period of a few weeks a slowly increasing dyspnea hitherto attributed to asthma. When at last suspected, the tumor is finally diagnosed when the obstruction is almost complete. A single session suffices for an endoscopic resection and the restoring of respiratory comfort. Radio or chemotherapy can then be started.

2.   Terminal cancers, already operated on or inoperable, irradiated or treated by chemotherapy, but recurring and unmanageable. In cases in which the growth of the tumor will inevitably lead to death by asphyxia (as, for example, when a recurrence occurs on the suture line of a pneumonectomy) a laser session may ensure a few months' survival. Another resection, followed by others, unfortunately at shorter intervals, will be necessary, but in many cases, each entailing merely two days of hospitalization.

These laser sessions for the treatment of cancers involve, or may involve, all sorts of technical problems, such as bleeding, massive secretions plugged beyond the obstruction, the size and extension of the lesion, which is sometimes sited in the trachea and the two main bronchi, but above all the asphyxic conditions of the patient. Here again, the rigid bronchoscope and jet ventilation under general anesthesia make these high risk deobstructions possible. The more so since it is absolutely essential to obtain an

effective desobstruction in a single session. An over timid resection can only result in a worsening of the situation.

Iatrogenic Stenoses

They entail far fewer technical problems. Restoration of the
diameter to 80 or 90% of its normal size is nearly always easily
achieved. If necessary, when the caliber appears insufficient,
dilation can be obtained by introducing the bronchoscope in the same
way as a bougie.

The severity of the structure makes little difference. It is
often the most severe stenoses that are easiest to destroy. Stenoses
leaving a lumen of less than 3 to 4 mm are a major indication.

In acute conditions the laser provides an outstanding resusci-
tation technique. In our hands, it has replaced all the other emergency procedures: whether the emergency surgery, dreaded by all alike,
problematic intubation using small bore tubes or tracheostomy resulting in an additional lesion. It is clean, rapid and always effi-
cacious method which restores an almost normal airway; it is safe and
requires no canulation. The patient immediately recovers a normal
ventilation and can get rid of the secretions plugged beyond the
stenosis.

Medium to long term results, on the other hand, are disappointing. Although the laser enables the state of asphyxia to be initially resolved, the stenosis often recurs 6 to 8 weeks later.

The prognosis depends of the anatomy of the stenosis. Two
opposite types of stenotic lesions may be found: a fibrous diaphragm
which leaves the tracheal wall intact, and, on the contrary, a destruction and collapse of the tracheal wall. These lesions however
often show a mixture of the two. During endoscopy, it is frequently
very difficult to determine the role played by each component and to
evaluate the damage done to the cartilagineous wall. A diaphragm, or
in extreme cases a funnel, will nearly always be observed, but the
tracheal wall lesion is impossible to assess. We have therefore
decided to classify the lesions in terms of the length of the stenosis, which accurately reflects its severity. For stenosis of 15 mm
we have obtained exactly 50% of definite successes. For those over
30 mm recurrence is inevitable.

Needless to say, when the stenosis recurs, a further attempt
with laser resection is possible, but the likelihood of a definitive
success is reduced. This nevertheless enables major tracheal surgery
to be delayed in patients who cannot be operated on immediately and
especially in the case of intensive care patients with a complex
pathology, the trachea being dealt with last of all. Finally, re-

calibration can be usefully performed before sleeve resection since it facilitates the tasks of both the anesthetist and the surgeon. When such a resection is impossible, this recalibration allows a Montgomery tube to be placed in position as a stent.

For granulomas which have developed as a result of contact of a tube with the trachea owing to a prolonged intubation, the laser is again an excellent indication. The destruction of these, sometimes profuse or voluminous granulomas may prevent the secondary formation of an ultimate fibrous stenosis. It is difficult to prove this, and even more to account for it. It however, seems to us that it is very likely true.

TECHNICAL PROBLEMS

The technical problems raised by laser resections in bronchology have not so far been completely solved. The laser generator, the fiber and the endoscope have not yet reached their final stage of development.

The Laser Generator

The one we use in our discipline is the Nd YAG only. The transmission of the beam along a flexible fiber, and its power, make this choice compulsory. In most indications, and in particular for cancers, its parameters are adequate. On the other hand, the disappointing results observed in stenosis would militate in favor of the choosing of a different wavelength. In this definite group the CO_2 perhaps would give better long term results. Its vaporization certainly affects tissues less than the carbonization of the YAG. This mode of destruction is probably responsible for a new iatrogenic lesion, scar of which can, in its turn, cause a stenosis.

The Fiber

It is, in daily practice, the source of most incidents. The cost of such a fragile component needs to be cut down and on the spot maintenance should be possible. The caliber should likewise be reduced so as to improve the available power.

The Endoscope

Our choice of the endoscope may appear paradoxical in this era of fibroscopy. We chose the old steel bronchoscope from the outset. Our experience has born out this choice for two crucial reasons: convenience and security.

The easiness in handling highly plead in favor of the large bore bronchoscope: the incomparable superior quality of vision through the rigid optic enables a far more accurate resection to be performed, flush with the tissue to be spared.

The cleansing of the optic glass, which needs to be done so frequently and is so taxing, can be achieved in a few seconds, permitting an immediate resumption of the procedure. An air jet can be added to clean continuously the tip of the optic.

The sucking of secretions and blood is permanently achieved through a catheter extending beyond the target to keep it dry.

Tumoral fragments can be extracted with large biopsy forceps.

Finally, the accuracy of the aim is ensured by the absolutely immobility of the tube within the laryngo tracheal tree.

The security is, obviously the decisive argument. Indeed, these endoscopic resections are, in the majority of the cases we have to treat, more or less fraught with danger, owing to the lesion to be excised or to the patient's condition. Very often, these two risks are found in combination since the major interest of the laser in our speciality is to restore the patency of a severly compromised airway. In over 200 cases, the procedure was carried out under condition of extreme emergency, on patients in acute asphyxia and sometimes in coma. The rigid bronchoscope alone made this possible. Obviously, some accidents occur, but none of these patients could have been treated using a fiberscope.

The main advantage is that adequate ventilation can be assured under general anesthesia as soon as the bronchoscope has been introduced. In most dramatic cases, when a patient has to be anesthetized while sitting up, the respiratory condition is improved immediately as soon as the bronchoscope has been inserted and the jet ventilation started. The type of jet ventilation routinely used can be adjusted at any moment to suit the patient's respiratory condition and to meet the surgical requirements. The anesthetist can, for instance, adjust the insufflation rate and volume so that is does not interfere with the surgeon's work and allows him to achieve an accurate aim.

During the past few months, we have tried high frequency jet ventilation, which ensures an almost complete immobility of the bronchial tree. It cannot be used in all cases, but it has made most of our resections much easier.

The hemorrhagic risk is the second decisive argument of security for the rigid bronchoscope. If it is moderate, it can be continuously sucked by the catheter pushed beyond the target. If the bleeding is more important swabs soaked with a vasoconstrictor can be

used. But, quickly, a moderate hemorrhage can endanger the life.
That is the particular problem of resections in bronchology. For us,
an hemorrhage kills by anoxia much before killing by blood loss. It
may be necessary to use a rescue procedure: pass the obstruction at
all costs, pushing the tube into a free main bronchus. This will
ensure, after suction of blood and clots, a sufficient ventilation
and the bronchoscope acts as a stopper by blocking until a clot can
form.

However, many teams, both in France and abroad, disagree with us
and continue to use the fiberscope[4-5]. Over the last few months,
there has nevertheless been a definite swing in favor of the rigid
tube. Prof. Dumon's team, who use now a rigid bronchoscope, had like
use realized a high number of resections[6]. Most of the other teams
have dealt with problems that they overcame by an oral intubation,
the fiberscope being inserted through the tube. It would have been
so much simpler to have a recourse to a rigid bronchoscope! Others
limit their indications to minor lesions, or are obliged to resort to
repeated sessions in order to secure a desobstruction. But a radical
cure ought to be achieved in a single session.

The old bronchoscope is admittedly old fashioned and that is its
principle fault. It is possible to modify a classical tube for this
new purpose, including three inner channels to ensure supply of
light, suction and jet ventilation. Such a tube will soon be available on the market.

It seems clear that, in the future, the indications for thermic
laser applications will not change. They are specific and limited in
number. Progress is possible with a view to achieving better
results. This will imply the need for various types of laser gener-
ators, perhaps with adjustable wavelengths; better adapted endoscopes; stronger and cheaper fibers. But the place of the Nd YAG
laser is in any case already well established in bronchology.

REFERENCES

1. L. Toty, Cl. Personne, P. Hertzog and A. Colchen, Rev.Fr.Mal.
      Resp., 7:57-60, 475-482 (1979).
2. Cl. Personne, A. Colchen, L. Toty and G. Vourc'h, "Tracheal
      surgery under Nd YAG Laser. Results on 521 cases, 905
      sessions." in Laser Tokyo 81 (4th Congress of the Inter-
      national Society for Laser Surgery), 20,II:21-24 (1981).
3. L. Toty, Cl. Personne, A. Colchen and G. Vourc'h, Thorax,
      36:175-178 (1981).
4. K. Oho and T. Ohtani, Laser surgery in the trachea and bronchus
      via the fiberoptic bronchoscope, in: "Laser Tokyo," (4th
      Congress of the International Society for Laser Surgery),
      16-9 (1981).
5. R. R. Hetzel and F. J. C. Millard, Br.Med.J., 286:12-16 (1983).
6. J. F. Dumon and E. Reboud, Chest, 81:278-284 (1982).

EARLY EXPERIENCES WITH CO_2 LASER FOR THE TREATMENT OF SEPTIC PATHOLOGY IN NEUROSURGERY: EXPERIMENTAL MODELS WITH OTHER SOURCES

V. A. Fasano, G. F. Lombard, P. Martinetto*,
S. Tealdi and R. Tealdi

Institute of Neurosurgery
*Institute of Microbiology
University of Turin, Italy

INTRODUCTION

At the first time, the CO_2 laser was utilized in infective neurosurgical pathology as a surgical cutting instrument to remove inflammatory pseudomembranes in chronic osteomyelitis, and as a vaporizing instrument on the dura mater surface.

Successively, the instrument, defocussed and at a low power, was used for a prolonged and diffuse photocoagulation of the surgical cavity, particularly, of the dural surface and of the osteomyelitic bone edges, with the aim to sterilize tissues.

So, we saw a shortening of the average time of wound healing and a lack of recurrence of the septic pathology.

Then, we have treated, with CO_2 laser, intracranial infective pathology: i.e. primary abscesses, capsulated or not, circumscribed purulent encephalitis, secondary abscesses in surgical cavities (patient operated for intracranial hematomas and tumors). In these cases we have obtained a lack of septic recurrences and an improvement of neurological post-operative course.

Thanks to these results, we have continue to use the laser in infective pathology; for giving an experimental support to these results we have carried on researches in vivo (on the experimental animal) to see the interation between the laser and inflammatory tissue, and in vitro (on bacterial culture: in solid and liquid media) to see the laser effect on the bacterial cell. The bacterial cell has been also sensibilized to the photodynamic effect of the

219

laser (Argon, He-Ne), with hematoporphyrin. The goal of these experiments is to understand the role of the thermal, photochemical, and
mechanic resonance laser effects in the interaction between laser
radiation and bacterial cell.

CLINICAL

Cases

The cases treated with laser have been divided as follows:

- Superficial, extracerebral, infective pathology (infection of
  cutaneous flap, osteomyelitis of the bone, dural infection and also
  one case of superficial circumscribed meningo-encephalitis): 9
  cases.
- Intracerebral, infective pathology (primary abscesses, capsulated
  or not, and secondary infections of surgical cavities, i.e. intracerebral hematomas and tumors): primary 6 cases, secondary 4 cases.

Surgical techniques

The CO_2 laser in superficial infections and osteomyelitis has
been used, at high power (40-50 W), to cut old inflammatory adhesions
and pseudomembranes and to photocoagulate the infiltrated bone edges;
at low power (10-20 W) and defocussed, on dura mater; and at a very
lower power (5 W), in superficial meningo-encephalitis.

All the surgical field and the internal surface of the cutaneous
flap have been photocoagulated several times, removing the eschars.
In deep abscesses the laser has been used (at high power) to cut the
capsule and to perform a lobectomy in encephalitis; then, (defocussed
and at a low power: 5 W) to photocoagulate the contaminated cavity.

Results

In cases with extracerebral infective pathology, the wound
healing, in 90% (8 cases), occurred within 7 days after operation;
in 10% (1 case) there was a delayed wound healing (10-15 days) without recurrences of inflammatory process.

In cases operated with traditional methods, a recurrence of
infective pathology with a secondary wound healing (after 20 days)
occurred in 44% (4 cases); a delayed wound healing (10-15 days)
occurred in 33% (3 cases); only in 22% (2 cases) the wound healing
occurred in 6-7 days.

In cases with intracerebral infective pathology, no septic recurrence occurred in the first group of cases operated with laser (10 cases); in cases operated with traditional methods, recurrence of septic process occurred in 40% (4 cases out of 10 operated).

The mortality in intracerebral infective pathology is the same, either in cases operated with laser or operated with traditional methods. In the first group mortality was not due to septic recurrence but to a cardiac infarction (1 case) and to the neurological pre-operative status (1 case operated in a deep coma). In the second group the mortality was due to septic recurrence only (2 cases) there was a worsening of pre-operative neurological status, due to a septic recurrence: a second operation has been necessary.

EXPERIMENTAL

In Animals (Guinea Pigs) Experiments

We have used a Staphylococcus aureus culture, in Nutrient Broth (BBL), held at 37°C for 24 hours. After having obtained transmittance 90 of this brothculture, we have diluted it 1:100 with the same cultural mean.

- Superficial infection. It is reproduced only in acute phase (it is not possible to obtain a chronic osteomyelitis): results of laser irradiation are the same obtained in clinic.

- Deep infection. It is not possible to reproduce a deep capsulated abscess but only a diffuse purulent encephalitis: for this reason the survival in animals is very low.

In Vitro Experiments

We have used a Staphylococcus Aureus culture with the same preparation used in animal. The irradiation has been made on cultures in solid and liquid medium.

We have used several laser sources: CO_2, Nd:YAG, Argon, He-Ne. After irradiation we controlled the successive phases of bacterial growth. In solid medium, the bacterial extinction that we saw is not significant because a thermal effect is evident and it is not possible to evaluate a photodynamic effect.

Because of the exponential extinction of laser radiation in a high liquid column (4,6 cm. in tube), the culture irradiation has been made with a tube in movement. At first time, laser irradiation has been made for several minutes (no more than 5 minutes); then, we reduced irradiation time to no more than 60 seconds to avoid thermal effect. Temperature increase was evaluated with thermistors and never was more than 30°C.

In liquid medium a decrease of cellular growth was of about two exponential factors less than untreated controls. Minimum decrease has been obtained with CO_2 laser; the maximum with Nd:YAG.

In our experimental conditions there is not a true bacteriostatic effect (decrease of bacterial growth is only of two exponential factors), so, in further experiments we have sensibilized the bacterial culture to the photodynamic effect of laser radiation (Argon, Dye lasers), with hematoporphyrin.

In Vitro Experiments with Hematoporphyrin

Material and methods. We have used a Staphylococcus Aureus culture, in Nutrient Broth, held at 37°C for 24 hours. After having obtained transmittance 90 of this brothculture, we have diluted it 1:100 with the same cultural mean and put it into a glass tube, 12 mm in diameter, for a total volume of 1 ml.

Hematoporphyrin 2Cl at 0,05 mM and 0,025 mM concentration, melted in 0,1 ml ethyl alcohol and 0,9 ml of sterile distilled water was added to the solution. The spectrum of our hematoporphyrin in neutral solvents has a range of absorption between 5000 Å and 6300 Å. The source used was an Argon laser by Lexel (Palo Alto California, USA) Mod. 95-4 (From 0 to 5W. Peak energy: 5W, CW mode. Power density: maximum $3,3.10^3$ W/cm^2, minimum 10 W/cm^2).

The inoculum was 5.10^4 UFG/ml and the final volume was 2 ml. The tubes have been appropriately wrapped in an aluminium foil and the experiment was performed in darkness to avoid the fluid photo-activation.

Two controls, consisting of EMP 2Cl at the concentration of 0,05 mM, 0,025 mM and of the same bacterial inoculum above mentioned, were stored in the dark; two were exposed to the daylight for 30 seconds and after, incubated at 37° for 24 hours.

One same contents tube was irradiated 30 seconds with the laser without shaking, another, on the contrary, was subjected to a vertical movement during the irradiation during irradiation, for the same time.

Another operation was made putting the tube content into a plate, 5 cm in diameter, obtaining a fluid layer high 1,5 mm approximately. After the irradiation with Argon laser all tubes and plates were incubated at 37° for 24 hours.

Argon laser was used defocussed on plates at a power of 4 W and with a power density of 0,25 W/cm^2 and focussed in tubes at a power of 4 W and with a power density from 60 to 70 W/cm^2.

After these operations we counted colonies numbers, according to conventional microbiological techniques, in triplicate plates to obtain an efficient statistical evaluation.

Results

The results show St. Aureus culture growth, incubated at 37°C for 24 hours, after Argon laser treatment, power 4 W for 30 seconds. In Table 1 it is evident that EMP 2 Cl does not present bactericidal activity on the sample held in the dark, while a growth inhibition has been evaluated after day light exposition. But no inhibition occurred when the tube was not shaken, in relation to the fact that the monochromatic light is absorbed by the first liquid layers with exponential values.

The starting temperature resulted 22°C; during the laser action, rose of 2°C, after it went to 28°C, and then returned to normal condition. In the agitated tubes a subsequent increase of the temperature up to 30°C was observed. In plates the medium increase of temperature was 3°C. The following data demonstrate that the bacterial mortality has bound to a photochemical and not to a thermal effect.

When the experiments were performed in plates we noted that a certain degree of inhibition was present.

Table 1. Activity of Laser Argon in Various Experimental Conditions with Photoactivation Mediated by Hematoporphyrin $_2$Cl (EMP$_2$Cl): values expressed as UFC/ml.

St. Aureus Growth UFC/ml	Argon Laser in tube without light	Argon Laser in agitated tube without light	Argon Laser in plate without light	Day light in tube	In the dark in tube
EMP$_2$Cl: 0,05 mM	5.10^8	4.10^3	$1,2.10^2$	2.10^2	6.10^6
EMP$_2$Cl: 0,025 mM	5.10^8	$1,5.10^3$	2.10^2	5.10^2	2.10^6
EMP$_2$Cl: without	2.10^8				

Besides Argon laser we have also used in these experiments a
Dye laser (from green 488 to red 640 nm) by Lexel Mod.Aurora (Mod 150
Pump laser + Mod 600 Dye laser – Peak energy 1,5 W. Beam diameter
2.22 mm. Divergence 1.07 mrad. CW mode. Power density: maximum
3,9 W/cm^2).

A concentration of EMP 0,05 mM, with which we have obtained the
best results, was used.

We had the best results with the Argon on shaken cultures and
thin layer plates; so in the test with dye laser we used these two
parameters.

As it is possible to see in Table 2, the presence of EMP
increases the bacterial death-rate by comparison with controls without EMP, with logarithmic decreasing of 4-5 times; for this reason,
we can think that the final results are the same for Dye and Argon
laser. The absorption bands of the EMP Cl 2 could make possible the
photoactivation process in the presence of radiant sources of different wavelengths.

CONCLUSIONS

The surgical experience shows that laser is really useful in
superficial infections. In fact, we have observed a shortening of
the average time of wound healing of about 50%, and lack of recurrence of the septic pathology.

In intracerebral infective pathology treated with CO_2 laser we
have seen the lack of local complication and the lack of recurrence
of the infection. The use of the laser had not a great influence in
the mortality or in the recovery time because these patterns are
greatly influenced by the preoperative neurological status and by the
anatomic characteristics of the lesion (presence of a capsule, surrounding encephalitis, abscessual collection volume).

Table 2.

St. Aureus Growth UFC/ml	Dye Laser in agitated tube without light	Dye Laser in plate with light
EMP_2Cl 0,05 mM	$6,2.10^4$	9.10^3
EMP_2Cl without	5.10^8	5.10^8

The clinical good results of laser treatment in cerebral infections is probably due to several factors:

1.   Thermal effect
2.   Photodynamic effect
3.   Resonance effect
4.   Mechanic surgical effect

Thermal effect acts in surgical field and in experimental plates with solid medium, where bacterial colonies are destroyed by the increase of temperature.

Photodynamic effect (not demonstrated in vivo), in vitro, it seems not to have a real bacteriostatic effect by itself; while, supported by a photoactivating substance (EMP), it really inhibits bacterial growth. Future experiments will be performed with Hp D (hematoporphyrin derivative) that has an absorption spectrum varying from the near UV to red (Blum, 1941; Dougherty, 1979) and with other dye compounds.

Resonance effect seems to play a role when the wavelength of laser source (Nd:YAG) correspond to the bacterial body dimension.

Mechanic surgical effect is very important in operative field, where the surgeon removes layer by layer the eschars produced by the photocoagulation of inflammatory tissue.

So, the good surgical results of laser treatment of superficial infective pathology may depend from the combination of thermal and mechanic effects.

While, in vitro experiments suggest that thermal effect may be avoided if photodynamic effect could be improved by photosensibilization of bacterial cells.

REFERENCES

Blum, H. F., 1941, Photodynamic action and diseases caused by light, New York Reinhold Publishing Corporation, pp. 211-237.
Britt, R. H., Enzmann, D. R., and Yeager, A. S., 1981, J.Neurosur., 55:590-603.
Dougherty, T. J., Lawrence, G., Kaufman, J. H., Boyle, D., Weishaupt, K. R., and Goldfarb, A., 1979, JNCI, 62:231-237.
Laws, E. R., Cortese, D. A., Kinsey, J. H., Eagan, R. T., and Anderson, R. E., 1981, Neurosur., Dec. 9(6):672-678.

PLASTIC AND RECONSTRUCTIVE SURGERY BY LASER

R. Pariente

Institute of Plastic Surgery
"La Sapienza" University
Rome, Italy

Our experience in the use of CO_2 laser in reparative plastic surgery started in 1974, i.e. less than one year after the realization of the first CO_2 system for surgical use.

We think that the best way to give a clear panoramic vision of the possibilities offered by the use of the instrument in plastic surgery, is to group the operations on the basis of the type of the tissues being cut, in order to evaluate whether, and to what extent, the laser has demonstrated its usefulness in each specific instance.

The analysis of the results obtained in each group of operations on specific tissues will allow us to determine the real benefits attainable by using laser in the various fields of reparative surgery.

As we mentioned above, the laser that we have always used is the CO_2 laser, the only one having real properties of a surgical scalpel.

We have listed our operations as follows:

Table 1.

Laser Surgery on teguments
Laser Surgery on highly vascularized tissues
Laser Surgery on adipose tissue
Laser Surgery on the skeletal apparatus
Laser Surgery on the muscular apparatus

By cutting teguments with a laser, we obtain a very quick and neat incision, totally bloodless as everybody knows, because the laser beam, while cutting, also seals blood and lymphatic vessels. As is very clearly shown by microscopic examinations, the cut is neat: the cells are perfectly normal with no appreciable trace of tissue necrosis. The cut looks exactly like another one, obtained by means of a scalpel; in another cut, obtained by means of an electro-diatermo-scalpel, the necrosis of the edges is very visible.

The dissection obtained is therefore much more accurate. Generally speaking, the obliteration of blood and lymphatic vessels must be considered useful also in oncologic tegumentary surgery. Furthermore, the laser allows surgeons to effect the exeresis with the least possible squeezing; and many sustain that squeezing if corresponsible of metastasic intraoperative diffusion.

Notwithstanding every personal opinion, any scientific demonstration and every polemic, regarding the possibility to reduce the risk of intraoperatory diffusion of the tumors by using the laser, there is an irrecusable fact: any school of Surgery prescribes that, in the removal of a tumor, the techniques must be as traumaless as possible, thus avoiding as much as possible complex manipulations, squeezing of the tissues, compressions etc..... Many of these schools prescribe the substitution of the instruments that have been in contact with the tumor..... We would not discuss these prescriptions; however, one thing is certain: by using the laser, we are in a position to use a "non touch" technique much better than by any other instrument.

Moreover, the laser permits, before closing the surgical wound, to make an additional purification of the wound, because it destroys, by means of strokes of its beam, any neoplastic cell that, in spite of every caution, might be still present.

The proper question, whether or not by using the laser it is possible to reduce the incidence of local recurrencies and of post-operative metastasization, cannot certainly find an answer in the enthusiasm of laser surgery's pioneers; an answer will only follow clinico-statistical studies, made on a wide scale and for a very long time.

Years and years will pass, before the hope (based until now on some experimental research, and on the "feeling" of some surgeons), will find confirmation in the scientific research rigor.

However, the actual advices in oncologic fields seem to frame differently the problem of metastasization and multifocality of malignancies.

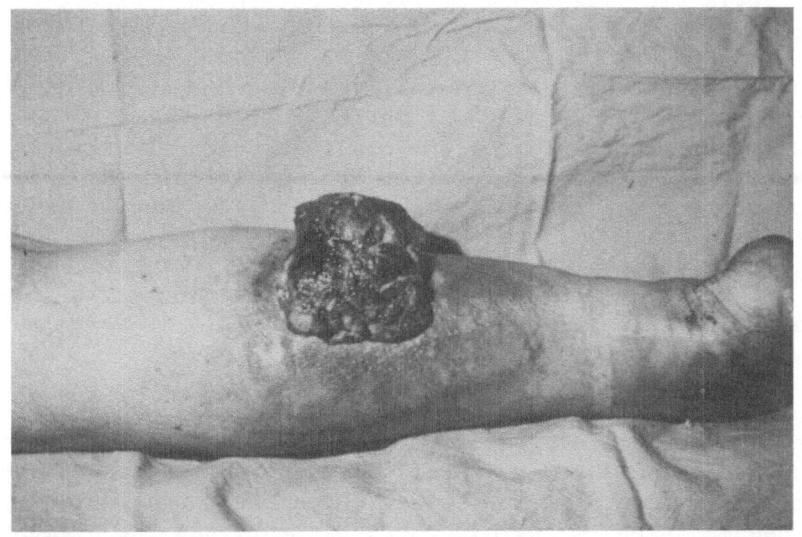

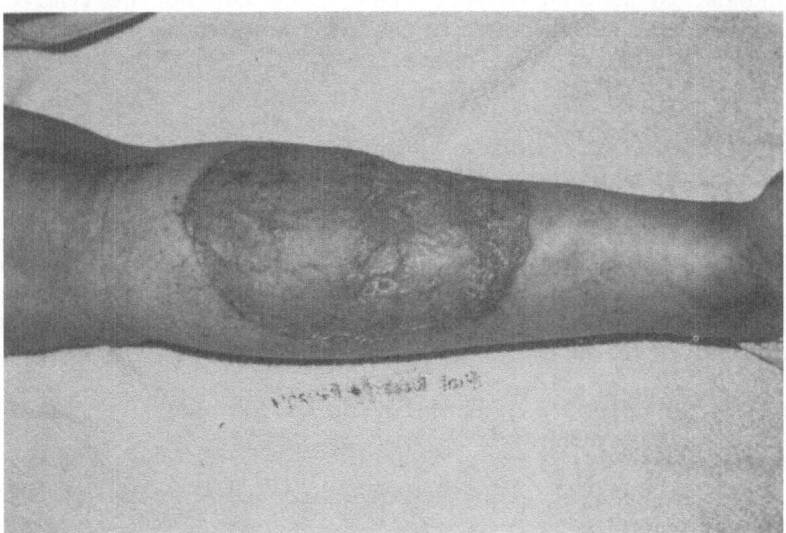

Fig. 1.

Due to the absence of necrosis on the cut's surfaces, we are in a position to obtain immediate repair with free grafts (Fig. 1a,b) or flaps.

The cicatrisation of a laser wound is little slower than that of a scalpel wound; a laser wound usually heals about two or three days later. The scars obtained by the use of a laser are hardly different from the scars obtained by the use of a scalpel: however, they are not almost invisible like those we try to obtain when effecting surgery for aesthetical purposes.

Before closing these brief notes regarding the action of the CO_2 laser on cutaneous tissue, we think useful to refer to a condition that, even if not important from the health's point of view, is nevertheless an important problem from the psychologic and social aspect, and whose solution was highly helped by the use of laser.

We are referring to tattoos, made for decorative purposes, or for magic and religious reasons, or as identification marks in some particular social groups. Removal of these tattoos is very difficult, and is nowadays in high demand.

Among all the methods proposed for this removal, the best results are given by an empiric dermo-abrasion, made using, like abrader, some crystals of NaCl. This very ancient method is useful in the removal of superficial tattoos, which are roughly made by a free hand needle, used to introduce the pigment, has the pigment granules deposed in so deep layers, that they can not be removed without damaging the germinative skin layer and without causing, consequently, the formation of scars.

We believed that it might be advisable to open some ways of penetration of the saline material; and we tried to do it with a CO_2 laser, to reduce to a minimum tissue damage in the treated area.

We used the laser beam in pulsed mode, with discharges duration between 5/100 and 1/10 of a second, and a power of about 9 watts. The beam, focalized, opens the skin some very small holes, going deep to reach the pigment granules. The tissue all around these holes remains safe, and will help the healing of these small lesions.

Through these holes, saline solution will reach the pigment granules, thus eliminating them.

Results obtained with such a method are quite satisfactory.

We shall now refer to surgical operations on angiomas, or any highly vascularized tissues.

I believe that in this field we may obtain the utmost advantages by using laser, and not only because we can save a lot of blood. To work in an almost bloodless field, or at least not to work under a blood pool, as it often happens in operations for angioma when using conventional equipment, means to work more neatly; it also very often means to be able to remove the formation "in toto", without the risk of leaving portions "in situ", which unavoidably will cause recidiv-ation, and at the same time means, on the other hand, to be able not to abduct other tissues which may be saved (Fig. 2a,b).

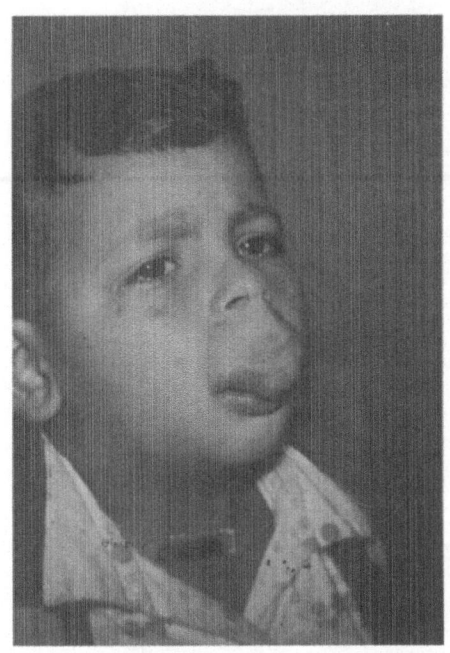

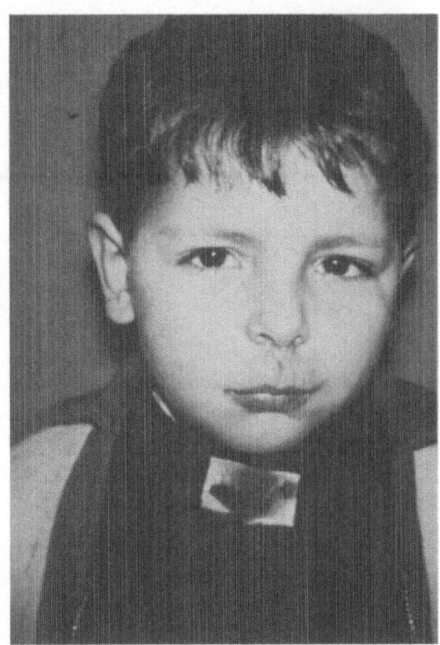

(a)   (b)

Fig. 2.

The situation is very different when we must operate on adipose tissue. Laser cuts fatty tissue with difficulty. It is possible to use the laser on fatty tissue: but unless this is required for specific reasons, why should we, at all costs, use the laser? We see for example that it is possible to effect a reductive mammaplasty by using the laser (Fig. 3a,b). The results are acceptable, neither better nor worse than those obtained by conventional surgery: but this operation is rather more lengthy.

If we have, for instance, to operate a polykistic mastopatia during its active phase: i.e. a precancerosic lesion where a mammectomy is required, the prevailing factor is the possibility of effecting the exeresis while avoiding tissue squeezing: therefore the use of a laser is advisable. After laser surgery, as everybody knows, prosthesic repair can be done immediately.

With regard to the bony tissue, we must admit that the use of the laser has not given results as satisfactory as on other tissues due to the low water content of the bony tissue, especially the cortical one.

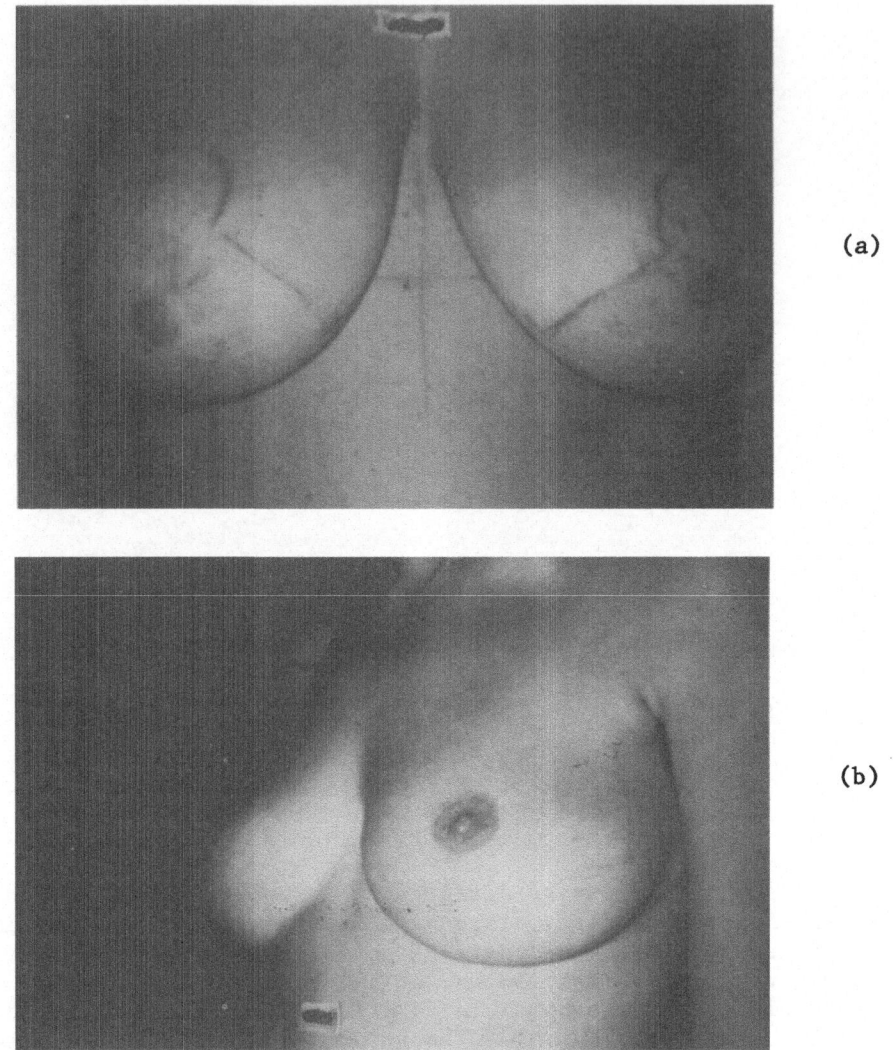

(a)

(b)

Fig. 3.

However, as the analysis of microscopic sections show, the bone sections obtained by mean of a laser does not show any necrosis, nor does the dissection obtained by means of a saw, while the necrotic area is quite large in the dissections obtained by means of an electro-diatermo-scalpel.

The most advantages are offered by the laser when operating on the muscular apparatus, especially when wide and deep undermining, tendineous disconnecting, and dissecting of muscular structures are required.

Also in these instances, the micropictures show the absences of tissue necrosis in muscular dissections obtained by the laser, while they show a considerable amount of necrosis and structural changes in dissections obtained by electro-diatermo-scalpel.

We have found a specific advantage, for example, in effecting a series of operations which nowadays are being carried out very frequently, it is those where myo-cutaneous peninsula shaped flaps are used.

As well known, the myo-cutaneous peninsula is obtained utilizing a wide muscle situated under the skin, and provided with an insertion pedicle through which the main vascular-nervous bundle gets in.

In the most frequent cases, that of the reconstruction of the mammary region demolished due to cancer of the breast, this muscle is the Latissimus Dorsi. The latissimus dorsi becomes vector of a skin dorsal insula, and as soon as it has been disconnected (with the only exception of its humeral insertion where the main vascular nervous pedicle enters) is being rotated forward below the prethoracic skin, so as to substitute the muscular masses (pectoralis major and minor) which where demolished during the Halstead operation (Fig. 4a).

This operation involves highly bleeding stages: in addition to subcutaneous undermining, which practically affects the whole hemithorax, it is necessary as a matter of fact to effect the subtotal disconnection of the muscle, which involves considerable bleeding even though not necessarily upon aponeurotic connections, but certainly upon the scapula and the muscles connected thereto, whereas furthermore the surgeon works below a deep subcutaneous tunnel.

In these instances, the use of the laser allows a considerable saving of time and blood, and a much neater result.

To complete the analysis we wish to point out that the mammary gland itself is replaced by a silicon prosthesis being placed below the muscle.

To the end of reducing unavoidable differences, a corrective controlateral mammaplasty, and reconstruction both of the mammary areola and of the nipple will complete the surgery.

For reconstructing the areola we use a very thin free graft originating from the healthy controlateral areola; for the nipple, we use a whole thickness graft taken from the fifth toe pulp (Fig. 4b)

Recently, with the diffusion of the theory considering cancer of the breast as a systemic illness with high risk of multifocality, we have extended, for prophylactic purpose, the indication to perform a

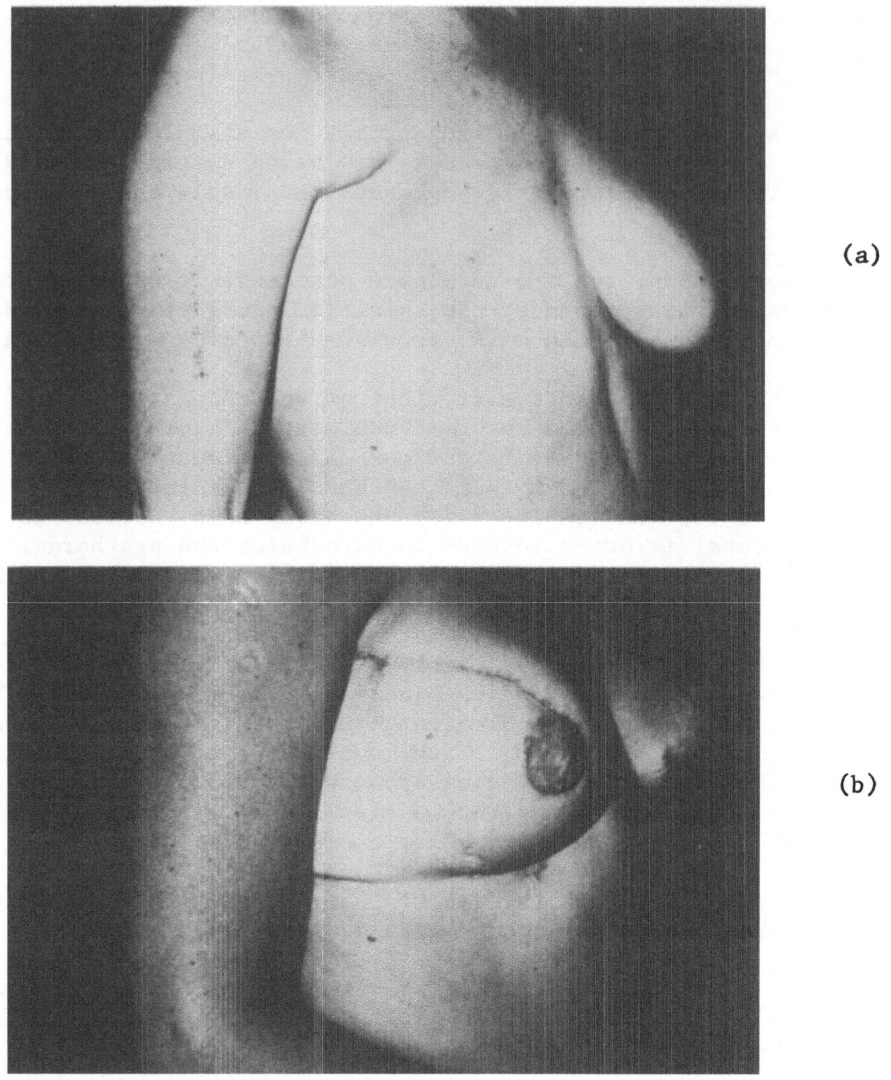

(a)

(b)

Fig. 4.

simple controlateral mammectomy, followed by immediate reconstruction
by means of a silicon prothesis.

I shall now mention briefly other cases where the use of the
laser gives wonderful results:

Table 2.

Operations carried out on infected areas
Operations on the hands
Operations performed on patients affected by
  hemocoagulation troubles

Thanks to the laser, we were able to suture wounds deriving from the excision of highly infected ulcerations and sores, and we obtained healing by first intention, while had we used conventional methods, the healing would certainly have occurred by second intention. This most likely is due to the fact that the laser beam does not spread the infection over the cutting surface as would the scalpel; not only that, but the laser beam helps organic defences destroying germs existing on the wound edges, thus speeding up the healing of the wound.

When operating the hands, it is always to be in mind that the hand does not contain dead spaces; each structure is essential, so that it is necessary to operate with utmost accuracy to avoid to damage unreplaceable structures: very often these structures to be preserved are very close to others on which we have to operate, because they are affected by a neoplastic process, or for any other reason. In these cases, very often, we draw advantage by the use of the surgical microscope, connected to the laser.

In these cases, the benefits given by the laser are such that we must overlook the undoubtful disadvantage of the rather more length duration of the operation. The possibility of working in a bloodless field (without the irrational and very dangerous Esmarch tourniquet) facilitates a very accurate dissection; the surgeon works as easily as on an anatomic table.

If we take into account how rich is the lymphatic network of the hand, it is obvious that conventional operations will leave numerous lymphatic capillaries open, which the laser instead scale. Therefore, in my opinion, we should give green light to laser for mostly all operations on the hand. A compensation to the inconvenience represented by the longer time required by the laser operations, is also the fact that the patients operated with the laser suffer less post-operative pains than those operated conventionally. Furthermore, the laser does not originate oedemas, and the functional recovery is speeded.

Following cases are typical ones where the use of the laser finds a particular indication: Professional Radiodermatitis of the hand, (Fig. 5a,b,c), Rheumatoid arthritis and new growth of the dorsum of the hand (Fig. 6a,b).

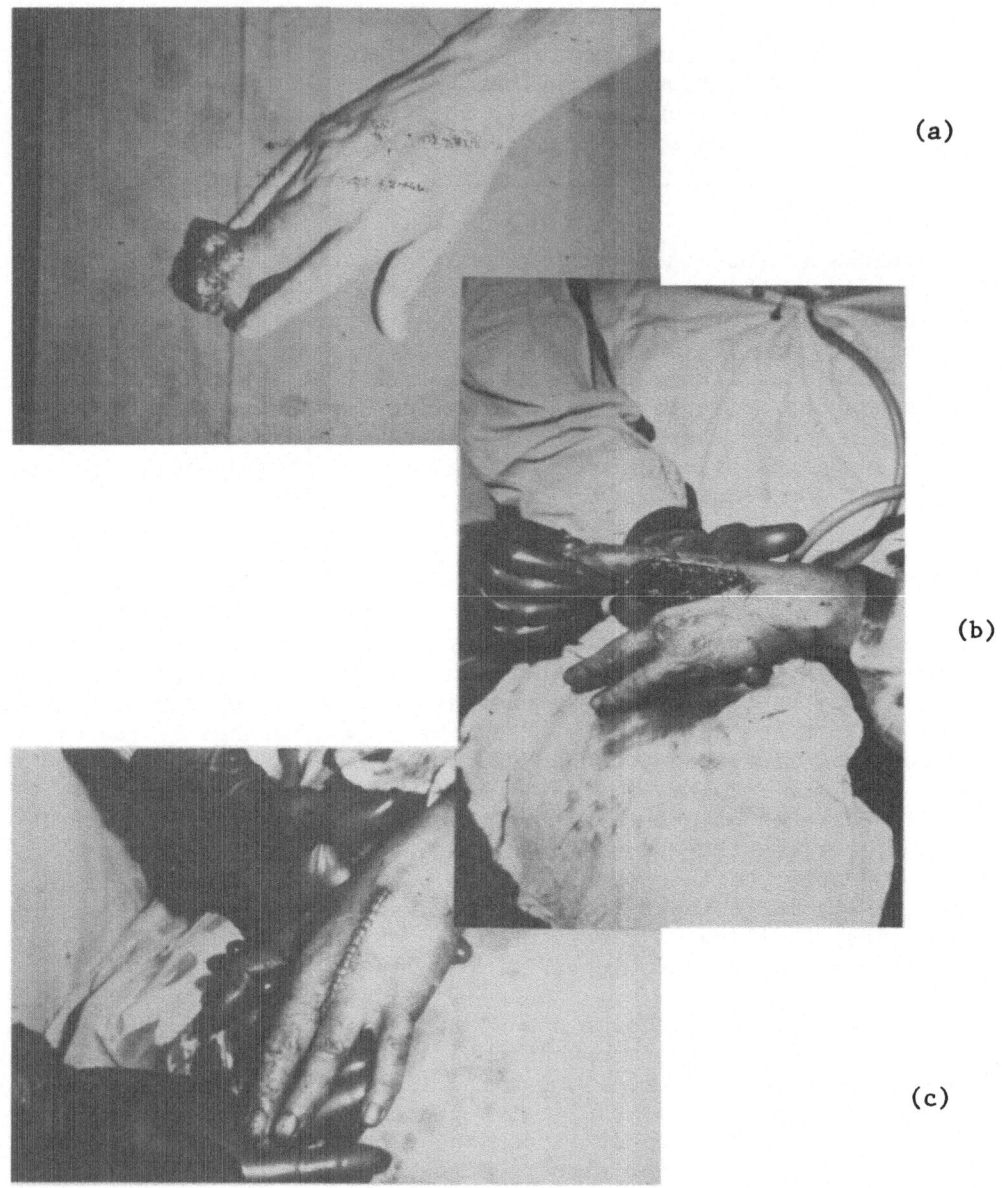

(a)

(b)

(c)

Fig. 5.

To conclude this rapid survey, I feel it is necessary to mention briefly the great advantage offered by the laser when operating hemophiliac patients. As is well known, these patients are subject to uncheckable bleeding. Only recently their bleeding has been

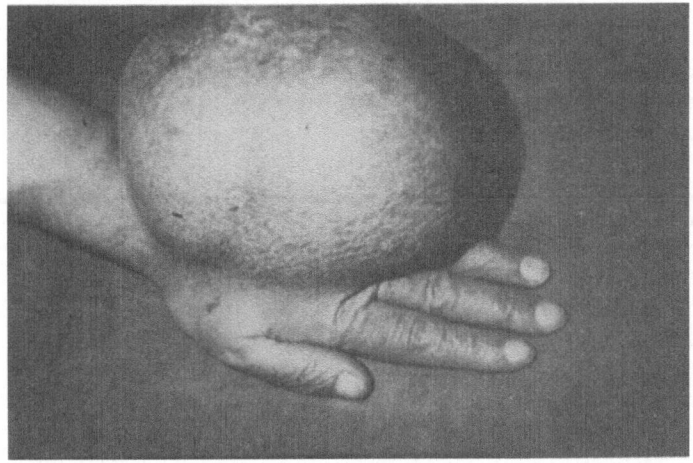

(a)

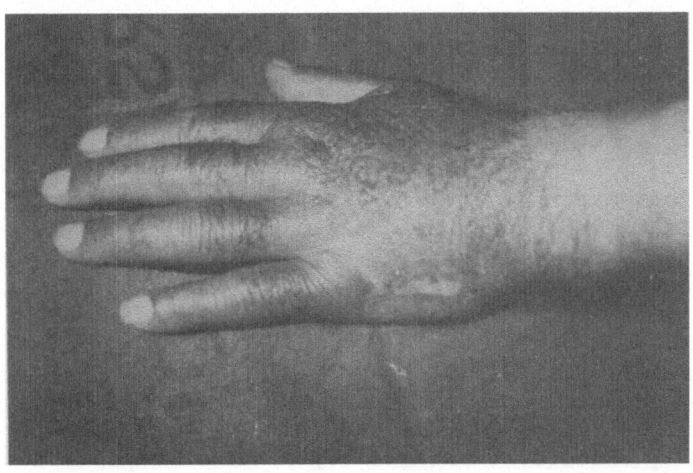

(b)

Fig. 6.

laboriously checked by means of highly expensive medical therapies, based upon hemoderivates products which so far as it has been ascertained up to date, quickly lose their efficacy and must therefore be administered in ever growing doses. In these instances the laser has proven very useful both when synoviectomy to fight hemartri was required, and when effecting other kinds of operations.

To conclude, in my opinion the laser, far from being the universal panacea as assumed by a few, is nonetheless a very important

factor contributing to the progress of surgery, a factor which day
after day shows new facets and new fields of applications.

Today we have at our disposal at least three different kinds of
lasers: CO_2 for surgery, Argon laser and Nd YAG laser for photocoagulation and endoscope surgery.

Other kinds of laser are now being studied; their technical
specifications are different, and thus also their biological effects.
We do hope that Medicine will soon be able to take advantage also of
these new discoveries.

REFERENCES

Apfelberg, D. B., and Maser, A., 1980, Lasers in Surg. & Medicine,
        Vol. I Nr. 1.
Apfelberg, D. B., 1982, Lasers in Surg. & Medicine Vol. X Nr. 2.
Aronoff, B., 1981, Lasers in Surg. & Medicine, Vol. I, Nr. 4.
Bandieramonte, G., 1983, Lasers in Surg. & Medicine, Vol. 2, Nr. 3.
Chavoin, J. P., 1982, Proceed. I Congr. E.L.A. Cannes.
Entin, M. A., and Daniel, R. K., 1975, Proceed. I Intern.Symp.Laser
        Surg. Jerusalem Acad.Press.
Giler, S., 1982, Proceed. I Congr. E.L.A. Cannes.
Haggai, T., 1977, Proceed. II Intern.Symp.Laser Surg. Dallas, (Texas)
        Jerusalem Acad.Press.
Henning, H., 1982, Lasers in Surg. & Medicine, Vol. 2.
Hulsbergen, H., 1983, Lasers in Surg. & Medicine, Vol. 2, Nr. 3.
Kaplan, I., and Pariente, R., 1974, Policlinico, Sez.Chir. Vol. 81.
Koebner, H. K., ed., 1980,"Laser in Medicine," John Wiley & Sons.
Labandter, H., and Kaplan, I., 1975, Proceed. I Internat.Symp.Laser
        Surg. Jerusalem Acad. Press.
McCarthy, W. H., and Stephens, F. O., 1979, Proceed. III Congress
        Laser Surg. Graz. Set.
Pariente, R., 1973, Proceed, XXIII Congr.Ital.Soc.Plastic Surg.
        Torino.
Pariente, R., 1973, Riv.Ital.Chir.Plast. Vol. V, 2.
Pariente, R., 1975, Proceed. I Internat.Symp.Laser Surg. Jerusalem
        Acad. Press.
Pariente, R., 1976, Acta Acad.Lancisiana, Rome, Vol. XX, I.
Pariente, R., and Kaplan, I., 1975, Trans. VI Intern.Congr.Plastic
        Surg. Paris.
Pariente, R., 1977, Trans.X Europ.Congr.Int.Coll.Surg.Milan.
Pariente, R., and Baiani, G., 1977, Proc. II Symp.Laser Surg. Dallas
        (Texas) Jerusalem Acad.Press.
Pariente, R., 1978, Policlinico, Sez.Chir. Vol. 85 Nr.I.
Pariente, R., ed., 1981, Trans. II Congr.Ital Soc.Laser Surg. Rome,
        Oct.
Pariente, R., 1982, Trans. World Congr. Maxillo Facial Surg.
        Sorrento, Naples.

Pariente, R., 1982, Proceed. I Congr. E.L.A. Cannes.
Pariente, R., 1982, "New Frontiers in Laser Medicine & Surg." Excepta
        Medica, Amsterdam.
Ryuraburo, T., and Mitsuhiro, O., 1979, Proceed. III
        Intern.Congr.Laser Surg. Graz, Set.
Toshio, O., 1980, Med.Laser Res.Co.Ltd., Fukui Print Co. Japan.
Toshio, O., 1981, Lasers in Surg. & Medicine, Vol. I, Nr. 4.
Simeon, C. M., 1982, Proceed. I Congr. E.L.A. Cannes.
Vershueren, R. G., and Oldhoff, J., 1975, Proceed.I Intern.Symp.Laser
        Surg. Jerusalem Acad.Press.

LASER SURGERY IN HEMOPHILIA

A. Musajo-Somma

Department of Plastic Surgery
University of Bari, Italy

Hereditary bleeding disorders present a deficiency of a single factor necessary for hemostasis and which is the result of a genetic mistake. We begin this article with an overview of the hemostatic mechanism; thereafter we especially consider the defects of hemophilias and the usefulness of laser surgery in those patients.

Primary events of hemostasis are really independent of the clotting mechanism. Immediately after a wound, and interaction starts between blood vessels and circulating platelets. Platelets come into contact with collagen resulting in a loose primary hemostatic plug; platelets release their contents resulting in retraction of the vessel around the plug. At his point bleeding stops and the intrinsic and extrinsic pathway of coagulation activated result in a firm clot consisting of a fibrin mesh, intertwining with the platelets.

Moreover, in hemophilias, a poor quality fibrin is formed and larger wounds have no sufficient clot retraction or definitive clot formation to permit any hemostasis at all. In hemophilias the defect is in the intrinsic system. This system is tested by the partial thromboplastin time. Hemophilia A, or classic hemophilia, results from a deficiency of Factor VIII coagulant activity. Factor VIII or antihemophilic globulin has a normal amount in the hemophiliac, but its function is not normal. Hemophilia B is a deficiency of Factor IX and the result again is the inefficient function of the intrinsic system in the activation of further steps of th normal coagulation cascade. In summary, hemophiliac disorders are sex-linked, genetic pathologies, resulting in the deficiency of specific plasma factor coagulant activity.

Even minor wounds may bleed severely in hemophiliacs, therefore any opportunity to stop bleeding must be searched for.

Specific therapy is available since plasma extraction of specific factors taken from blood donors was obtained about twenty years ago.

Since 1964, using simple method of cryoprecipitation, a concentrate is obtained which contains large amounts of the original factor VIII. The cryoprecipitate is very effective in controlling bleeding and can also be prepared in the form of lyophilized concentrates of factor VIII.

By giving substitute therapy of factor VIII the hope is to achieve a factor VIII plasmatic level, even if only for a short time, capable of obtaining hemostasis therapy is the development of antibodies to factor VIII in about 15% of severely affected patients. Moreover, hepatitis offers another adverse side effect in those patients who receive many infusions of factor VIII, extracted from a huge pool of plasma.

Even dental extraction is now a relatively safe operation which may be carried out in many instances on an out-patient basis, thanks to homologous factor VIII concentrates. But, even if factor VIII is adequate, there is an important reason for caring about local therapy after tooth removal: the tooth socket is like an open wound to be coped with until healing is achieved with normal mucosa bridging the alveolar gap.

Oral hemostasis in hemophiliacs is very difficult because any coagulant material (e.g. bovine thrombine, antifibrinolytic drugs, etc.) locally applied to the mouth is easily diluted and washed away by saliva; moreover, tissue fibrinolysis, that is very active in any mucosal area, enchances clot lysis[1]. Therefore, it is well explained why, until recently, many authors have advocated extensive replacement therapy and hospitalization before oral surgery and during the post-operative healing period. Even local surgical measures, such as electrodessication or diathermy, never work in oral bleeding because the eschar removal may itself result in severe bleeding later on (See Table 1).

A new tool in the surgeon's hands is the laser. The laser scalpel is an optical device using the power properties of light (laser is an acronym derived from light amplification by stimulated emission of radiation). A laser beam can be produced in over 200 different materials which may be in solid, semiliquid or gaseous states. Laser beam emission can be focused into a very fine spot at different wavelengths dependent on the nature of the utilized medium. Laser has different important properties that make it a unique tool for surgery: cutting power, penetration through water, hemostatic effect and a very small paraincisional necrotic zone.

Table 1. Local and General Measures to Stop Mucosal Bleeding in
Hemophiliacs.

	Local Measures	
	Effective	Non Effective
Non Surgical	Pressure and Vasopressor Agents	Packing Alone Absorable Dressing Chemocautery
Surgical	Laser Photoagulation	Galvanodiathermy
General Measures	– Substitutive Therapy – Tranexamic Acid	

Previous positive experience with lasers in reconstructive
surgery prompted us to find more specific medical indications for the
different laser models, under the pilot study of the Italian National
Research Council[2].

Therefore we selected four severe A hemophiliacs, aged from 7 to
31 years, with oral bleeding caused by inflamed gums or amovable
deciduous tooth.

Plasma factor VIII was in every case less than 1% and proper
laboratory investigations were performed on an out-patient basis.

Rationale for management was based on:

1. the necessity for surgical extraction
2. evaluation of clotting factor deficiency
3. checking a circulating inhibitor
4. available laser-technology

Two patients were severe hemophiliacs with high responder
inhibitors (75 and 80 Bethesda Units). Hemostasis by laser is
accomplished by the welding of lymphatic and capillary vessels rather
than by clot formation as with diathermocoagulation. Argon laser
offers an excellent coagulative action with a narrow parasurgical
necrotic zone and was selected to perform effective coagulation of
gums.

Argon laser used in this clinical study produces intense light
in the blue-green spectrum, with a spot of 1 mm, and pulsed power of
3 watts lasting 0,2 seconds each pulse; its radiation was flexibly
transported through a fiber-optic wavelength to the target area
(Fig 1).

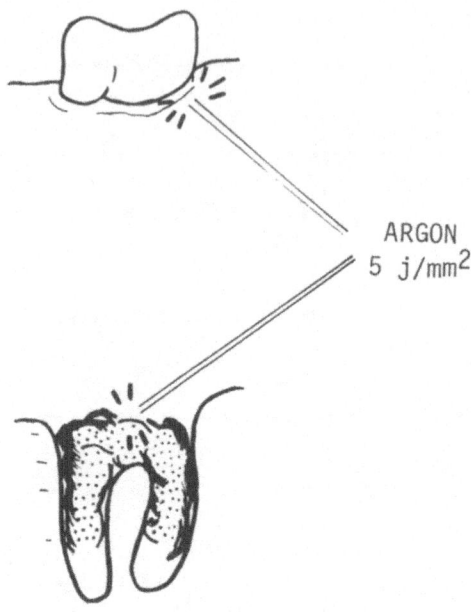

Fig. 1. Argon laser photocoagulation of the gengival area.

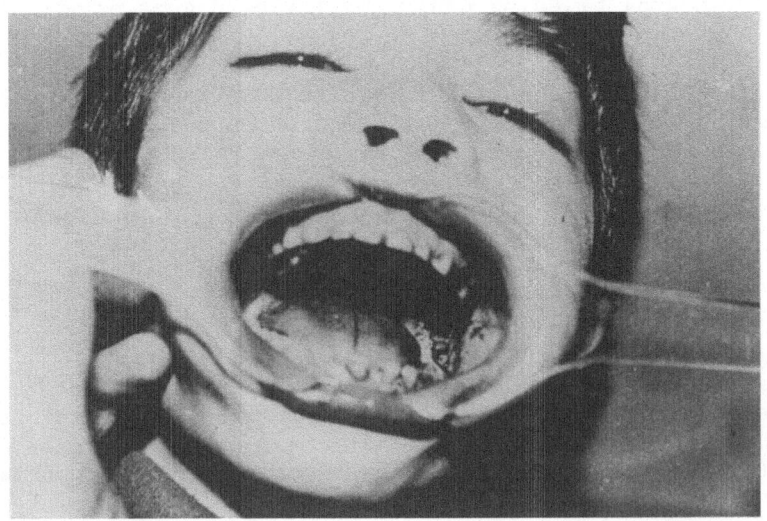

Fig. 2. Oral bleeding from deciduous teeth in a young hemophiliac.

Plan of treatment for single extractions (of removal three deciduous teeth and one root extraction in four patients) was as follows: each patient was treated under local anesthesia and/or under neuroleptoanalgesia on a day-hospital basis (Fig 2).

No substitutive therapy at all was given.

Surgery began using laser photocoagulation of the gengival area around the tooth to be removed (Fig 3); elevators for luxation and extraction forceps were then utilized to remove the loosened tooth; then if necessary, granulation tissue was curetted, and laser photocoagulation reapplied to the residual bleeding socket. Bleeding was always stopped by laser (Fig 4).

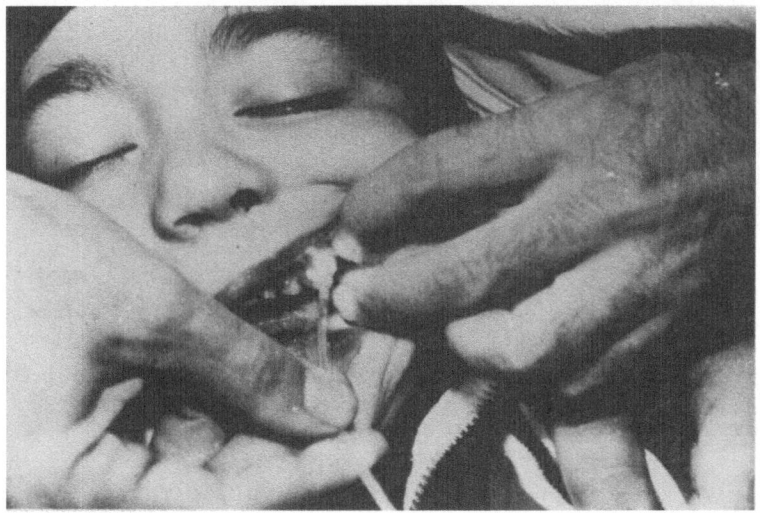

Fig. 3. Argon laser lift through a fiber-optic waveguide.

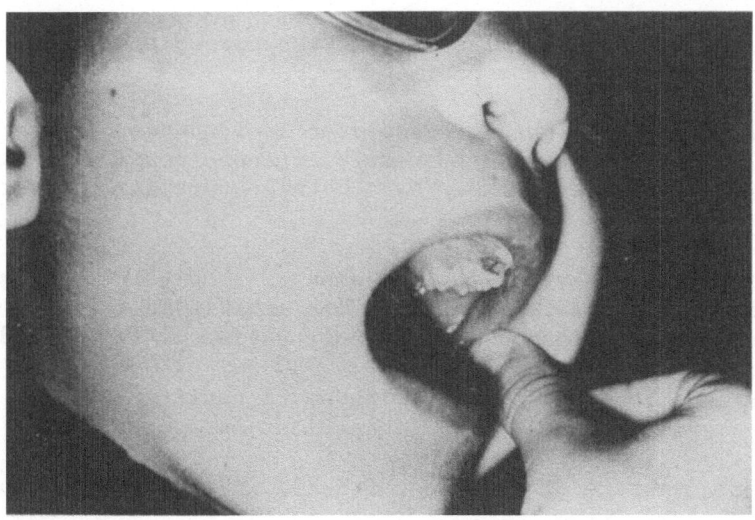

Fig. 4. Mucosal area healed.

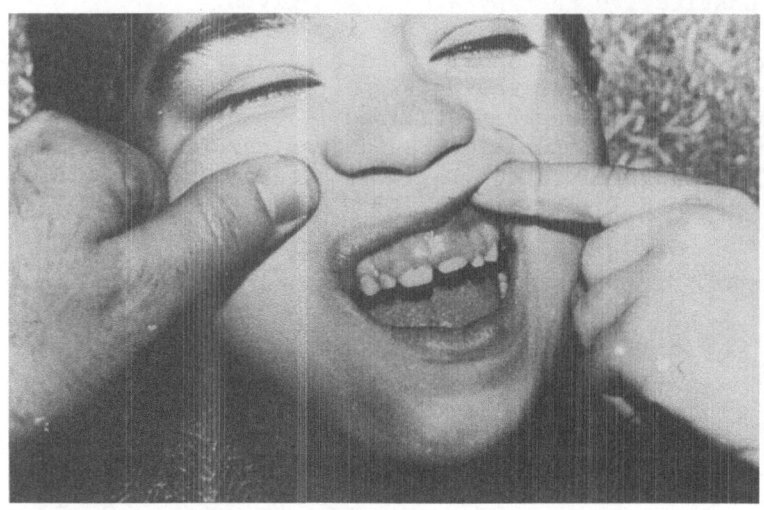

Fig. 5. Permanent tooth in the treated area.

Antifibrinolytic therapy was given (tranexamic acid 1.5 gr/250 mls saline i.v.) immediately after surgery and then, orally, every three hours for a week (0.5 gr each dose).

Wet packs with tranexamic acid were applied to the raw area and changed immediately before patient discharge.

Patients were checked on an out-patient basis, everyday until healing was completed (Fig 5).

No patient developed bleeding in the post-op phase and healing was perfect between 4th-7th post-op day.

Laser surgery has been suggested to avoid dangerous over-treatment with blood products and to reduce extended hospitalization. Therefore our first results may have been encouraging for several reasons:

1.   in saving time hospitalization when oral surgery is necessary in hemophiliacs, even in those with inhibitors
2.   no need to give factor replacement, saving money and reducing risks
3.   quick and bloodless surgery with sound healing
4.   psychological benefit both for patient and relatives.

The disadvantage of this technique is inherent to the cost of the laser system.

REFERENCES

1. E. Cavalcanti, M. Zacchino, and T. Ripa, Odontoiatrica Practica, 15:253 (1980).
2. A. Musajo Somma, N. Ciavarella, A. Scaraggi, and T. Ripa, Hemostasis 10(Suppl. 1):229 (1981).

REFERENCES

1.
2.

MICROSURGERY BY MEANS OF A Nd-YAG LASER

C. Balacco-Gabrieli

Institute of Opthalmology
Bari University, Italy

INTRODUCTION

In January 1983, a Nd-YAG pulsed laser was made available for clinical tests carried out in Bari by the author (C. Balacco-Gabrieli) within the CNR national research program "High Power Lasers". At that time it was the first laser system to work in Italy in a public institution and the third in Europe after the mode-loched laser (Aron-Rosa, in Paris) and the Q-switched laser (Fankhauser, in Berne).

After a period of experimentation on animals' eyes we have treated about 300 patients affected by various pathologies, often of a complicated nature.

Since the beginning, the results have been entirely satisfactory and beyond our initial expectations.

The Sirius microruptor II was installed by Lasag AG of Berne with the collaboration of Prof. Fankhauser; it is extremely versatile and may enlarge the field of ophthalmologic microsurgery, being an alternative and often a substitute to the traditional microsurgery.

DESCRIPTION OF THE INSTRUMENT

The microruptor II is composed of a laser source, an optical system of dispatch and a hinged optical arm which connects the source with the dispatch system. The most important parts of the instrument are the contact lenses which are used for irradiating the various ocular structures.

The source is a Nd-YAG laser(λ = 1.06μ) pulsed by a Pockel's cell (electronic shutter). According to whether the Pockel's cell is functioning or not, one can work in Q-switched or in free-running mode.

The laser beam passes across a system of mirrors and a hinged optical arm, to the dispatch system, consisting of a special optical head and a standard slit-lamp. The observation system consists of a Zeiss stereoscope.

The laser beam has an angular aperture of 16° with a focal diameter of about 70μ and a focal length of 150 mm; it is focussed on the center of rotation of the slit lamp.

To focus the invisible Nd-YAG laser beam, a small, visible, low power, continuous He-Ne laser is used; this is inserted into the optical head with two beams which exactly surround the Nd-YAG beam. The point of intersection of the two He-Ne beams marks the focus of the Nd-YAG.

It is possible to move the focus YAG in front of that of He-Ne by about 3 mm. This allows an increase in the diameter of the YAG focal spot (maximum 1 mm).

The operator is protected from the light reflected from the laser by a filter incorporated in the observation system which absorbs radiation of 1.06μ (anti-YAG filter).

There is no laser emission if the control at the operator's side have not been programmed and if the patient is not in the correct position (the control system blocks if the patient has not placed his forehead correctly).

The YAG contact lenses which are used for irradiation are of a particular crystal (BK-7) rather than of metaacrylate; the latter does not support the high intensity which is involved when working near the cornea.

The lenses give a high quality image and do not distort the focal spot; what is more, they increase the angular aperture of the laser beam (maximum 24°) and reduce the diameter of the spot. This produces an increase between the transversal selection of the laser beam at the level of the cornea, and the diameter of the focal spot.

The contact lenses are a highly important part of the optical system because they do not only protect the crossed dioptric means, but they fulfill a special function of connecting the extra-ocular and the intra-ocular trajectory of the laser beam.

There are three types of contact lenses: one for irradiation of
the iris (CGI 1); one for irradiation of the pupillary region and
retro-pupillary structures (CGP 1); and two lenses for the anterior
chamber and angular structures (CGA 7- CGA 8, according to the
corneal curve). For the posterior vitreous and the retina, Goldmann
lenses can be used.

The system works either in Q-switched (mechanical effect) or in
free-running mode (thermic effect). The laser operated in the free-
running mode is destroying the tissues. This destruction is brought
about by the optical breakdown and the wave of impact from which it
derives, and it determines the level of the focal spot.

When high intensities of energy are emitted in a very short
time, and are concentrated on a very small area, owing to the high
electric fields the breakdown happens; this consists of the ionization of tissue with formation of a plasma, which absorbs most of the
emitted energy. At the level of the optical breakdown acoustic waves
are produced from the impact. The effect of the optical breakdown is
limited to the focal point and the surrounding area, about 200µ in
depth, and 0,5 mm sideways.

In Q-switched mode the pulse lasts for 12 ns with energy from
0.35 to 80 mJ (recharging time of about 2.5 sec).

Both single and bursts of pulses (maximum 9) can be used in
Q-switched and in free-running mode. When working with free-running
in 10 msec, only single pulse can be used.

SECURITY OF THE SYSTEM

Even though the microruptor II is a highly powerful instrument
capable of destroying any ocular structure it is a very safe and
reliable instrument; but, at the same time it is necessary that the
operator must be very careful and obey a few fundamental rules. It
is important that he chooses his target precisely and the best possible focalization; it is easy to imagine the damages caused by an
unwanted emission. However, it must be added that it is difficult to
miss the focalization or to damage the surrounding tissues with the
laser beam.

To focus the instrument, two continuous low power He-Ne revolv-
ing beams are sent exactly around the Nd-YAG beam; this has an
angular aperture of 16° which means that the depth is very small.
The focussing is laborious, but at the same time very precise, so
that the emission can be seem at the lowest level.

Although the two rotating He-Ne beams may cause a slight disturbance during treatment, they have the important function of con-

trolling the YAG laser trajectory beam. When there is an obstacle blocking this trajectory beam (for instance the iris when one works on the retina or in the vitreous) the beams begin to flash.

The contact lenses which connect the extra-ocular trajectory and the intra-ocular laser beam provide a quite good immobilization of the eye being treated; their main function is to increase the angular aperture of the laser beam to 24°, thus increasing also the ratio between the diameter of the laser and the focal spot.

The greater this ratio, the greater the power at the focal spot with a better utilization of the breakdown; besides the density of laser power at the level of surrounding tissue is less, and the work can be done with less risk to such structures and with greater power. By increasing the aperture of the beam one increases also the precision of the spot.

CLINICAL EXPERIENCES

To summarize, the microruptor II can be used either in Q-switched (mechanical effect) or in free-running mode (thermic effect). If they are correctly employed, both methods can produce excellent results. When used together, then the machine is even more effective,

In a 7 month period, we have treated over 300 patients affected by various pathologies which are listed in the following:

Acute and chronic angle block glaucoma	60 cases
Pupillary block glaucoma	29 cases
Chronic simple glaucoma	19 cases
Ectopia pupillae	3 cases
Pupillary membranes of various nature	73 cases
Iris-corneal adherences	42 cases
Iris-capsule adherences	18 cases
Vitreous strands	39 cases

Every treatment has been conducted by the surgery, and the average length of time for each patient was 5-10 minutes.

In some cases it was necessary to have two sittings, when the transparency of the aqueous did not allow us to continue treatment. In these cases the energy loss along the trayectory of the laser beam is very low, as the light is not absorbed or scattered. We preferred to follow different treatment in two or more sittings when the acqueous has become cloudy due to the presence of deposits from tissue destruction.

In most cases only a local anesthetic was used; when the single
contact lenses did not work to obtain a sufficient immobilization of
the eye, than a retrobulbar injection of xylocaine was used.

When working on the iris or the angled structures, miotics were
used, while we have used mydriatics for pupils or retropupillary
structures.

The patients did not suffer any pain; but they just felt a light
humming from the acoustic waves produced from the optical burst,
inducing no feeling of apprehension.

PATHOLOGICAL RESULTS

Acute and Chronic Angle Block Glaucoma

In angle block glaucoma (both acute and chronic) we have performed excellent iridotomies with the Q-switched YAG laser. Iris CGI
1 lenses were used for all treatments. In most cases miotics drops
were inserted before treatment to obtain a stretching of the iris; in
some cases glycerol drops were used so as to have a sufficient transparency of the cornea.

When the iris was particularly thick and when there was neovas-
cularization then Q-switched treatment was preceded by free-running,
with energy of 400-600 mJ in 10 msec, on the area already chosen for
the iridotomia.

In Q-switched we used a burst of 4 pulses with energy varying
from 8 to 19 mJ per pulse, according to the thickness of the iris,
independent of its colors. In every case, perfect iridotomy have
been performed.

One single sitting was sufficient in most cases, and often only
one pulse was necessary to obtain sufficiently wide iridotomie.

In phakic eyes, it is necessary to perform an iridotomy on the
outline so as not to run the risk of damaging the lens.

Undesirable and collateral effects were not experienced. Only
in a very small number of cases there were small hemorrhages due to
broken iris vessels; however, by the following day there were no more
traces of blood in the anterior chamber.

Pupillary Block Glaucoma

In cases of pupillary block glaucoma in phakic eyes, we performed an iridotomy in the same way as for the angle block glaucoma;

in aphakic eyes, we attempted to directly remove the pupillary block, and the treatment was the same as that for pupillary membranes.

Chronic Simple Glaucoma

We treated some open angled glaucoma using the laser in free-running for trabeculotomy according to Wise.

The angled structures were irradiated with 10 ms pulses and energy from 800 to 1.100 mJ. The results were good, as were those obtained using the Argon laser.

However, some difficulties were experienced during treatment for the focalization of the spot, which seemed to us difficult to direct; and for the time required to recharge to instrument (working in free-running with pulses of 10 ms the recharging time is about 20 sec).

Ectopia Pupillae

We performed 3 iridotomies on three subjects with ectopia pupillae caused by over-work of the iris after cataract extraction.

With 10 msec pulses of 600 mJ (free-running) we initially marked out the outlines of the new pupil; then using Q-switched, with a burst of 4 pulses and energy per pulse of between 12 and 16 mJ, we performed an ample iridotomie. Iris CGI 1 lenses were used throughout.

In all three cases, treatment was carried out in two sittings at a distance of one week. This is because after the initial impact of Q-switched, the aqueous became cloudy due to material from the destruction of the iris. A light increase in ocular tone was noted, well controlled within two or three days with the use of timolol locally and acethazolamide generally.

Inflammation was quickly decreased with the use of anti-inflammatories used both locally and generally.

Two of the 3 patients have a 10/10 visus, the third 6/10 due to retinal problems.

Pupillary Membranes

Numerous pupillary membranes were treated, almost always in aphakic eyes; we do not yet have any experience of membranotomies in pseudophakic eyes.

For most cases, irradiation lenses CGI 1 were used; only in very few cases no lenses were used. In these last cases we noted a great endotelial damage with big stromal oedema.

In all cases visual acuteness increased often dramatically except where other retinal pathologies of various natures were present.

The membranes were usually the consequence of cataract surgery (posterior capsules, posterior capsules with cortical residue and cataractous masses) or of a perforating wound.

For slender membranes in tension, like opaque posterior capsules, after extracapsular extraction, bursts of 4 pulses, with energy per pulse of 4-6 mJ, were sufficient.

For thick membranes (capsules with cortical residues of cataractous masses) and membranes not in tension, bursts of 6-8 pulses, with energy per pulse up to 23 mJ, were used.

One sitting and a low number of impacts were sufficient for thin capsules in tension, for thick membranes, capsules with cortical residue or cataractous masses, a high number of impacts and more than one sitting was necessary for the debris to be reabsorbed.

If we intended to perform a membranotomy then just the visual axis was liberated; when a deep observation was necessary, then we tried to liberate the whole pupillary field.

In almost all cases, we had a light increase of the optical tone, resolved with timolol and acethazolamide in very few days, and signs of inflammation were quickly checked with anti-inflammatories. It should be said that these conditions were present only in a very few cases, when it was decided to increase the number of sittings (especially when there was cortical residue).

Iris-corneal Adherences

We treated many iris-corneal adherences, both as a consequence of cataractous intervention and secondary to perforating injuries.

Treatment in all cases was straight forward and exempt from dangerous collateral effects.

Contact lenses (CGA 7-CGA 8) were used for the irradiation of the angle structures which easily allowed us to reach adherences even near the center of the cornea.

Bursts of 4 pulses, with 8-10 mJ energy per pulse, were used in Q-switched mode.

When there was a neovascularition of the iris, then free-running preceded Q-switching, with 10 msec pulses and 600-800 mJ energy. No damage to the corneal endothelium, already destroyed by adherence neither post-operational problems were noted.

Using this method we have obtained excellent results with far less trauma than with traditional microsurgery.

Iris-capsule Adherences

Very few iris-capsule adherences were treated owing to the inconvenience of usually having to perform an anterior capsulotomy which results in an opaque crystal. Even when low energy is used, it is still possible to damage the lens.

Vitreal Strands

Many vitreal strands were treated in anterior chamber, at the level of pupil and in the vitreous.

In the first case, we found little difficulty; angled contact lenses (CGA 7 - CGA 8) were used and bursts of 4-6 pulses with energy per pulse usually between 10-12 mJ in Q-switched mode.

Neither was there great difficulty in treating strands in the pupillary area; for these, we used contact lenses (CGP 1) for the pupillary region, and bursts of 4 pulses with energy 10-12 mJ energy per pulse in Q-switched mode.

With no doubt it has been most difficult to treat the strands at medium and posterior levels of the vitreous; in these cases we used Goldmann's contact lenses, and bursts of 4-6-8 pulses, according to the state of tension of the strands with energy per pulse between 10-12 mJ in Q-switched mode.

In the case of a vascularized strand we preferred to use the pre-treatment in free-running with 10 msec single pulse of 600-800 mJ energy. For the strands very close to the retina (less than 4 mm) we preferred to irradiate the retinal area near the strand, with 800 mJ pulses in 10 msec in free-running; this helps to avoid retinal hemorrhages and other retinal damage.

Despite the difficulties experienced, the results were good. We cut off strands which impeded vision, and those that produced tractions on the retina.

A sufficient period of control has not yet passed, but up until
the present we have not had any undesirable collateral effects or
post-operative problems.

This method is similar to classic vitrectomy, although it can
never be a substitute: Q-switched mode destroys while the vitrectomy
removes. The two different methods are not competitive but compli-
mentary; for instance, the laser can treat vitreous strands and
preretinal membranes which are difficult to reach and treat with
classic surgery, then a vitrectomy can be performed, having been made
easier by the former laser treatment.

Using Goldmann 3 mirrors lens it is easy to direct the laser
beam towards structures which are not easy to reach or are completely
unavailable using the vitrectomy. It is less traumatic using the
laser and, most important, when one does not obtain the required
results with laser, then vitrectomy can be used; it is not possible
to do vice-versa.

CONCLUSIONS

At the moment, the most important principle of the Nd-YAG Q-
switched laser is the destruction of structures inside the eye which
disturb vision. An example of this is the destruction of pupillary
membranes of various natures and of vitreous strands, which disturb
vision and can determine a retinal detachment owing to traction.

The problem of the strands remains wide open as today we do not
have a sufficient number of cases enabling us to give a definite
answer; in any case, the problem of vitreous strands will be examined
in the near future, both for better treatment, and the removal of
possible damages sustained during treatment.

Another important point about the Nd-YAG Q-switched laser is the
iridotomy both for optical as for prophylaxis for acute glaucoma, and
in some therapeutic cases too.

The iris-jaloid adherences and iris-corneal can be treated very
satisfactorily without danger to nearby structures as the spot
focalizes well and the treatment does not require high laser energy.

We do not have experience of either anterior or posterior capsulotomy in pseudophakic eyes, but we intend to carry out experiments
in the near future.

The subject of free-running remains completely open. We have
experience in the trabeculotomy according to Wise, and in some pre-
treatment on the iris and retina only.

Experimentation using free-running has yet to be used on the retina, choroid and in the macular area; also using the Nd-YAG laser to compare the hemostatic effect with that of the Argon laser. As yet, we do not have sufficient experience, but we have already prepared a program of experimentation using animals.

In our brief experience we have been able to appreciate the advantages of the Nd-YAG laser system and the various perspectives it offers in the treatment of ocular diseases.

We believe however, that we can give a valid contribution towards the improvement and perfection of this already excellent instrument.

It would be very useful to put the laser source with a different wavelength (for example, the visible Argon wavelength) on the Nd-YAG, as at present it seems impossible to substitute the Argon laser for certain types of treatment; in any case, the two lasers compliment each other in the results of our experiences.

To conclude, this new Nd-YAG laser system, in treating many pathologies of the anterior segment and of the vitreous, opens a new way. The relative simplicity of operation which requires a certain experience and knowledge of the operator, places in the hands of the opthalmologist the possibility of resolving numerous complicated ocular pathologies.

There is no necessity for the patient to stay in hospital; and many of the pathologies considered resolvable only with surgical operation of high specialization in narcosis using operational microscopes, sophisticated instruments and selected surgical equipment, can now be treated in a far simpler fashion using the laser.

REFERENCES

1. F. Fankhauser et al., Int.Ophthal., 3,3:129-139 (1981).
2. F. Fankhauser et al., Int.Ophthal., 5:15-32 (1982).
3. R. H. Keates, S. Fry, and W. J. Link, "Neodinium-YAG Ophthalmic Laser," Monography (1982).
4. D. Riquin, F. Fankhauser, and J. Lortscher "Contact Glasses for Use with High Power Lasers," Bern, Switzerland.
5. P. Roussel and F. Fankhauser, "Contact Glass for Use with High Power Lasers - Geometrical and Optical," Bern, Switzerland.
6. C. Balacco-Gabrieli, G. Avolio, G. Mastrandrea, and O. Stefania, "Prime Esperienze con il Laser Nd-YAG Sulla Patologia del Segmento Anteriore Dell'occhio," Congresso Nazionale della Società Italiana di Laser Chirurgia e Applicazioni Biomediche, Milano, 23-25 Maggio (1983).

LASER SAFETY

LASER DOSIMETRY AND DAMAGE MECHANISMS:

UNRESOLVED PROBLEMS

D. H. Sliney

US Army Environmental Hygiene Agency
Aberdeen Proving Ground, MD 21010, USA

INTRODUCTION

The use of any physical or chemical modality in science or in
medicine and surgery requires the application of dosimetric concepts,
terms, quantities, and units. The choice of any domestic quantities
requires certain assumptions relating to the nature of the interaction of the chemical agent (or drug) or the physical agent (e.g.
acoustic radiation, light, or ambient heat or cold). If these underlying assumptions are seriously in error, the wrong physical quantities may be chosen, with the result that empirical observations may
not fit theory, the development of new theory will be hindered and
experimental results will not be readily comparable between different
laboratories. The use of standard terminology is also very important. Fortunately, the CIE (Commission International d'Eclairage) and
other international organizations have established a standardized
system of radiometric quantities and units[1,2].

RADIOMETRIC QUANTITIES AND UNITS

Two systems of terms and units are used to describe optical
radiation. One has a physical basis: the radiometric system. The
other, the photometric system, attempts to describe the optical
radiation in terms of its ability to elicit the sensation of light by
the eye. Figure 1 gives the most commonly used radiometric terms and
the "preferred" units for each system. There are generally analogous
units in each of the two systems, and Table 1 is arranged to show
these similarities. Although the radiometric system of units can be
used across the entire spectrum, the photometric system is limited to

a. DESCRIBING A SOURCE

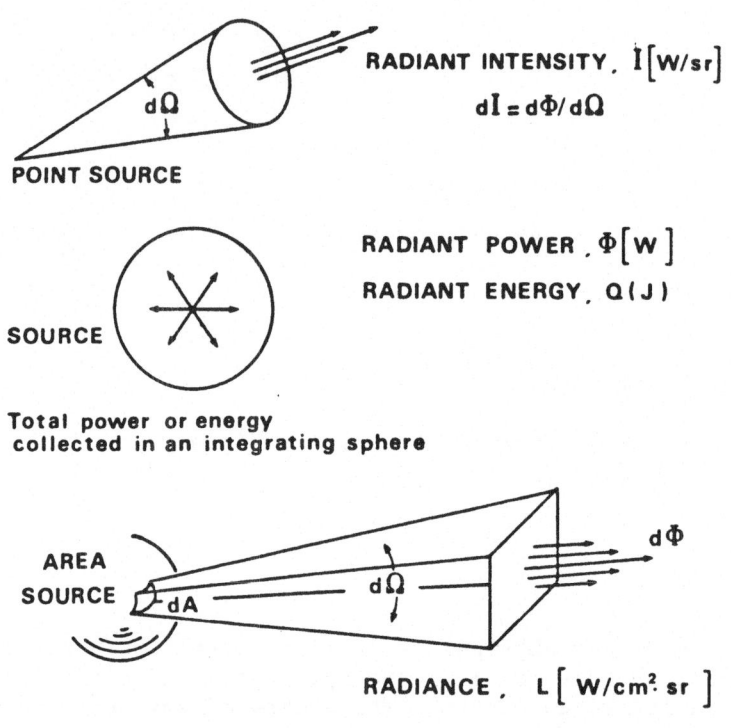

RADIANT INTENSITY, I [W/sr]

$$dI = d\Phi/d\Omega$$

POINT SOURCE

RADIANT POWER, Φ [W]

RADIANT ENERGY, Q(J)

SOURCE

Total power or energy
collected in an integrating sphere

RADIANCE, L [W/cm² sr]

$$dL = d\Phi/dA \cdot d\Omega$$

b. DESCRIBING AN IRRADIATED SURFACE

Continuous source   Pulsed source

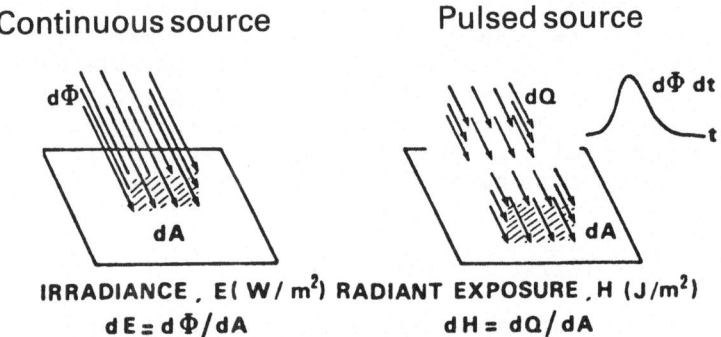

IRRADIANCE, E(W/m²)   RADIANT EXPOSURE, H (J/m²)
$$dE = d\Phi/dA$$ $$dH = dQ/dA$$

Fig. 1. Simple conceptual drawings of some radiometric terms.

describing light (i.e. electromagnetic radiation that is visible)
from approximately 300–400 nm to 760–780 nm.

Of all the terms given in the figure, the following will be
encountered most often: optical energy (Q), optical power (Φ, or

Table 1. Useful CIE Radiometric and Photometric Terms and Units[a,b].

		Photometric	
Term	*Symbol*	*Defining Equation*	*SI Units and Abbreviation*
Quantity of light	Q_ν	$Q_\nu = \int \Phi_\nu dt$	Lumen-second (lm · s) (talbot)
Luminous energy density	W_ν	$W_\nu = \dfrac{dQ_\nu}{dV}$	Talbot per cubic meter (lm · s · m^{-3})
Luminous flux	Φ_ν	$\Phi_\nu = 680 \int \dfrac{d\Phi_e}{d\lambda} V(\lambda) d\lambda$	Lumen (lm)
Luminous exitance	M_ν	$M_\nu = \dfrac{d\Phi_\nu}{dA} = \int L_\nu \cdot \cos\theta \cdot d\Omega$	Lumen per square meter (1m · m^{-2})
Illuminance (luminous density)	E_ν	$E_\nu = \dfrac{d\Phi_\nu}{dA}$	Lumen per square meter (1m · m^{-2})
Luminous intensity (candlepower)	I_ν	$I_\nu = \dfrac{d\Phi_\nu}{dr}$	Lumen per steradian (1m·sr) or candela (cd)
Luminance[c]	L_ν	$L_\nu = \dfrac{d^2\Phi_e}{dr \cdot dA \cdot \cos\theta}$	Candela per square meter (cd·m^{-1})
Light exposure	H_ν	$H_\nu = \dfrac{dQ_\nu}{dA} = \int E_\nu dt$	Lux-second (1x·s)

		Radiometric	
Radiant energy	Q_e		Joule (J)
Radiant energy density	W_e	$W_e = \dfrac{dQ_e}{dV}$	Joule per cubic meter (J·m^{-3})
Radiant power (radiant flux)	$\Phi_e P$	$\Phi_e = \dfrac{dQ_e}{dt}$	Watt (W)
Radiant exitance	M_e	$M_e = \dfrac{d\Phi_e}{dA} = \int L_e \cdot \cos\theta \cdot d\Omega$	Watt per square meter (W·m^{-2})
Irradiance or radiant flux density (dose rate in photobiology)	E_e	$E_e = \dfrac{d\Phi_e}{dA}$	Watt per square meter (W·m^{-2})
Radiant intensity	I_e	$I_e = \dfrac{d\Phi_e}{d\Omega}$	Watt per steradian (W·sr^{-1})
Radiance[c]	L_e	$L_e = \dfrac{d^2\Phi_e}{d\Omega \cdot dA \cdot \cos\theta}$	Watt per steradian and per square meter (W·sr^{-1}·m^{-2})
Radiant exposure (dose, in photobiology)	H_e	$H_e = \dfrac{dQ_e}{dA}$	Joule per square meter (J·m^{-2})

Footnote to Table 1.
_____

a The units may be altered to refer to narrow spectral bands in which
case the term is preceded by the work spectral, and the unit is
then per wavelength and the symbol has a subscript λ. For example,
spectral irradiance H_λ has units of $W \cdot m^{-2} \cdot m^{-1}$ or more often
$W \cdot cm^{-2} \cdot nm^{-1}$.
b While the meter is the preferred unit of length, the centimeter is
still the most commonly used unit of length for many of the above
terms and the nm or μm are most commonly used to express wavelength.
c At the source, $L = \dfrac{dM}{d\Omega \cdot \cos\theta}$; at a receptor, $L = \dfrac{dE}{d\Omega \cdot \cos\theta}$

sometimes P), and the optical energy or power per unit are of absorbing source, which are known as irradiance (E), and radiant exposure
(H) respectively. Most of these units have an area contribution.
The preferred unit of area in the System International (SI) is the
meter^2 (m^2) or some 3-decade multiple or sub-multiple thereof (e.g.
mm^2, μm^2, nm^2). However, since the cm^2 is very nearly the size of
some laser beams and only slightly larger than the dilated pupil of
the eye, as well as a reasonable detector of aperture size, the unit
of cm^2 is very popular.

LASER OUTPUT

     Peak power or total energy emitted per pulse are often used in
describing a laser's output. On the other hand, the biological
effect of radiant energy incident upon tissue is clearly not dependant upon the total energy but upon the energy and power absorbed per
unit area of the absorbing tissue surface. The biological effect is
usually also a function of duration. The terms of irradiance (W/cm^2)
and radiant exposure (J/cm^2) are used.

IRRADIATION OF A SURFACE

     Optical radiation is normally absorbed in a thin tissue depth.
Therefore, to describe effects of radiant energy (Q in joules) upon
tissue, most investigators have used power or energy per unit area
(e.g. W/cm^2 or J/cm^2) and in early years termed these "power density"
or "energy density". However, the standard CIE nomenclature power
density and energy density refer to power or energy per unit volume
and not per unit area, and therefore their erroneous use should be
discouraged. Rather than use radiant energy as a general term, the
term "radiation" is preferable. One other term once widely used to
connote both irradiance and radiant exposure is "beam intensity". By
CIE standard terminology, "radiant density", I, is power per unit
solid angle (watts per steradian) from a point source. This term has
little value in laser bioeffect studies.

BRIGHTNESS

The terms "radiance" (and "luminance") are used to describe the rate of light or radiant energy leaving the source per solid angle, i.e. the "brightness" of the source. Alternatively, their terms "irradiance" and "illuminance" are used to describe the level arriving at a given point in space, i.e. the level measured by detector which measures the number of photons falling upon a unit of surface. The concept of radiance is a very important in the evaluation of light sources, since radiance is conserved during focusing[1].

UNRESOLVED PROBLEMS: MECHANISMS

Other lectures in this School have covered the basic concepts of interaction mechanisms and thermal models. I would like to briefly review four puzzling findings which do not appear to be adequately predicted by thermal models - or for that matter - by any model[2,3,4].

Most of us would agree that there are three to four distinctively different laser interaction mechanisms with biological tissue: photochemical, thermal, thermomechanical, and optical breakdown. The latter two mechanisms are limited to short pulse and ultrashort pulsed lasers. Photochemical damage mechanisms are characterized by the reciprocity of irradiance and exposure duration, and by a long-wavelength cutoff where individual photon energy is insufficient to break or change a critical molecular bond. Figure 2 shows how the time dependence of injury threshold for a photochemical event differs from that application to the rate-process of thermal injury. This leads us to our first unresolved problem.

Problem #1: The Pulse Duration Dependence of Retinal Injury

Simple models of thermal injury predict that for a very short exposure durations where heat flow would be insignificant, that reciprocity (of time and irradiance) should also exist for a thermal injury mechanism. Threshold injury data of skin, cornea and retina do not show this. The pulsed injury thresholds for the retina have been the most carefully studied[2-6].

The rate-constants for thermal coagulation of proteins are such that thermal models that include a rate-process dependant term predict that injury from a 20-ns exposure should occur before heat flow becomes significant (i.e. in less than 100 μs)[7]. The temperature rise in the outer receptor layer where irreversible damage occurs is 10 times less for a 20-ns pulse than for a 1 ms pulse, but the time-temperature history in this layer should be the same for either pulse, since heating occurs in this layer only from conduction of

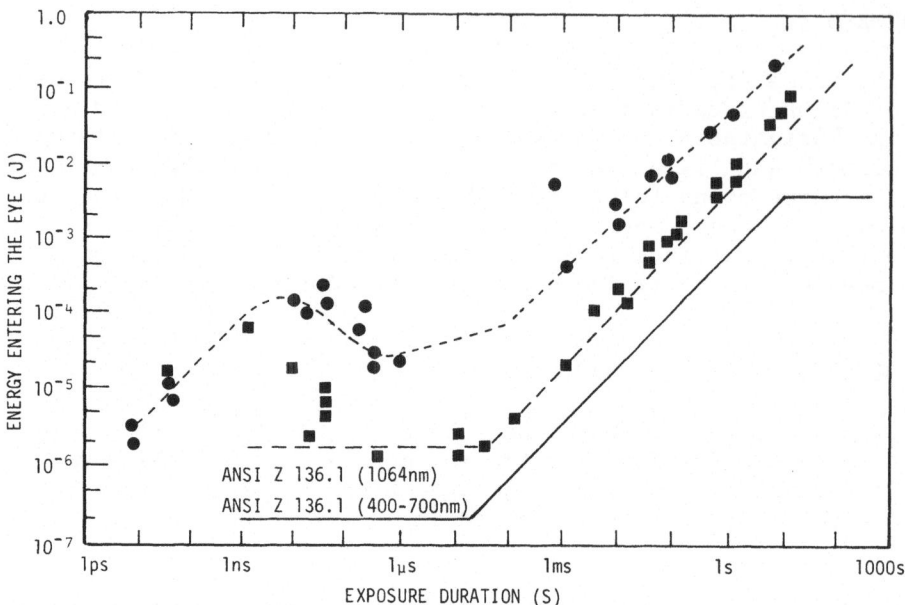

Fig. 2. The dependence of retinal injury threshold upon pulse
         duration. The injury thresholds in the rhesus monkey are
         plotted as a function of total laser energy entering the eye
         vs exposure duration. The dashed line shows the laser
         safety limits (ANSI Z-136.1). Note the anomalous increase
         in thresholds for both 1064-nm and visible light at 10 and
         20 ns when compared to thresholds at 10-20 μs. Safety
         limits do not exist for mode-locked pulse durations.
         Adapted from refs. 2 and 8.

heat from the adjacent RPE. The rate-process models do predict one
scaling relation: the decrease in threshold radiant exposure (J/cm²)
as a function of the pulse duration raised to the 3/4 power for
exposure durations ranging from 20 μs to 10 s as shown in Figure
3[6].

     None of the retinal thermal models predict the anomaly seen in
Figure 3 for thresholds between 1 μs and 500 ns to 10 μs exposures
for the same image size and same wavelength[2]. This is a curious
phenomenon; and many more data points between 10 ns and 1 μs that are
not shown offer solid evidence that this temporal dependence of
threshold for such short exposures is real and nor an experimental
artifact. What is the explanation? Two possibilities appear reasonable to me. As the pulse duration decreases, more energy could go
into a noninjurious dissipative mechanism such as an acoustic transient at threshold. Of course this very mechanism is known to be the
principal cause of severe, almost explosive, damage in the retina for
suprathreshold exposure levels. An alternate, and perhaps more

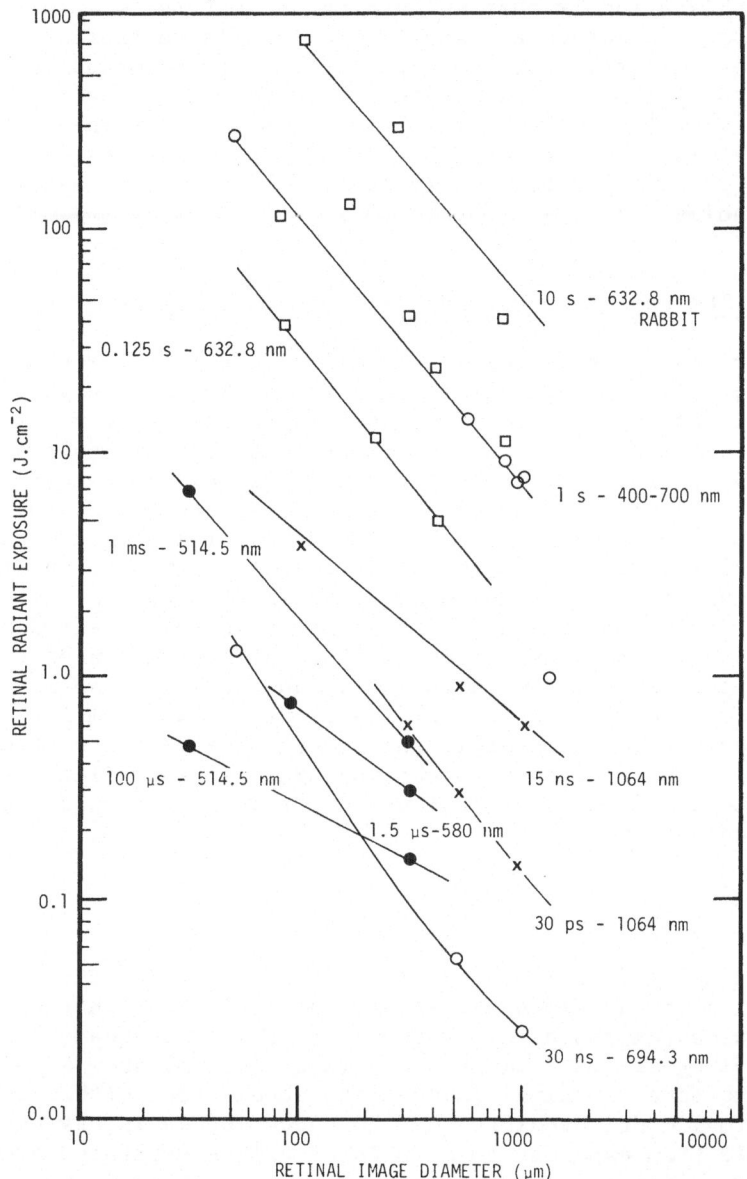

Fig. 3. Scaling relation for retinal injury threshold with spot
         size[2]. While the reduced exposure threshold for large
         spot sizes (d_r) is expected for longer exposure durations
         (due to heat condution), the same dependence of
         approximately $1/d_r$ is surprising for 15-ns and 30-ps pulses.

appealing explanation is that melanin may be saturable absorbed like some other organic molecules, and as the irradiance increases for shorter exposure durations, the amount of energy absorbed in the granules becomes less. I have tried to detect such behavior by irradiating a vial of melanin in a water suspension with a Q-switched pulsed laser beam, but was unable to achieve a clear result because I could not prepare a sufficiently dense and uniform suspension of melanin granules.

Problem #2: The Spot Size Dependence of Retinal Injury

Another scaling relation is seen in Figure 3: the retinal injury threshold for all exposure durations, from 30 ps to 10 s varies approximately as the reciprocal of the image diameter for image sizes of 20 nm to at least 1 mm. All of the thermal models would predict that for exposures less than approximately 100 μs the temperature rise would be so great prior to any significant heat flow that there should be no difference in injury thresholds for exposure durations less than 0.1–1.0 ms, and there should also be no variation of threshold with image size. The melanin-granule models suggest some decrease in damage thresholds for the RPE cells themselves for the shorter exposure durations, since the incandescent granules could destroy adjacent structures in the cell, but certainly no image size dependence of injury could be explained. Heat flow from large images during 1-s exposures is insufficient to cool the center of the image for the same irradiance that causes no injury in a smaller image. This leads to the image size dependence predicted for exposure durations exceeding 1 ms.

Problem #3: The Repetitive Pulse Dilemma

Perhaps the most curious contradiction of the thermal model is the collection of experimental retinal threshold data that clearly show that pulsed exposures separated by durations adequate for almost complete cooling of small image sized (e.g. 100 ms) show an additivity of exposure. Minimal image sizes should cool within 1–10 ms and the thermal model predicts that no reduction in injury threshold should result from exposure to a train of pulses separated by that duration, even if the pulsed exposures are superimposed over the same retinal area. A large collection of data suggests that one can empirically predict that the threshold exposure of a single-pulse in the train is decreased proportionally to $N^{-1/4}$ where N is the number of pulses in the train[9]. This relation holds for 10-ns pulses, 10-μs pulses, and 10-ms pulses over a wide range of pulse separations and exposure durations for small image sizes. According to one study it appears not to occur for large image sizes[8]. But then another study with large image sizes does tend to show the additive effect [10]. One phenomenon - biological repair - may repair these and the other anomalies[2].

Problem #4: Biostimulation

Perhaps no area of laser bioeffects studies is more controversial than that termed "biostimulation". I would include myself among those who are very skeptical that many of these low-irradiance He-Ne and Ga-As laser effects really exists. However, I would not completely discount them. I find it difficult to conceive of a mechanism either microscopic or macroscopic in scale to explain many of the reports which are termed biostimulation (e.g. wound healing, cell growth, pain relief, etc). Without a theory for the mechanism it is difficult to know what radiometric parameters should be used. For example, let us hypothesize that it could be the result of laser speckle. In this case irradiance or radiance exposure, power or energy do not adequately describe speckle.

Why should I mention speckle? This phenomenon of coherent light occurs due to interference effects following diffuse scattering. The result at the nearly microscopic level is that localized irradiance values can vary more than 100-to-1.

Still further research is also necessary to understand the effects of speckle upon retinal function at very low, chronic exposure durations as reported by Zwick[11]. It may well be that is we force ourselves to stare at a large-area diffuse laser reflection, the speckle which produces a myriad of areas of very high contrast borders at the retina may affect those neutral mechanisms that detect borders in our normal world. Fortunately, most of us find it uncomfortable to stare at speckle for long periods, and this effect is hopefully just a laboratory curiosity. Clearly there is still much to learn vis-a-vis laser irradiation of the eye.

REFERENCES

1. Commission International de l'Eclairage, CIE: International Commission on Illumination, "International Lighting Vocabulary," Paris, CIE No. 17(E-1.1) (1970).
2. D. H. Sliney and M. L. Wolbarsht, "Safety with Lasers and Other Optical Sources," Plenum Publishing Corp., New York (1980).
3. D. H. Sliney, Kvant Elektron., 7(10):2271-2281 (1980) and Soviet J.Quant.Electr., 10(10):1278-1284 in translation.
4. D. H. Sliney, Interaction mechanisms of laser radiation with ocular tissues, Proceedings of the 11th Symposium on Laser Induced Damage in Optical Materials, NBS Special Publication, US Department of Commerce, National Bureau of Standards, Boulder, CO (1982).
5. W. T. Ham, J. J. Ruffolo, Jr., H. A. Mueller, and D. Guerry, Vision Res., 20(19):1105-1111 (1980).
6. R. G. Allen, Retinal thermal injury; Ocular effects of non-ionizing radiation, Proc.Soc.Photo-Opt.Eng., 229:80-86 (1980).

7.   C. Hu and F. S. Barnes, IEE Trans.Biomed.Eng., BME-17:220-229
     (1970).
8.   G. A. Greiss, M. F. Blankenstein, and G. G. Williford, Health
     Physics., 39(6):921-927 (1980).
9.   B. E. Stuck, D. J. Lund, and E. S. Beatrice, "Repetitive Pulse
     Laser Data and Permissible Exposure Limits," Report No. 58,
     Letterman Army Institute of Research, Presidio of San
     Francisco, San Francisco, CA (1978).
10. J. E. Walkenbach, Determination of retinal lesion threshold
     energies of pulse repetition Nd^{+3}:YAG laser in the rhesus
     monkey, Masters Thesis, Medical College of Virginia,
     Richmond, VA, June (1972).
11. H. Zwick and D. L. Jenkins, Coherency effects on retinal neural
     processes in Pseudemys, Technical Note No. 81-20 TN, Dec. 82,
     Division of Ocular Hazards, Letterman Army Institute of
     Research, San Francisco (1982).

SAFETY ASPECTS OF ANTERIOR SEGMENT

LASER PHOTODISRUPTERS

H. Sliney

US Army Environmental Hygiene Agency
Aberdeen Proving Ground, MD 21010, USA

INTRODUCTION

During the past two years the world of ophthalmic surgery has been alive with the discussion of the latest surgical technique using lasers. Two European ophthalmologists pioneered this laser application: Prof. Danielle Aron-Roas in Paris and Prof. F. Fankhauser in Bern[1-2]. In essence this application consists of focussing a very high-peak-power, pulsed Nd:YAG laser beam into the anterior segment of the eye to cut membranes by vaporizing tissue within a very small focal zone. In this focal zone the beam irradiance is so enormous (of the order of 10^{10} W/cm^2) that optical breakdown occurs in ordinarily "transparent tissue", forming a high temperature microplasma ($\sim$ 0.2-0.3 mm across) to vaporize and thereby cut tissue[3-4]. Two types of Nd:YAG pulsed lasers are used: Q-switched (8-20 ns) and mode-locked (7-10 pulses in a train of 30 ps pulses). The optical breakdown phenomena responsible for igniting the microplasma differ with each pulse form, but the surgical results are basically the same[3-4]. In this lecture we shall explore the safety aspects of this procedure without delving deeply into the details of the mechanisms of microplasma formation (which are reported elsewhere). The safety aspects of anterior segment laser surgery with Nd:YAG lasers are centered upon two aspects: one must consider hazards to the patient's eye and also the protection of the surgeon and any bystanders.

RETINAL SAFETY

The beam cone angle, total beam energy, wavelength, distance of the microplasma from the retina, plasma shielding, and clarity of media all are important factors. Each of these shall be considered in the following paragraphs.

    The beam cone angle plays a very important role. A large cone
angle reduces the retinal irradiation for the same laser energy.
However, a large cone angle encounters more spherical aberrations of
the cornea and the beam may intercept points where pre-focal absorption could occur (e.g. at corneal scars). In theory, a larger cone
angle should also produce a smaller focal spot, reducing the threshold for plasma production.

    The status of the cornea, or corneal clarity, is also important.
Even a slight haze in the cornea or along the focal path will blur
the focal spot size resulting in greater energy required to produce a
plasma for cutting, while the retinal exposure area remains basically
the same. A failure of plasma production will produce a higher
retinal exposure level since plasma "shielding" will not take place.

    Of course total beam energy also influences the risk of adverse
effects. The risk of damage to adjacent structures by acoustic
transients (shock waves), by thermal injury, and by other effects
increased with an increase in pulse energy. Repeated, carefully
aimed shots of lower energy are safest. However, repetitive pulsing
can cause problems if the patient moves only slightly during a salvo
of pulses, with the result that other structures along the line of
motion can be cut.

    With the present system there is little choice in wavelength.
The 1064-nm wavelength of the Nd:YAG laser is a good choice because
retinal absorption is low compared to the retinal absorption at
shorter wavelengths. Longer wavelengths, e.g. 1220 nm also penetrate
the ocular media, but far less reaches the retina. Only extensive
experience will show whether the greater absorption in the anterior
segment is safer, since pre-focal loading is greater and structures
anterior to the focal point may absorb too much energy.

    The most important factor with regard to retinal safety appears
to be the distance of the focal spot from the retina. Surgery of the
anterior segment does not expose the retina to damaging irradiances.
At distances greater than 5 mm from the retina there is little chance
that retinal exposure levels will result in retinal injury[3].
Mode-locked retinal injury thresholds are lower than Q-switched
thresholds for smaller image sizes, but not greatly less for the
larger image areas characteristic of this surgical technique[3,5].

    There has been a great deal of discussion in the literature on
the nature of plasma shielding in this technique. While this effect
is greater in the mode-locked pulse form, both Q-switched pulses and
mode-locked pulse trains exhibit this phenomenon which provides a
substantial factor of protection to the retina and to other structures posterior of the plasma. From 50% to more than 90% of the
laser pulse energy is absorbed and scattered by the plasma, resulting
in far less exposure of the retina. The initial plasma also limits

the growth of the plasma from the focal point to points posterior of it[4]. The experienced surgeon places the plasma center just behind the membrane so that the plasma grows into the membrane. The plasma may be thought of as making the initial microincision; and the shock wave that follows separates the membrane, opening the incision.

No discussion of retinal surgery is complete without a consideration of retinal injury mechanisms. The surgeon should keep in mind that injury to the retina by ultrashort pulses differs from photocoagulation from a 0.1-s argon laser burn. The acoustic component of injury can disrupt retinal membranes, etc. A valid benefit-vs-risk judgements when working near the retina depends upon this understanding.

SAFETY OF THE SURGEON AND BYSTANDERS

The surgeon sees only the red He-Ne (632.8 nm) aiming beam; the longer wavelength is filtered to protect his or her eye. Thus we can be assured that the operating surgeon is not at risk during the procedure.

The safe viewing distance for bystanders is typically greater for an argon laser photocoagulator than for an infrared laser photodisrupter. This finding is a result of the high collimation of the argon laser beam. Safe viewing distances may be as close as 30-50 cm for diverging reflections from the cornea or contact lens. Nevertheless, for medical-legal reasons there is some merit in having onlookers unfamiliar with the procedure to wear laser eye protectors.

Safety switches and other system safeguards are also desirable. In the United States, the Food and Drug Administration Laser Product Performance Standard requires certain safety features[5]. These include protective enclosures, a remote connector to permit the use of door interlocks, a protective filter in the viewing optics, warning labels, a beam attenuator and an energy monitor.

REFERENCES

1. D. Aron-Rosa, J. C. Griesemann, and J. J. Aron, Ophthalmic Surg. 12:496-499 (1981).
2. F. Fankhauser, P. Roussel, and J. Steffen, Int.Ophthalmol., 3:129-139 (1981).
3. D. H. Sliney, YAG Laser Safety, in: "YAG Laser Ophthalmic Microsurgery," S.K. Trokel, ed., East Norwalk, Appleton-Century-Crofts (1983).
4. M. A. Mainster, D. H. Sliney, and C. D. Belcher, and S. M. Buzney, Ophthalmology, 90(8):973-991 (August 1983).
5. D. H. Sliney and M. L. Wolbarsht, "Safety with Lasers and Other Optical Sources, A Comprehensive Handbook," Plenum Publishing Corp., New York (1980).

LASER SAFETY IN LASER SURGERY

D. H. Sliney

US Army Environmental Hygiene Agency
Aberdeen Proving Ground, MD 21010, USA

INTRODUCTION

Lasers used for surgery and for some other biomedical applications can pose potential hazards to both the patient and the laser
operating personnel. Because the laser beam is normally in the open
when emitted from a surgical laser, special precautions are necessary. Unlike many industrial application of lasers, the very nature
of most laser surgical procedures require both a flexible and often
an open beam and the use of administrative controls with protective
eyewear rather than engineering controls such as beam enclosures,
baffles, etc. The potential for hazardous exposure to laser radiation can therefore be quite high for some surgical personnel[1,2,3].

LASER SAFETY STANDARDS

There are two major areas for standardization in laser safety:
Exposure Limits (EL's) and Codes of Practice[4]. As it turns out,
to have workable codes of practice, it became necessary to develop a
laser hazard classification scheme which provides still one more area
for standardization: Accessible Emission Limits (AEL's). In 1983,
Technical Committee TC76 of the International Electrotechnical Commission (IEC) approved an international standard on the radiation
safety of all laser products[5]. National standards are largely in
agreement with the IEC standard.

HAZARD ANALYSIS AND LASER CLASSIFICATION OF A NEW LASER SYSTEM

Any evaluation of laser hazards must consider three major
aspects: (a) the laser's emission; (b) the environment, and (c) the

275

persons who are potentially exposed and the laser operators. The
first aspect is covered by a laser hazard classification scheme.
The last two aspects rely on the user's judgement. Laser safety
standards aid considerably in hazard analysis and setting controls.
Control measures vary with laser class.

The basic concept of the laser hazard classification scheme is
to place all lasers into one of at least four classes to indicate the
degree of hazard. Class 1 laser products are essentially "safe" and
are typically totally enclosed so as not to emit hazardous levels.
Class 2 laser products are limited to visible lasers with an output
power of less than one milliwatt which are safe for momentary viewing. Class 3 laser products are not safe even for momentary viewing;
procedural controls and protective equipment are normally required in
their use. Class 4 laser products are normally considered much more
hazardous than Class 3 lasers since they may represent a significant
fire hazard or skin hazard and may also produce hazardous diffuse
reflections. General concepts of laser safety are dealt with quite
extensively in the aforementioned standard and elsewhere (e.g. Sliney
and Wolbarsht, 1980)[1,4], and will not be discussed further here.
Most surgical lasers are in Class 4 since their output power exceeds
0.5 W. The upper power limit of a class is termed the accessible
emission limit (AEL).

EXPOSURE LIMITS

In general, laser classification limits (AEL's), are probably
used far more frequently than exposure limits (EL's) for the eye and
skin. In most applications, there is no intentional exposure of
persons and EL's are seldom applied. However, in many diagnostic
applications, intentional exposure is the norm and EL's must be
determined.

The applicability of EL's to the patient must be considered.
In non-diagnostic application, the EL's may not apply to the patient.
The EL's were drived with the promise that a potentially exposed
individual was awake, unrestrained, and not under medication or
anesthesia. A factor of three to ten is typical between the EL and
the level which produces a clearly visible reaction is tissue. The
limits were designed to be sufficiently low to preclude any adverse
functional or delayed effects. However, the limits cannot be considered as fine lines between safe and dangerous conditions. If
exposures of patients occur at levels above the EL, a benefit-vs-risk
judgement is in order. Should the eye be exposed to laser radiation
under conditions where ordinary pupillary reactions and eye movements
would not take place, the EL's would not be sufficiently conservative
to preclude a risk of injury in a diagnostic. Likewise, if tissues
are exposed to ultraviolet or visible radiation at levels at or even
below the applicable EL following administration of photosensitizing

drugs, the EL may not be sufficiently conservative. To analyze the
risk of unintentional injury of tissue under these conditions, it
may be necessary to go back to the literature on laser bioeffects
studies[1,2,4].

SURGICAL LASER SAFETY

With the development of modern, portable surgical lasers, the
designation of a specific or for laser surgery has grown less common.
During early developments of surgical lasers, specialized facilities
were often designed[1,2]. Today, three general types of lasers are
most commonly employed in surgical procedures which utilize lasers;
the argon laser (488 nm and 514.5 nm), the neodymium:YAG laser (1064
nm), and the carbon-dioxide laser (10.6 μm). From a safety standpoint, it is emphasized that of these three types of lasers, the
carbon-dioxide laser's far-infrared-wave-length radiation does not
penetrate the ocular media of the eye and is, therefore, far less an
eye hazard than the other two types of lasers for the same power
level.

Hazardous laser reflections are a primary safety concern in
laser radiation safety. For example, where endoscopic delivery
systems are used, the concern for reflections is less, but one must
still be concerned with backflash[6]. The reflective hazards associated with a surgical laser beam in the open are minimized when using
curved surface instruments. The calculated reflected beam irradiance
from a flat metal surface is far greater than from a curved surface.
Reflections from a curved-metal surface are typified in stainless
steel surgical instruments, and are minimized by a short-focal-length
lens in the beam delivery system. Although not always feasible, this
feature dramatically reduces the probability of a hazardous reflection. High-numerical-aperture fiber optical light guides will also
produce a rapidly diverging beam of laser radiation and reduce the
chance of unintended hazardous exposure. Positive-action exposure
switches, beam shutters, and key-lock master control switches are all
features that provide additional safety for the user and the patient.
It is generally agreed that only Class 1 or Class 2 lasers should be
used in diagnostic instruments unless the actual patient exposure is
otherwise insured to be below the applicable EL's.

Ancillary hazards from fires, improper ventilation[8,9], and
improper electrical grounding must not be ignored. To date the most
serious accidents in laser surgery have occurred as a result of
oxygen supported fires in flammable airways through the use of COH^2
lasers. The proper choice of flame-resistant airway and anesthetic
gas mixture[9] is critical in endotracheal surgery.

LASER SURGERY FACILITIES

Class 3 or Class 4 laser operations should be carried out in an enclosed facility need not be "light tight", but there should be no direct beam path outside of the facility. Diffuse baffles, door interlocks, and window shades are called for in most situations. It is not necessary to seal light leaks at the bottom of doors or at vents well above eye level. Glass windows are acceptable for CO_2 lasers. Appropriate control are generally accepted when the hazards are properly understood[1,2].

Quite often surgeons have available a fire-resistance target such as alumina, or silicon-carbide fire brick to focus a hand-held laser "scalpel". This inanimate target aids the surgeon in developing a "feel" for the instrument and provides a visible indication of the size and location of the focal spot. Metallic surfaces must not be used for this practice.

LASER ACCIDENTS

While there have been a number of minor skin burns or burns on OR clothing from laser beam reflections of CO_2 lasers, the only major accidents relate to fires caused by laser ignition of plastic airways during otolaryngological procedures. With oxygen used in connection with anesthetic gases, an intense combustion is supported with severe or fatal consequences. Snow et al.[9] recommended that combustible gases be avoided during such procedures. In opthalmic surgery the initiation of retinal vessel hemorrhaging with retinal laser photo-coagulators occasionally has been reported.

Most standards urge the drafting and posting of safe use procedures. These help remind laser users of the hazards of the laser and emphasize safe practices.

The only serious safety problem encountered with opthalmic argon laser photocoagulators has been the reflected secondary beam from a contact lens[1,7].

CONCLUSIONS

Safety training is critically important for all users of surgical lasers. Also, designers of laser equipment should be knowledgeable regarding the adverse effects of laser radiation and the need for appropriate hazard controls. Laser users should remember that EL's were developed for normal individuals, and the EL's may not always apply to photosensitive patients or to a patient in certain medications of under anesthesia.

REFERENCES

1. D. H. Sliney and M. L. Wolbarsht, 1980, "Safety with Lasers and
    Other Optical Sources," Plenum Publishing Corp., New York.
2. D. H. Sliney, 1981, Laser Safety, in: "New Developments of
    Lasers," L. Goldman, ed., Springer-Verlag.
3. L. Goldman and R. J. Rockwell, 1971, "Lasers in Medicine,"
    Gordon and Breach, New York.
4. World Health Organization, 1982, "Environmental Health Criteria
    Document No. 23, Lasers and Optical Radiation," World Health
    Organization, Geneva.
5. International Electrotechnical Commission (IEC), 1983,
    "Radiation Safety of Laser Products Equipment Classification,
    Requirements and Users Guide, Document TC76," IEC, Geneva.
6. C. Gulacsik, D. C. Auth, and F. E. Silverstein, 1979, Appl.Opt.,
    18(11):1816-1823.
7. D. L. Jenkins, 1979, "Non-ionizing Radiation Protection Special
    Study No. 25-42-0310-79, Hazard Evaluation of the Coherent
    Model 900 Photocoagulator Laser System, Jan-Feb, US Army
    Environmental Hygiene Agency, Aberdeen Proving Ground, MD
    (NTIS No. ADA068713).
8. K. W. Marich, J. B. Orenberg, W. J. Treytl, and D. Glick, 1972,
    Am.Ind.Hyg.Assn.J., 33(7):488-491.
9. J. C. Snow, B. J. Kripke, M. S. Strong, G. J. Jako, M. R. Meyer,
    and C. W. Vaughan, 1974, Anesth.Analg., 53:507-512.

LASER SAFETY IN THE USE OF LASER SYSTEMS

E. Righi

INFN
Frascati, Italy

INTRODUCTION

This Course has confirmed the great possibilities which laser systems offer to medicine, including some very interesting aspects of oncological therapy. Nevertheless laser light should be considered as a possible cause of injury when used incorrectly. Lasers are used for a great number of purposes, i.e. in medicine, research, telecommunications, industry and in show business, hence the time has come to single out all the relative risks entailed by the use of lasers and define a set of protectionistic measures. Laser safety has been dealt with in a competent and interesting way by Sliney in his lectures; moreover, a great deal of documentation of fundamental importance[1,2] is available to all those who operate in the field of laser safety as well as in the field of photoradioprotection.

I wish to contribute to this documentation by pointing out some characteristics which are indispensable to qualify the risks entailed by the use of lasers and the vulnerability of "critical" organs, emphasizing,, whenever possible, the experience that we have accumulated in this field[3,4].

RISKS, CONNECTED WITH LASER LIGHT AND CRITICAL ORGANS

Laser light is constituted by a beam of monochromatic, coherent and well collimated light. The primary risk related to the use of laser light lies precisely in the fact that laser light can carry even very high amounts of energy. Another aspect to deal with when evaluating the photobiological behavior of critical organs is the wave length interval of lasers which range between 100 nm and 1 mm.

Wave length is especially important in defining the topography of eye injuries, as shown clearly by the diagram of the transmission of the optic dioptric system as function of wave length, experimentally demonstrated by Geeraets and Berry[5].

The pulse length of laser light ranges from 10^{-12}s (1 ps) to more than 0,1 s (continuous wave lasers).

Pulsed lasers fall into four categories, depending on the operative techniques applies: pulsed non Q-switched (P non QS); pulsed Q-switched (PQS), modelocked (ML); Ultra Short (US).

The correlation between the energy output of the beam and the duration of the impulse determines the degree of potential damage. For example, a QS neodymium laser with an energy output of 50J in 30 ns will emit a beam of 1,7 GW of power; the same laser having an inferior energy output, let us say of 10J, but a duration of approximately 10 ps, and with ML techniques applied to it will reach a power of approximately 1 TW.

Another characteristic of laser light lies in its high degree of radiance, even lighter than the most intense conventional luminous sources. A mercury vapor lamp with a power output of 1000 W has a radiance of 10^5 W/m² steradiant or the surface of the sun has a radiance of 10^7 W/m² steradiant, whilst a helium-neon laser with a power output of 1 mW or a ruby laser with a power output of 100 KW can reach levels of radiance as high as 2.10^9 and 6.10^{14} W/m² steradiant respectively[6].

This peculiarity of lasers, combined with correct focalization of the beams, can produce a truly remarkable density of energy, such to enable the fusion of even the hardest materials (i.e. steel, alumina).

Due to its anatomical structure and to its optic behavior the eye is the organ most vulnerable by laser light, hence it can be defined, using a radioprotectionist terminology, as "critical" organ. Owing to the different absorption capacities of the various components of the eye, ocular pathology varies according to the wave length. With lasers in the ultraviolet and in the distant infra-red the lesions hit the more superficial ocular structures (cornea, iris, lens) whilst when applying lasers which emit in the visible and in the near infra-red the vulnerability of the eye lies in the potential transmission of laser beams through the transparent structures of the eye and in the refraction power of the ocular dioptric system itself, thus ensuing, by means of focalization, a considerable concentration of incident energy on small areas of the retina. The degree of retinal damage depends upon the quantity of incident energy and on the type of exposure. The damage can vary from a small burn, which clinically has very little importance, to serious lesions of the

macula, which entail deterioration of visual acuity up to massive
hemorrhage, or even extrusion of tissues in the vitreous humor, in
short the loss of the eye[7]. Moreover, continuous exposure of the
retina, creating small condensations of pigment (melanomata), dis-
cromies of the macula, related to forms of macula photoretinite.

In addition to eye injuries one should also take into account
the possible effects on the skin. Laser light can cause superficial
or deep burns. The degree of damage depends on the area affected,
the intensity and wave length of the laser beam and on the pigmentation of the affected area. Due to the stratiform structure of skin
having disomogeneous optic characteristics, the incident beam will be
partly directly reflected, partly absorbed, and partly undergo scattering phenomena. Experimental observations have shown percentages
of reflection which go from 23,5% for argon lasers (454,5 nm) to
66,5% for ruby lasers (694,3 nm)[9]. It is therefore obvious that
as far as lesions and therapeutic efficiency is concerned the wave
length adopted is of great importance.

BIOSTIMULATION

Together with the destructive effects which are used for surgical applications, laser beams at low density of energy, can produce
a repair effect, the so called "biostimulation effect"[10].

The destructive effects of laser light are due to three fundamental actions: a thermic action, a thermo-acoustic action and a
photochemical action.

As far as biostimulation is concerned the trigger mechanism have
not been delineated quite so precisely, yet the laser light has shown
to have an antiphlogistic and a vasodilatant action as well as a
stimulating action as regards the regeneration of tissues. Although
the use of laser light in biostimulation has been advocated for many
different indications, the therapeutic field for which it has been
employed is the treatment of some cutaneous lesions of the ulcerative
that show a torpid behavior and that often resist traditional
therapies.

The principal dermatological indications proposed by the various
experts concern essentially all forms of ulcers, no matter what their
etiopathogenesis may be (vascular, dysmetabolic, traumatic, infec-
tious, neurological etc...) and more specifically, ulcers of the
legs, and sores of the bed-ridden.

As for the type of laser employed there is a tendency to choose
the helium-neon ones (He-Ne). The irradiation modalities are essentially two, namely "point by point" and "scanning" (sometimes also
known as "sweeping"). In the former case the cutaneous lesion is

irradiated along the edges and inside for approximately 3-5 minutes, with the source of the beam situated at a distance of 50-70 cm. The power used for this kind of therapy is in the order of 2,5-5 mW (up to 15 mW). In the case of direct application by means of optical fibers however, the power has to be appropriately modified.

When applying the scanning mode, irradiation is done with He-Ne laser of 25 mW, guided by an automatic scanning device which moves the beam continuously over the entire injured area. The point by point method probably exerts a more incisive biostimulating action on the periferical epithelial spots of the injured area, but has the disadvantage of needing longer treatment times and the therapeutical commitments are undoubtedly far greater.

Laser therapy in biostimulation should also be taken into consideration for other purposes, especially for application to physiotherapy (orthopedics, traumatology and rheumatology), to reflex therapy up to acupuncture ("light acupuncture"). In addition to He-Ne, other types of lasers, in the low power range, such as CO_2 and diode lasers are used for those therapies.

At the INFN National Laboratories Franscati an automatic computerized pin-pointing device has been elaborated suitable for low powered laser sources, which can be used, among other things, in the point by point laser therapy for biostimulation (the device can also be applied to scanning irradiation when a correct choice of times and location memorization is made).

The device is easy to handle, hence a statitics can be compiled in a relatively short period of time, such that the efficiency of laser therapy in biostimulation by means of the point method can be verified on a far more rigorous manner. Moreover, the device has the great advantage of guaranteeing better safety conditions for the practitioner and the patient.

The growing interest shown vis-a-vis this kind of laser therapy in at this stage encouraged primarily by clinical results according to which the relationship between the treatment and the healing effects cannot have occurred simply by pure chance.

PRIMARY AND ASSOCIATED RISKS

In order to define the protectionist measures useful to prevent the dangers caused by laser light, all useful information relative both to the primary and the associated risks should be carefully recorded.

The parameters related to primary risks enable both a protectionist characterization of the laser system in question and a dosimetric elaboration by means of proteximetric calculations or derivated nomograms. These parameters, must not be underestimated because of realizing for laser radiation a "monitoring" instrumental dosimetry, similar to that used for ionizing radiations.

As far as the safety aspects are concerned, the equations proposed by Burnett are to be kept in mind[11]. As an example I wish to recall an accident which was brought to the attention of our Medical Service and for the reconstruction of which the above mentioned safety calculations were employed[12].

As an experiment of saturable absorbers was being carried out, the saturable absorbers being the target of a Q.S. pulsed ruby laser having an approximate power of 7 MW. The technician in charge proceeded to the alignment of the laser cavity by means of an external collimator placed at a distance of approximately 2 m from the cavity (Fig 1).

This kind of laser emits beams of light having a wave length of 694,3 nm, an energy of 0,1J and a duration of 15 ns. During this operation the remote trigger of the electrode was turned on to "manual", while the bench of condensers was charged and the ground safety device disconnected. The technician brought his right eye up to the eye-piece of the collimator when, due to the closing up of the trigger, probably caused by the discharge of the triggering tyratron or by some other casual disturbance, he was dazzled by a sudden and intense red light. After a momentary blinding, he noticed that his right eye had altered his perception of colors, as if everything had suddenly turned green. This disturbance lasted approximately 2-3 hours.

The small copper reticle of the collimator, placed at a distance of 1 cm from the ocular lens, was destroyed. The absence of the laser radiation ocular lesions stirred up some doubts among the observers. The clinical report of this event could in fact hardly

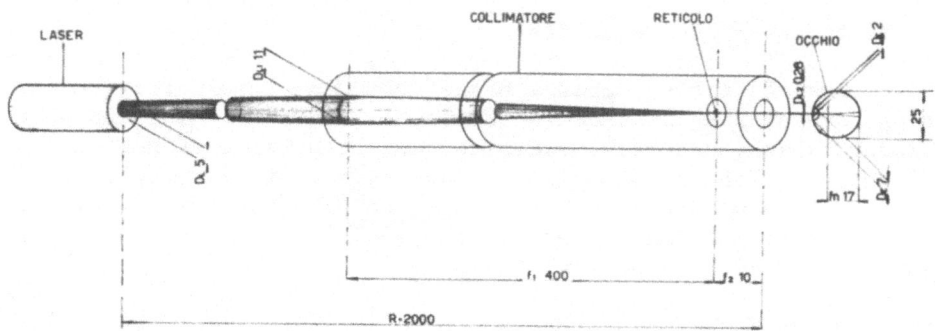

Fig. 1.

correspond to the so far proven destructive powers of laser energy, or to the objective disturbances acknowledged by the technician in question (dazzling and temporary alteration of chromatic sensibility). It was therefore decided to carry out a dosimetric study to establish, at least in an orientative fashion, the causes which could have brought about this situation.

In order to evaluate the density of incident energy on the retina, various exposure conditions were thought of, some of which being entirely theoretical, but nevertheless useful to appreciate the degree of danger entailed by the accident. The first condition admitted an exposure of the eye placed at a distance of 2 m from the laser source, without an interposition of the collimator. On the basis of the above said hypothesis the values of energy density on the retina resulted as follows:

$$I = 188 \text{ J/cm}^2 \text{ with pupillary diameter} = 2 \text{ mm}$$
$$I = 2300 \text{ J/cm}^2 \text{ with pupillary diameter} = 7 \text{ mm}$$

In the second hypothesis the distance between the eye and the laser source remained the same by the interposition of the undamaged collimator was, as opposed to the first hypothetical condition, taken into account.

The calculation acknowledged that the laser beam, emitted through the eye-piece, had completely penetrated the eye through the pupillary hole. The hypothesis could have been acceptable in view of the eye alignment, moreover confirmed by the dazzling and by the dimensions of the optic system in question.

The result, on the basis of the assumed condition and in consideration of Fraunhofer's diffraction was the following:

$$I = 880 \text{ J/cm}^2$$

The levels of energy resulting from the above calculations are extremely destructive and demonstrate the considerable degree of potential danger of the accident.

In actual fact the accident was affected by some purely fortuitous phenomena, such as the destruction of the copper reticle of the collimator, which undoubtly determined an attenuation of the energy on the target. The incident laser beam on the copper reticle could have induced, in view of the brief duration of exposure (15 ns) and of the entity of the energy transmitted (0,1J), a passage towards a state of plasma of the material hit.

The energy which this process could have subtracted, was estimated to be roughly of the same order of magnitude as the energy originally emitted by the laser (0,1J).

Therefore, we can presume that the energy of the beam was entirely, or at least partly, dispersed in the destruction of the copper reticle, hence explaining the absence of lesions, which otherwise would have been very serious.

As far as associated risks are concerned, "volatile toxic substance" withhold particular importance. High powered continuous lasers, such as CO_2 lasers, have a considerable capacity to pollute the environment.

At our Laboratory in Frascati a CO_2 laser cutter is used which has presented some interesting problems of personnel protection and of environmental hygiene.

ENVIRONMENTAL AND INDIVIDUAL CONTROLS

In our country no law has yet been defined for protection against the risks associated to laser light. Nevertheless I wish to point out that in 1978 the Frascati division of CNEN (ENEA), elaborated laser safety norms based on internationally accepted criteria. These rules have been adopted in our Laboratory.

Among other things, these norms provide safety values useful in the classification of laser systems in accordance with the degree of risk ensued by each one[13]. Moreover, the Italian Association for Protection against Radiation (AIRP), through its Permanent Committee for Non-Ionizing Radiations, has formulated some criteria of medical supervision as regards risks connected to non-ionizing electromagnetic radiations, which include laser light[14].

The medical supervision essentially entails oculistic and dermatological examinations, as well as examinations carried out in accordance with preventive medicine criteria to establish the overall physical condition of the individual.

The oculistic examinations comprises, among other things, a visual acuity test, a chromatic sensibility test, and a biomicroscopy test with a slit lamp.

Retinography, especially if in color, represents an interesting component of the clinical documentation. As far as the dermatological examinations are concerned, a photographic survey in black and white, in color or even with the use of Wood's light is useful.

Staff who use high powered laser apparatuses only occasionally are obviously more prone to danger than the staff who are familiar with the various apparatuses. Thus it is advisable to adopt the same medical supervision for both the occasional and the habitual operators.

CONCLUSIONS

I wish to draw the attention of the readers to the fact that whilst rigorous safety measures are being adopted in the use of high powered lasers, there is a tendency to ignore safety measures when dealing with small and more commonly used laser systems, such He-Ne ones.

We are today witnessing a considerable proliferation of available laser equipment, albeit at energy levels that do not produce photocoagulation, and this raises the problem of diffusing and, if necessary, imposing rules and criteria for protecting patients and practitioners, this in analogy with what has already been done with positive results in other sectors. I am particularly thinking of the risk of causing damage to the eyes, a risk that users of continuous He-Ne lasers are often induced to underestimated on account of the small size of the equipment and the ease with which it is used.

REFERENCES

1. A. Mallow and L. Chabot, "Laser Safety Handbook," Van Nostrand Reinhold Co., New York (1978).
2. D. Sliney and M. L. Wolbarsht, "Safety with Lasers and Other Optical Sources: A Comprehensive Handbook," Plenum Publishing Corp., New York (1980).
3. E. Righi and S. Martellucci, Problemi di protezione e sicurezza per apparecchiature laser presso il Centro di Frascati del CNEN. Atti del 33^Congresso Nazionale di Medicina del Lavoro, Cagliari, 23-26 sett. (1970).
4. M. Di Pofi and E. Righi, Problemi di protezione dalla radiazione ottica, Ann.Ist.Super.Sanità., 16:443-476 (1980).
5. S. J. Geeraets and E. R. Berry, Ocular spectral characteristics as related hazards from lasers and other light sources, Am.J.Ophthalmol., 66:15-21 (1968).
6. G. E. Frigerio, "I laser, Rassegna di Applicazioni Mediche, Tecnologiche e Didattiche," Casa Editrice Ambrosiana, Milano (1976).
7. W. T. Ham, A. M. Clarke, W. J. Geeraets, S. F. Cleary, H. A. Mueller, and R. C. William, The eye problem in laser safety, Arch.Environ.Health, 20:156-160 (1970).
8. M. M. Zaret, Examinations for laser workers, J.Occ.Med., 14:255-259 (1972).
9. L. Goldman and R. J. Rockwell, "Lasers in Medicine," Gordon and Breach, eds., Science Publishers Inc., New York (1977).
10. E. Righi and M. Di Pofi, Laser therapy in biostimulation, Ellettromedicali, 4:13-17.
11. W. D. Burnett, Laser eye and skin hazard evaluation, Report SC-RR-68-174, Industrial Hygiene Division, pp. 3311-3321 Sandia Laboratory, Albuquerque, New Mexico (1968).

12. A. Apollonio and E. Righi, Considerazioni su di un incidente con laser, <u>Minerva Fisiconucl</u>, 14:182-193 (1970).

13. CNEN Centro di Frascati, Norme di sicurezza per l'uso dei laser (gennaio 1978).

14. P. Bernardi, M. Boggio, A. Checcucci, M. Grandolfo, E. Righi, C. Tamburello, and R. Zannoli, (Comitato Permanente per le Radiazioni Non Ionizzanti dell'AIRP), Raccomandazioni in tema di sorveglianza medica per rischio da radiazione elettromagnetica non ionizzante: <u>II - Dispositivi laser.Med. Lavoro</u>, 1:71-73 (1981).

DIAGNOSTICS AND TECHNOLOGICAL ASPECTS

POTENTIAL USE OF INCOHERENT AND COHERENT LIGHT-

EMITTING-DIODES (LEDs) IN PHOTOMEDICINE

R. Pratesi

Istituto di Fisica Superiore dell'Universitá and
Istituto di Elettronica Quantistica del CNR
Via Panciatichi, 56/30 Firenze, Italy

PRINCIPLE OF OPERATION OF LIGHT EMITTING DIODES

Light emitting diodes (LED) are semiconductor diodes which emit electromagnetic radiation when operated in forward direction. Contrary to that of a filament lamp, the radiation spectrum is limited to a narrow wave band. The wavelength of the emitted radiation is essentially determined by the semiconductor material used.

As is known, an extrinsic semiconductor with donor (acceptor) impurities is referred to as n-type (p-type) semiconductor since the conduction of current through it is predominantly by negative (positive) charges (i.e. electrons ("holes")).

A p-n junction diode is made by joining together pieces of p-type and n-type semiconductor materials. On making contact, electrons diffuse from the n-region to the p-region because of the concentration gradient which has been established. This depletes the n-material close to the metallurgical junction of free electrons leaving ionized donors which form a positively charged depletion layer. Similarly, "holes" diffuse from p-region to the n-region creating a negatively charged depletion layer in the p-material next to the junction. Figure 1 shows the conduction and valence band scheme after junction is formed.

Through a pn junction operated in forward direction, electrons from the n-region are injected into the neutral n and p region: in those semiconductors, such as GaAs, where direct radiative recombination of electrons with holes is likely, recombination in the depletion layer results in the emission of photons of energy $h\nu \simeq E_g$ (E_g = energy gap between conduction and valence band; ν = frequency

293

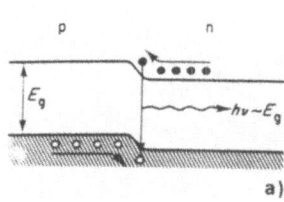 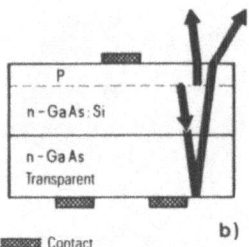

Fig. 1. a) Recombination of electrons (·) and holes (∘) at a
         forward-biased junction to produce photons; b) Structure of
         the diode body of an infrared (IR) LED.

of photon; h = Planck's constant). A high degree of crystal perfection is needed for the creation of effectively radiant recombination
as crystal defects act as center for non-radiating recombination.
Moreover, the junction should be shallow to minimize re-absorption of
the emitted light.

In GaAs LEDs, where the energy gap Eg ≃ 1.43 eV, the peak emission wavelength is ∿ 900 nm. When aluminum is added to gallium arsen-
ide (GaAlAs) the output wavelength can be controlled to peak within
the spectral range of 800 to 900 nm. By doping GaAs with silicon the
emission wavelength can be shifted to ∿ 950 nm. When indium is added
(InGaAs), the peak wavelength may be shifted to 1060 nm. Moreover,
by using mixed crystals of GaAs and GaP (Eg = 2.26 eV), that is
$GaAs_{1-x}P_x$ (x = 0-0.46), the energy gap can be increased and radiation
in the visible part of the spectrum can be produced. The colors
red-orange (γ_p ≃ 645 nm) and yellow (λ_p ≃ 590 nm) are realized with
nitrogen-doped $GaAs_{1-x}P_x$. The wavelength range can be further increased by suitably doping of GaP, for example GaP diodes doped with
sulphur and zinc emit in the green (λ_p ≃ 560 nm), while when doped
with zinc and oxygen they emit in the red (λ_p ≃ 700 nm).

Only semiconductors with an energy gap of > 2.6 eV can be considered for creating LEDs which produce blue light. Little research
has been done on these semiconductors, in comparison with GaAsP, and
great difficulty still exists in mastering these technologically.
However, it has already been possible to create SiC LEDs which emit
blue light. Nevertheless, economical mass production is still not
foreseeable.

Figure 1b shows the schematic of the diode body. By means of
liquid phase epitaxy (LPE), the active layer with a high degree of
crystal perfection can be grown onto a GaAs substrate. Due to the
amphoteric characteristic of the silicon impurity, the pn junction
forms automatically during the process of epitaxy. Part of the
radiation leaves the diode body on a direct path through the near

surface. However, radiation emitted in the direction of the substrate is also useful. For this purpose, the rear of the diode body
is mirrored and serves as a reflection surface.

     IR LEDs with a high radiation density are based on hetero-
structure consisting of GaAs and GaAlAs. This consists of several
semiconductors layers, located one above the other, with differing
compositions and differing energy gaps. In the case of dual hetero-
structure (DH structure) the thin ($\sim$ 0.2-1 µm) active region is
enclosed by two layers with a larger energy gap. In this way the
injected carriers are limited to a narrow band by potential barriers.
In comparison with the simple homostructure, the DH structure of an
IR LED provides a considerable increase in the quantum efficiency for
the radiant recombination. Moreover, limitation of the active recombination region by a semiconductor with a higher energy gap reduces absorption losses upon the exit of the radiation. In this way,
the recombination region can be placed near to the heat sink in order
to increase the thermal load capacity.

     Fig. 2a, b illustrates two typical mountings of IR LED which
allow the collection of the light emitted from both the edge and the
upper surface of the emitter. Figure 2c shows another typical mount-
ing of IR and visible LEDs (surface emitters): here, the transparent
plastic lens, in addition to the required protective and mechanical
supporting function, increases the coupling out of the beam and
ensures a certain degree of collimation.

Edge Emitters

     Some applications, such as optical communications, require
extremely small sources to ensure efficient coupling to single fiber
optical waveguides. Small sources sizes are achieved by sawing the
diode into relatively narrow widths and using only the emission from
a single face, in the recombination region of the junction. The
lateral size of the radiation area is usually defined by etching an
opening in an oxide insulating layer and forming an ohmic contact by
depositing a metal film in the open contact region. The edge emission structure (Figure 2d) has an oxide isolated metallization stripe
constricting the current flow through the recombination region to
the area of the junction directly below the stripe contact.

     When the edge emitting structure is combined with the DH struc-
ture epitaxial process, it becomes possible to restrict the emission
angle of the radiation pattern to a relatively narrow angle in the
plane perpendicular to the plane of the junction. This accomplish-
ment is due to the waveguide characteristics of the thin recombination layer.

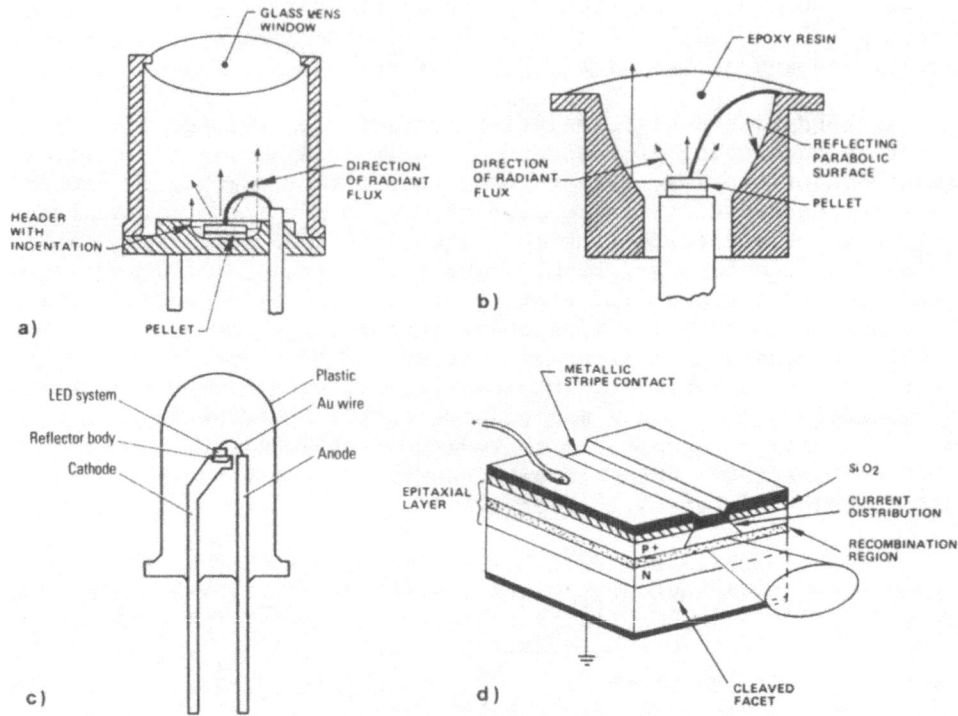

Fig. 2.   a,b) Schematic arrangements of typical IR-LEDs structures
          (combination edge and surface emitters); c) Typical struc-
          ture of an encapsulated LED; d) Schematic representation of
          a typical edge-emitting diode with directed emission.

Laser Diodes

Laser action from an edge emitting diode can be obtained when
enough carriers are injected into the combination region to provide
the optical gain needed to overcome the cavity's internal and external losses. Inverted population is built up and a photon produced
by electron-hole recombination induces further radiant recombi-
nations.

The optical resonator of the laser is usually formed by cleaving
the opposite ends of the diode to form partially reflecting optical
surfaces.

The adjacent sides are sawed to suppress lateral modes
(Figure 3).

Up to the threshold current, the laser diode behaves in the same
way as a conventional IR LED. As from the threshold value onwards

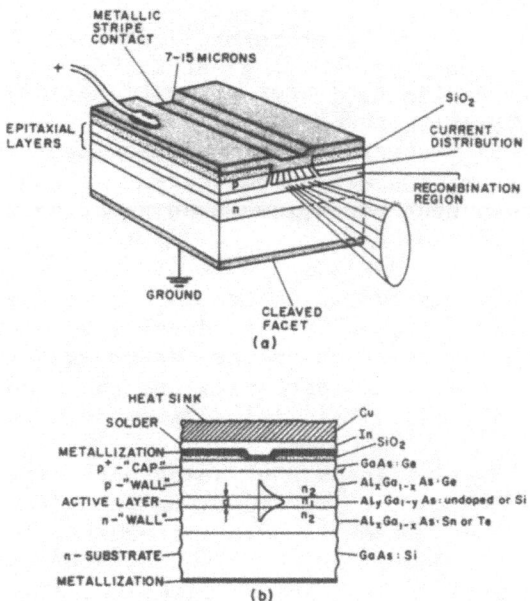

Fig. 3. a) Typical injection laser diode structure with dual
         heterostructure and strip geometry; b) cross-section of a
         DH-junction AlGaAs laser: the active layer of thickness d
         (0.05 to 0.2 μm) is placed 2 to 3 μm away from the contact
         stripe[1].

coherent radiation sets in, and the radiation current characteristic
rises steeply and linearly.

For continuous wave (cw) operation at room temperature, low
lasing-threshold currents are required, which implies a thin recombination region. In turn, the lasing transverse spot-size (0.5-1.0 μm)
provides a very divergent beam (when compared with other lasers) of
30-40° angular width in the plane perpendicular to the junction. In
the plane of the junction, light confinement is controlled by the
current flow, and thus the lasing spot is comparable in size to the
stripe-contact width. The laser emission is then elliptical, with
typical transverse and lateral bandwidths of 40° and 10°, respectively.

High-Power Diode Lasers

In recent years the need for IR diode lasers has been limited to
optical communication applications at relatively low power levels
(3-5 mW/facet). Now there is an ever increasing demand for high-

power (10-50 mW/facet) (single-mode) diode lasers to be used in
applications such as optical recording, long-distance high-data-rate
optical communications, high-speed printing, etc. During the last
two years a number of single emitter conventional diode lasers have
been described as emitting powers in excess of 50 mW cw from a single
facet. Very recently cw operation of multi-emitter phase-locked
arrays of diode lasers has been demonstrated. These devices have
been operated at cw output power levels as great as 2.5W from a
single facet[2].

The maximum power extractable form a cw diode laser is limited
by the mode spot-size. Typical powers densities which produce permanent damage of the output-faces of the laser are 6 MW/cm^2(cw). For
low-power devices (0.6x3.5 µm spot-size) the fundamental-mode power
that can be obtained reliably is limited to ∿ 5 mW/facet.

To overcome this limitation the lasing spot-size, both perpen-
dicularly and parallel to the junction, has been increased, and a
mode-dependent loss mechanism has been introduced to discriminate
against high order-mode oscillation. The most successful device to
date is the constricted double -heterojunction large-optical-cavity
(CDH-LOC) laser (Figure 4)[3]. The active layer is thinned from the
standard value of 0.15 µm to 0.05 µm, then causing the mode transverse spot-size to approximately double (from 0.6 µm to 1 µm). An
additional layer of index intermediate between that of the active
layer and of the n-AlGaAs layer allows the optical mode to propagate
mostly in the guide layer while obtaining optical gain from the
active layer. Mode transverse sizes close to 2 µm can thus be
achieved.

The discovery that LPE growth is strongly affected by top-
ographical features etched into the substrate has been used to produce thickness variations of the layers, thereby controlling the
laser-modal properties. As shown in Figure 4 an optical cavity is

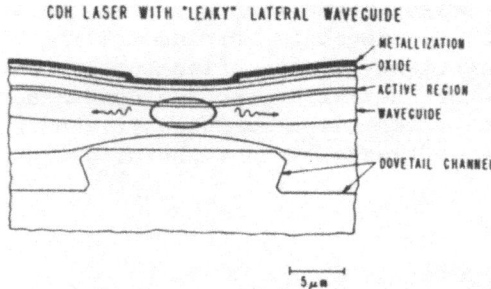

Fig. 4.   Schematic representation of high-power single-mode laser
          with constricted double-heterojunction (CDH) large-optical-
          cavity (LOC)[3].

formed above the mesaprofile of the substrate: a convex-lens-shaped
active layer atop a concave-lens-shaped guide layer. The combination
of the two provides a large spot-size (1.7 x 6.5 μm) and discrimi-
nation against high-order mode oscillation. Single-mode cw emission
in excess of 50 mW/facet has been reported for this laser[3].

LED Arrays

Diode Laser Arrays: The need of high recording data rates has
led to the development of multichannel optical recording systems
using monolithic arrays of CDE-LOC diode lasers[4] (Figure 5).

The arrays are mounted either to a common electrode or to a BeO
submount with separate electrodes for individual addressability. The
maximum output powers from a ten-unit array achieved in cw and pulsed
(100 ns wide pulses at 1 kHz repetition rate) operation are 320 mW
and 3.6 W, respectively.

A different approach for high-power laser arrays is shown in
Figure 6. It consists of 10 grain-guided 3.5 μm wide stripe injection lasers which are periodically spaced on 10 μm centers[5].
Optical coupling, resulting from evanescent wave overlap between the
adjacent closely spaced emitters, ensures a high degree of phase
locking and lateral transverse mode stability. 585 W/facet cw at
room temperature has been reported for this laser. The maximum
overall power conversion efficiency (electrical to optical) is 22%.
This value is much higher than that of He-Ne and argon lasers (0.5%),
Nd-YAG lasers (2%), and approaches that of CO_2 gas lasers (30%).

The increase of the number of emitters from 10 to 40 has led to
an output power as great as 2.5 W/facet[2].

Because the emitters of the laser array are mutually coherent,
the laser light can be focused into a single diffraction-limited
spot. From a high-power (200 mW/facet) phased array laser (10
emitters) operated at 770 nm, over 90 mW have been focused into a 2.5
μm diameter spot[6]. Efficient coupling with single-mode optical
fibers can be envisioned.

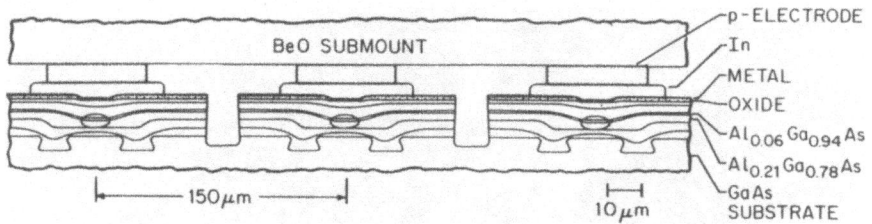

Fig. 5.   Schematic representation of monolithic CDH-LOC laser
          array[4].

    <u>Incoherent LED arrays</u>: Straightforward hybrid-circuit fabrication techniques should permit the assembly of single LED chips into tightly-packed arrays capable of producing the irradiance needed for PT, as already proposed by Epstein et al.[7]. Integrated optical systems could then provide the suitable radiation pattern or efficient coupling to optical fiber delivery systems. Figure 7 shows a picture of a 2 x 2 array of four IR LEDs already available commercially.

Optical-Fiber Delivery Systems

    The typical dimensions of the LED active area are very small: 300 x 400 μm and 13 x 62.5 μm for surface and edge emitters, re-

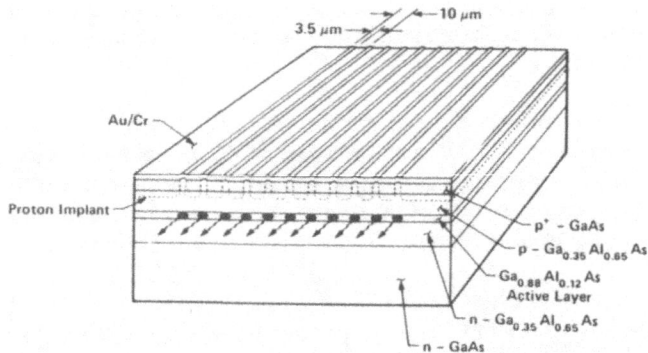

Fig. 6. Schematic of a phase-locked laser array[5].

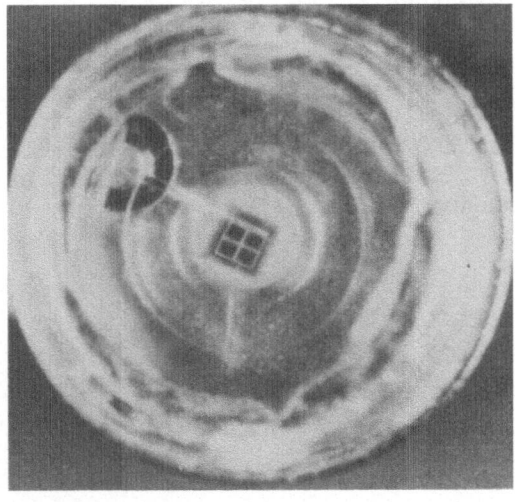

Fig. 7. Picture of a 2 x 2 array of IR incoherent LEDs.

spectively. The high irradiance of a multimilliwatt LED makes coupling to individual optical fiber an attractive possibility for biomedical applications of incoherent LEDs. Moreover, high coupling
efficiency and lower costs are now possible with new LED mountings
where in place of the plastic/glass collimating capsule a metal
windowed package containing an internal microlensing system is used
to project a uniform spot on the can's surface. This structure
provides efficient light coupling to optical fibers without the need
for expensive precision connectors required with the previous LED-
fiber coupling systems.

Coherent LEDs have much better directionality characteristics
and very efficient coupling to small-core optical-fibers is possible.
Mutually coherent laser arrays have been already efficiently coupled
with single-mode optical fibers.

MEDICAL APPLICATIONS OF INCOHERENT AND COHERENT LEDs

The availability of high-power incoherent and coherent LEDs in
the IR and visible (VIS) range will permit important changes and
progresses in the therapeutical applications of light.

The overall electrical-to-optical conversion efficiency of an
incoherent LED source is typically ∿ 10% in the IR and ∿ 5% in the
red. Coherent IR LED arrays have been operated up to 20% of overall
efficiency. These values are much higher than those of the best
laser sources, in the VIS and near IR, used today (argon-laser-pumped
dye laser: 0.05%; flash-pumped dye laser: 27%; Nd-YAG laser: ∿ 2%;
metal-vapor lasers: ∿ 1%).

The higher efficiency and the extremely small dimensions of
solid-state semiconductor sources can bring about great reductions in
the physical size, and in the cost of the systems for phototherapy.

Phototherapy with Incoherent Light

At present there are two important therapeutical applications of
incoherent light[9]: 1) photochemotherapy of psoriasis, the so-called
PUVA therapy, which utilizes the synergistic action of a drug (8-MOP)
and of UV-A radiation; 2) phototherapy of hyperbilirubinemia of the
neonate[10]. Fluorescent lamps are used as light sources in both
cases.

A new light therapy that is receiving an ever increasing
interest is the photochemotherapy of tumors[11]: here, again, the
combined effect of a sensitizer (hematoporphyrin = Hp) and of VIS
light is used to destroy selectively the tumor. Incoherent light has
been used at the beginning, later replaced by laser light.

Photomedical Use of Lasers

The widest use of laser light in medicine is for photosurgical and photocoagulative applications[12]. The CO_2 laser ($\lambda = 10.6$ μm) is the best optical knife for unpigmented tissues due to the extremely high absorption by water at that wavelength. VIS and near IR lasers (such as argon, dye, and Nd-YAG lasers) are used as photocoagulators. Tissue pigments (melanin, hemoglobin) absorb strongly the VIS light of argon and dye lasers, and, at a much smaller extent, the IR light ($\lambda = 1.06$ μm) of Nd-YAG laser (Figure 8).

VIS and IR lasers, operated at low power densities, have been advertised to cure everything from paralysis to psoriasis ("biostimulation" therapy). Scientifically controlled studies are still lacking. Research now centers on two major application areas: the relief of pain through acupuncture-type treatment, and the speeding

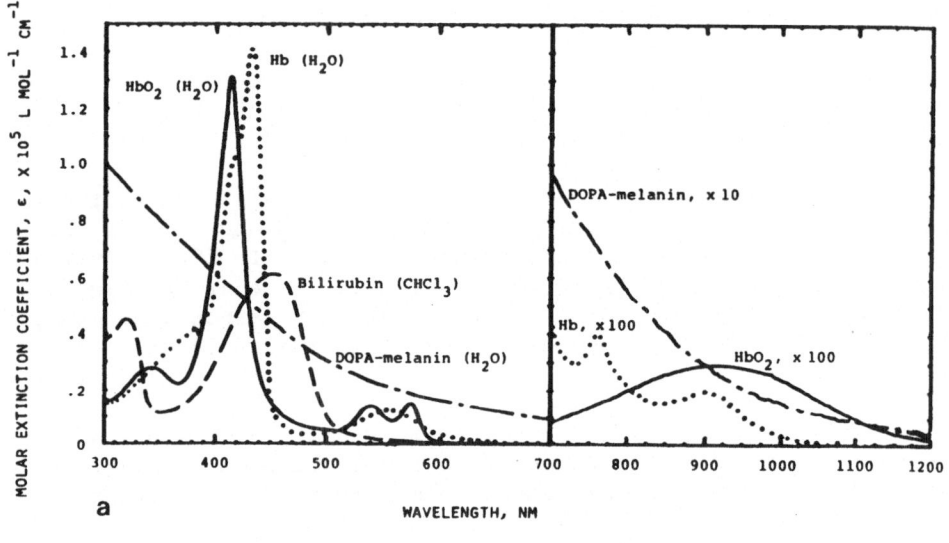

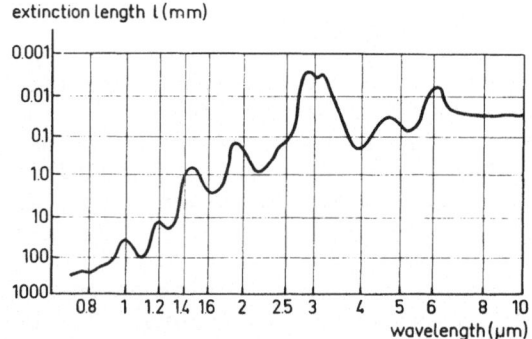

Fig. 8.   a) Molar extinction coefficients of main skin pigments;
          b) extinction length of water.

of wound or injury healing. Double-blind investigation of He-Ne
laser therapy of venous leg ulcers has proved no advantage with
respect to the controls[13].

"Coherence or Not Coherence"

The medical applications of lasers do not use in a direct way
any of the coherence properties of laser light. The spatial coherence is important to obtain highly collimated beams, which in turn
allow the concentration of almost all the power emitted by the source
into very small focal spots, and the efficient coupling to optical
fibers. Therapeutical effects connected to the high degree of
spatial and temporal correlation of laser light have not been reported to date. In principle cooperative phenomena at a biological
level (such as long-range correlation of enzyme activity induced by
laser light) may be expected[14-18], and it would be interesting to
devise experiments to evidentiate these correlation effects "in
vitro" and "in vivo". However, the possibility of taking advantage
of these potential effects depending in a direct way on the temporal
and spatial coherence of laser light is reduced by the severe
scattering properties of living tissues, which degrade quickly the
coherence of laser light as it propagates through them.

Therefore, when illumination of large external areas of the body
is needed, incoherent sources may be conveniently used in place of
lasers, if enough useful power is at disposal.

LED Phototherapy

In this section we will briefly discuss a few possible applications of LEDs sources to phototherapy.

Incoherent LEDS: At present the most powerful LEDs emit in the
near IR (800-1000 nm). IR LED sources with 100 mW capability are now
available at very low price. IR edge emitters with output cw power
as high as 1.2 W are already commercially available. Visible LEDs
are much less intense. High-efficiency red LEDs (λ_p = 660 nm, $\Delta\lambda$ =
60 nm) have maximum cw power of 5 mW, but a 4-fold increase has been
announced for these LEDs. LEDS emitting at $\lambda_p \simeq 635$ nm (i.e. at the
output wavelength of a He-Ne laser) are $\sim 1/3$ less intense. Green
LEDs have still lower output powers, and blue LEDs are available as
prototypes with extremely low emission.

Phototherapy of jaundice: The light sources currently used for
PT are fluorescent lamps emitting narrow-band blue (or green) light
or broad-band white light. These lamps are efficient and cheap.
However, the intense Hg-lines are always present superposed on the
continuous fluorescent emission of the phosphor. In particular, the
toxic and mutagenic UVA-line (365 nm) and violet-lines (405-408

nm)[19] are not completely cut-off by the plastic shield of the
illuminator, and by the incubator. Broad-band tungsten-halogen lamps
are also employed in PT. They require filters to select the useful
spectral band, so to avoid loading of the infant with heating and
toxic radiation.

The future availability of high-power and low-cost blue LEDs
would permit the construction of multi-LED or LED-array panels capable of producing the irradiance needed for PT (1-3 mW/cm^2 at the
body surface), without the presence of hazardous spectral
components[20].

Figure 9 shows the output spectrum of a SiC blue LED; the emission spectrum of the special-blue fluorescent lamp recommended for PT
is also shown for comparison. The peak emission is at $\lambda_p \approx 480$ nm;
the short-wave intensity drops sharply, and very little light is
emitted below $\lambda = 430$ nm.

According to our analysis[10,21] on the filtering action of the
skin reported in this Volume, the spectral profile of this LED
appears to be very suitable for PT. In fact, the blue-light emission
occurs mainly in the long-wave half-width of the bilirubin absorption
spectrum, and produces bilirubin absorption rates that approach the

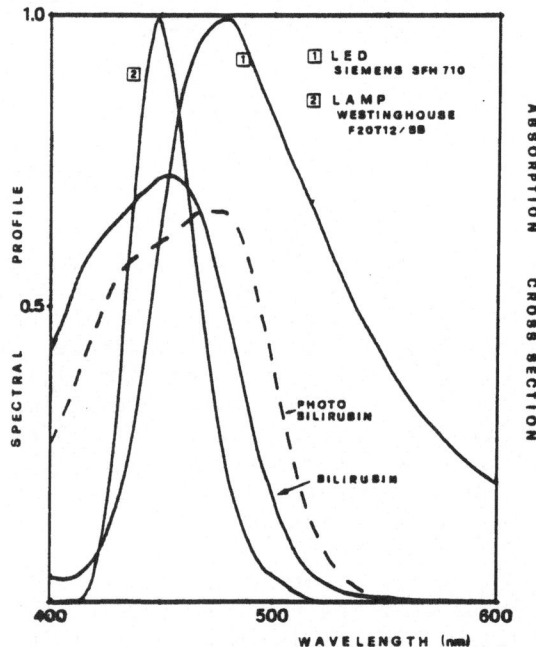

Fig. 9. Spectral profile of the light from a blue-emitting LED
         (curve 1). The spectrum of special-blue fluorescent lamp
         is shown for comparison (curve 2) (for details see Ref. 10).

corresponding rates of a special-blue lamp for increasing skin thickness (d) and blood content (γ)[10]. The following Table shows the bilirubin absorption rates of the Siemens LED (SFH-710) referred to the corresponding absorption rates produced by the Westinghouse special-blue lamp (F20T12/BB):

Source	"In vitro"	d_D = 200 μm			d_D = 400 μm		
		γ = 0	10^{-5}	10^{-4}	γ = 0	10^{-5}	10^{-4}
Blue-LED (Siemens SFH-710)	55%	61%	65%	97%	65%	69%	103%

(for the definition of the symbols see the paper by R. Pratesi in this Volume).

Finally, owing to the absence of short-wave components with $\lambda <$ 430 nm, and to the red-shift of the peak emission wavelength ($\lambda \simeq$ 480 nm) the light emitted by this blue-LED will be safer than that of fluorescent blue and white lamps[22].

Photoradiation therapy of tumors: Photoradiation therapy of tumors consists of the irradiation with VIS light of tumors previously sensitized with hematoporphyrin (Hp). Red light peaked at the longest wavelength secondary absorption maximum of Hp ($\lambda \simeq$ 630 nm) is usually preferred to ensure deepest penetration into the tumor mass. The main PT systems used to date are: a) filtered 5 KW Xenon arc lamps; b1) argon-laser-or b2) flash-lamp-pumped dye lasers; c) filtered 250 W quartz halogen lamps; d) gold-vapor lasers. Overall efficiencies are about 0.05% for a) and b1); 0.2% for c); 1-2% for b2) and d). Typical irradiances used for therapy are 25-100 mW/cm^2 for radiation in the 610-640 nm range.

High efficiency (5%) red LEDs emitting 5 mW (maximum) - power are available at low cost, and 20 mW LEDs are becoming available at moderate cost. Multi-LED or integrated LED-array panels could provide the suitable irradiation pattern and irradiance for PT of superficial tumors. Coupling of single LED chip to individual optical fiber represents a realistic possibility for multiple fiber implantation into tumor masses, and for irradiation of internal tumors with large bore endoscopes (as in gynecology). By reducing the LED temperature to liquid nitrogen temperature, the emission maximum at 660 nm can be brought into coincidence with the 630 nm peak of HpD. Moreover, the total light intensity increases at least by one order of magnitude.

The possibility of utilizing a multi-LED system as light source for photoradiation therapy has been successfully tested by following the Hp-sensitized photokilling of HeLa cells[11].

Power densities in excess of 50 mW/cm^2 have been obtained with 106 LEDs inserted into closely-packed, radially oriented holes in a metallic hemisphere with 20 mm inner radius.

Biostimulation phototherapy: At present, clinical successes of "biostimulation" therapies are claimed for laser operating over a wide spectral range (0.6-10.6 μm). When controlled studies will have proved the clinical efficacy of low-power irradiation in some selected pathology and determined its action spectrum, incoherent LEDs, if available with sufficient power outputs in the useful spectral range (as for red and near IR diodes), could represent a good alternative to laser sources.

Coherent LEDs

Near IR LEDs: In the wavelength emission range of the most powerful IR diode-lasers available (800-950 nm) the absorption by body tissues is relatively low (we are in the long-wave tail of hemoglobin and melanin absorption spectra, and in short-wavelength tail of water absorption (see Figure 8)). However, the absorption by melanin for a 820 nm diode laser is ∿ 5 times greater than that of a Nd-YAG laser (λ = 1060 nm). Moreover, oxyhemoglobin has a residual absorption that peaks at 900-950 nm. Thus a future diode laser with cw or average power output in the 2-10 W range may be used for microsurgical and microphotocoagulative applications on highly pigmented tissues. Efficient coupling to optical fiber delivery systems will permit precise microendoscopical treatments in microsurgery.

VIS LEDs: The development of high-power diode lasers in the visible is mainly a technological and economic problem, and it may be triggered by specific and important applications.

Phased array LED lasers have already been operated at 200 mW/facet cw output power at 770nm. A further shift to shorter wavelengths and power increase to 1 W level will open the entire field of applications covered at present by dye and argon lasers. Efficient coupling to small-core fibers has already been demonstrated.

Extention to longer wavelengths: Recent progress in GaAsP technology has been announced, resulting in longer-wavelength operation (1.3-1.6μm) and in high lasing performance, such as cw single-mode operation over 40 mW at 1.3μm[23]. Moreover, the development of led-salts materials is leading to diode lasers with longwave emission with λ > 2 μm[24].

As is well known, a laser with good cutting and photocoagulative properties is still lacking. Carbon dioxide lasers are excellent optical knives, but not suitable to coagulate blood vessels exceeding 0.5 mm diameter. In turn, Nd-YAG lasers have low cutting capability, and are mainly used as photocoagulators. This depends on the large difference between the water absorption coefficients at the corresponding wavelengths. According to Figure 8b, CO_2 laser radiation at 10.6 μm is absorbed completely within a tissue thickness of 0.1 mm, while Nd-YAG light penetrates several centimeters into body tissues.

As can be seen from 8b, a wavelength increase from 1.06 μm to 1.3 μm produces one order of magnitude reduction of the extinction length. Safer photocoagulative procedures would be possible at this wavelength, which allows better control of coagulative depths.

A further increase of wavelength to the range 1.4-2.5 μm reduces the extinction length to values expected for an "ideal" knife, that is a laser with good cutting and hemostatic characteristics at the same time. Diode lasers with sufficient output powers in this range will find interesting applications in microsurgery. Moreover, since the output wavelength of a diode laser can be turned almost continuously by varying the doping and the constituents of the diode, the selection of the most suitable wavelength for a given application (i.e. more cutting than photocoagulative power or viceversa) becomes feasible. In this spectral range quartz fibers are still sufficiently transparent, and endoscopic microsurgery and/or microphotocoagulation will become possible.

In addition to reduction of size and cost the introduction of diode-lasers will represent a noticeable simplification of the handling of laser medical equipments: built -in optical cavity and optical-fiber coupler will make the laser source assembly very compact and stable, thus releasing the need of dedicated laser technicians for constant aligning, peaking, cleaning, and, in case of dye lasers, messy dye changes, as usual with traditional medical lasers.

REFERENCES

1. D. Botez and J. C. Connolly, RCA Review, 44:64 (1983).
2. D. R. Scifres, R. D. Burham, C. Lindstrom, W. Streifer and T. L. Paoli, High-Power Diode Lasers, Paper TUC5, C.L.E.O., May 17-20, Baltimore, USA (1983).
3. D. Botez, Appl.Phys.Lett., 36:190 (1980).
4. D. Botez, J. C. Connolly, D. B. Gilbert, M. G. Harvey and M. Ettenberg, Appl.Phys.Lett., 41:1040 (1982).
5. D. R. Scifres, R. D. Burnham, and W. Streifer, App.Phys.Lett., 41:1030 (1982).
6. D. R. Scifres, R. A. Sprague, W. Streifer, and R. D. Burnham, Appl.Phys.Lett., 41:1121 (1982).
7. M. E. Mahric, M. Epstein, and R. V. Lobraico, A Proposal for Light Emitting Diode Arrays for Photoradiation Therapy, to be published.
8. T. Ormond, Fiber-Optics Components, EDN, 28:112 (1983).

9. J. D. Regan and J. A Parrish, (eds.), "The Science of
    Photomedicine," Plenum Press Ltd., New York (1982).
10. R. Pratesi, Phototherapy of Hyperbilirubinemia: Physical
    Aspects, in: this Volume.
11. R. Cubeddu and A. Andreoni, (eds.), "Porphyrins in Tumor
    Phototherapy," Plenum Press Ltd., New York, in press.
12. F. Hillenkamp, R. Pratesi and C. A. Sacchi, (eds.), "Lasers in
    Biology and Medicine," Plenum Press Ltd., New York (1980).
13. P. Santoianni, G. Monfrecola, D. Martellotta and F. Ayala,
    Effects of He-Ne Laser on Venous Leg Ulcers, Photodermatology,
    Clinical and Experimental., (in press).
14. N. Kollias and W. R. Melander, Phys.Lett., 57A:102 (1976).
15. J. P. Biscar, Bull.Mathem.Biol., 38:29 (1976).
16. S. J. Webb, M. E. Stoneham, and H. Fröhlich, Phys.Lett., 63A:407
    (1977).
17. E. J. Brändas and L. J. Dunne, Chem.Phys.Lett., 64:329 (1979).
18. S. A. Moskalenko, M. F. Miglei, P. I. Khadshi, E. P. Pokatilov,
    and E. S. Kiselyova, Phys.Lett., 76A:197 (1980).
19. R. Parshad, K. K. Sanford, W. G. Taylor, R. E. Tarone, G. N.
    Jones and A. E. Baek, Photochem.Photobiol., 29:971 (1979).
20. R. Pratesi and M. Scalvini, Biol.Med.Environ., 11:467 (1983).
21. R. Pratesi, L. Ronchi, and G. Cecchi, Skin Optics and
    Phototherapy of Jaundice, Photochem.Photobiol., in press.
22. B. S. Rosenstein and J. M. Ducore, Photochem.Photobiol., 38:51
    (1983).
23. T. Ikegami, Progress in Long-Wavelength Laser Diodes, Paper
    TuL1, Conference on Optical Fiber Communication, New Orleans
    Jan. 23-25, (1984).
24. W. Lo and D. L. Partin, Long-Wavelength ($\lambda > 2$ μm) Diode-Laser
    Sources, Paper TuL5, Conference on Optical Fiber Communi-
    cation, New Orleans, Jan. 23-25 (1984).

CO_2 LASER RADIATION DELIVERY SYSTEMS

A. M. Scheggi

Istituto di Ricerca sulle Onde Elettromagnetiche of CNR
Florence, Italy

INTRODUCTION

The realization of flexible waveguides for CO_2 laser light is an
important problem in industrial and medical applications. In general
the guidance is carried out mainly by means of an articulated arm
consisting of a number of metallic pipes with a flat mirror at each
rotating joint, whose optical axis must be precisely aligned. The
main drawbacks for surgical use are:

- misalignement of the mirrors which give rise to instability
  and degradation of the beam
- large dimensions and poor handling capability

Some improvements have been proposed: the flat mirrors have been
replaced with concave mirrors of suitable curvature for less string-
ent mechanical and beam alignment tolerances[1]. Another modification of the conventional articulated arm has been reported by Patel
[2] who developed a CO_2 laser delivery system for ophtalmic surgery.
The articulated arm incorporates straight dielectric hollow sections
where the propagation of a non degraded beam is accomplished by
bouncing at grazing angle against the walls. The delivery probe
incorporates a quartz tapered section which is used to match the
articulated arm to the probe tip (0.5 mm i.d.). Such a tip termin-
ated with a polished diamond window can be introduced directly into
the eye. However such arms still remain not suitable for cavitational and endoscopic surgery. For these operations flexible
waveguides are necessary playing the same role as silica fibers for
Ar of NdYAG radiation transmission. Different types of waveguides
have been proposed which can be divided into two main groups: longer
wavelength I.R. fibers and hollow waveguides.

I.R. FIBERS

As is well known the conventional silica fibers can be used in
the visible and near I.R. region up to 2 μm; beyond such wavelength
the intrinsic optical attenuation of the material due to the I.R.
absorption edge prevents the transmission of the radiation. On the
basis of preceeding experience in I.R. techniques, crystalline alkali
or metallic halides, fluoride and chalcogenide glasses have been
considered as potential materials for such fibers[3]. The technologies necessary both for fabricating the bulk material and the
fibers differ considerably from those employed for conventional
fibers. The polycristalline fibers are fabricated by extrusion
through a die, single crystalline fibers are obtained by growth from
a melt zone, while fluoride and calcogenide fibers are obtained by
the usual drawing fashion. One of the main difficulty especially
with crystalline fibers is that of finding a suitable cladding while
polycristalline fibers may present an excess of loss due to scattering from the grain boundaries.

TlBr and TlBrI (KRS-5) have been the most investigated materials
for medium I. R. fibers since the first announcement by Pinnow et
al.,[4]. Unclad and loose tube cladded fibers have been made by
extrusion with losses of $\sim$ 0.5 dB/m at 10.6 μm. CO_2 power transmission of some tens of W has been reported. The disadvantages of
these fibers are their high toxicity, water solubility and brittle-
ness. A CO_2 flexible delivery system is already commercially available for maximum power transmission of 20 W (KRISTEN - Horiba Infrared Optical Fiber).

Several other kinds of polycristalline fibers were fabricated
such as KCl, AgCl, AgBr and AgCl–AgBR alloys for grain growth reduction[5] and losses of $\sim$ 5 dB/m have been reported. Ag halides
fibers are not toxic, not hygroscopic and quite plastic. AgCl cladding for AgBr core fibers has also been proposed due to the small
refractive index difference. Efforts have also been made for fabricating single crystal AgBr, CsBr, CsI fibers either by growth from
melt or by pulling down[6,7,8]. Their transmission loss at 10.6 μm
is 9 dB/m for AgBr and 5 and 13 dB/m for CsBr and CsI respectively.
A loose tube protection was proposed for mechanical and humidity
protection of Cs halides fibers.

Glass materials, which are in general the most suitable for
fabricating fibers, have also been considered and in particular
efforts have been successfully made on fluoride and chalcogenide
glasses for I.R. fibers. However as fluoride glasses are not transparent above 8 μm and hence cannot transmit CO_2 laser radiation, Se
and Te glasses seem to be the only ones convenient, but they are
toxic and rather brittle. Among varieties of selenide glasses
$Ge_{30}As_{15}Se_{55}$ glass was used for preparing[9,10] fibers which were
drawn in the classical fashion from rods synthesized in sealed silica

ampoules. The bare fibers were coated with a loose polyolefin plastic material plus mechanical reinforcement by a heat-shrinkable polyethylen tubing. Measured losses were reported of ∿ 20 dB/m at λ between 8 and 10.6 μm.

HOLLOW WAVEGUIDES

Hollow core waveguides seem to be suitable for practical use and several kinds of such waveguides have been proposed and fabricated ranging from the so called flexible infrared transmissive waveguides [11,12], helical circular waveguides[13], circular metallic waveguides[14], dielectric coated metallic waveguides and hollow core-glass clad waveguides[15,16,17]. At I.R.O.E. experimental investigations have been carried out on metal (brass or alluminium) rectangular and closed waveguides (Figure 1) which appear today to represent the most interesting solution for some application[18,19,20]. It turned out that the roughness and reflectivity of the two (wider) guiding plates are the most important parameters for a low attenuation. The best result has been obtained by using a brass waveguide 1x20 mm cross section whose walls were accurately polished: the transmission was measured to be of 76% per meter with 80 W input power corresponding to 1.2 dB loss. These losses increase to 2.1 and 2.5 dB in the presence of a 90° curvature and 90° twisting respectively. The coupling problem was also considered in order to use the metal waveguide in connection with a commercially available domestic CO$_2$ laser for medical applications (Valfivre LSS-100). The layout of the assembled laser and guide system is shown in Figure 2. It includes a specially designed coupler, with a mirror mounted on a rotating joint, a waveguide tapered input, and an handpiece with focusing output lens. The mechanical joint with the mirror allows the guide to rotate in the vertical plane, thus reducing the necessity of twisting the guide itself and avoiding the consequent losses. Such a rotation, of course, makes the laser polarization to be no more aligned parallel to the wider walls at the waveguide input; however, no effective additional loss has been measured for a rotation angle of the joint up to 30°. The focused output can provide a very narrow line of light which may be useful for distributing the laser energy over a large area or for cleaner and faster cuts over long sections. This system has been already used in experimental maxillo-facial surgery[21].

Another type of hollow waveguide under investigation at I.R.O.E. [22,23] is the hollow core-oxide glass clad fiber[19]. Many glasses exhibit an anomalous dispersion of the refractive index in the wavelength region around λ = 10 μm. More precisely these glasses show a very strong absorption near λ = 10 μm due to molecular vibrations, consequently the refractive index of the medium can be written as a complex number n = n_r + ik where k is related to the absorption coefficient of the material; when k is large enough n_r results < 1.

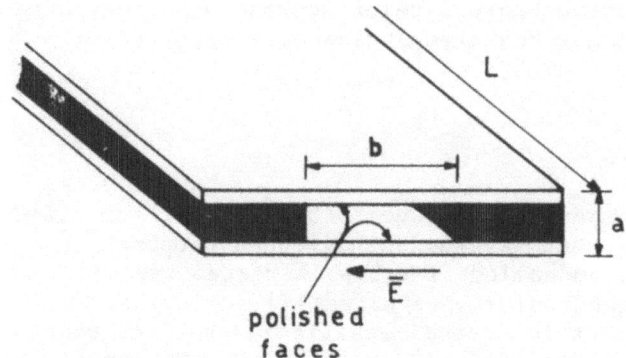

Fig. 1. Rectangular metallic waveguide for I.R. radiation. Typical
values are a = 0.5 ÷ 1 mm, b = 10 ÷ 20 mm while L ≥ 1 m.

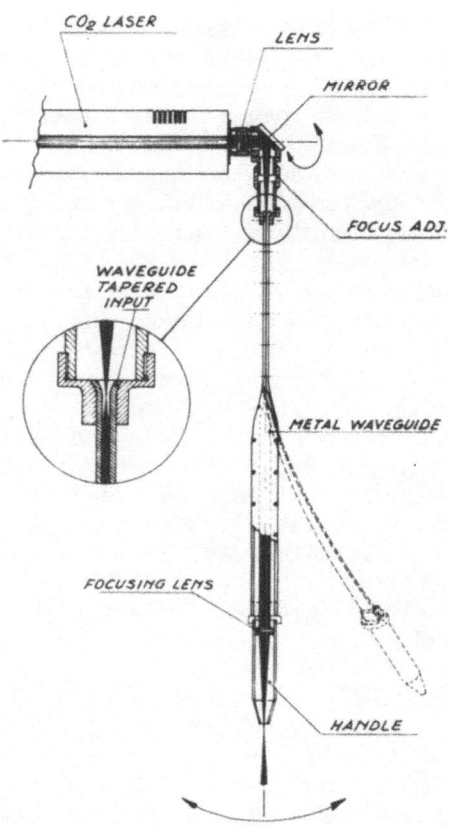

Fig. 2. Sketch of the Laser-metal guide system for medical
applications.

Accordingly one can think of utilizing this effect to realize a fiber working at this wavelength. The working principle is the same as for ordinary fibers: while for a step index fiber the propagation occurs by total reflection at the interface between two media of different indexes (for instance glass-air, $n_{air} < n_{glass}$) in this case we have the complementary situation: the light is totally reflected at the interface air-glass. In the first case the glass constitutes the fiber core and a second material can be chosen as cladding (with $n_{clad} < n_{core}$), in the second case the core is constituted by air ($n = 1$) while the cladding is formed by the glass with $n_r < 1$ (Figure 3). Consequently the I.R. radiation transmission occurs in air (no loss) and the multiple reflections at the interface give rise to losses due to the strong absorption of the glass. Accordingly it is necessary to choose a glass presenting $n_r <$ 1 but not too high value of k at the operating wavelength. In particular it has been shown that the presence in SiO$_2$ glass of heavy ions, such as Pb, shifts such operative wavelength towards $\lambda \approx$ 10 μm.

We have experimented on hollow fibers drawn from oxide glass capillaries with different Pb percentages which present a minimum value of n around λ = 9.3 μm and used three emission lines (λ = 10.6, 9.5, 9.24 μm) of a c.w. CO$_2$ IEQ-CNR laser in order to achieve a "tuning" of the laser emission on the minimum transmission losses of the fiber. The maximum power obtainable from the laser at each wavelength was 15 W due to the introduction of a grating placed at one end of the cavity for tuning the laser on the three emission lines. Figure 4 shows the attenuation per meter as a function of the wavelength measured on two 2 m long fibers having glass cladding with different Pb percentages. Note that these values include coupling losses which were typically of about 0.6 dB. These fibers have a 0.6 mm internal diameter, 1 mm external diameter and a Kynar protection coating (applied on line during the drawing process) of $\sim$ 20 μm. They result quite flexible with a minimum curvature radius of 30 cm before breaking. From the curves of Figure 4 it turns out that the most favorable situation occurs for a glass with 27% Pb percentage at λ = 9.24 μm. These results suggest that an efficient CO$_2$ laser-fiber system could be obtained by maximizing the laser power output on the

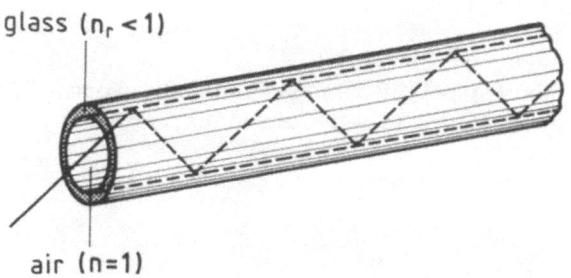

glass ($n_r < 1$)

air (n=1)

Fig. 3. Hollow core-glass fiber.

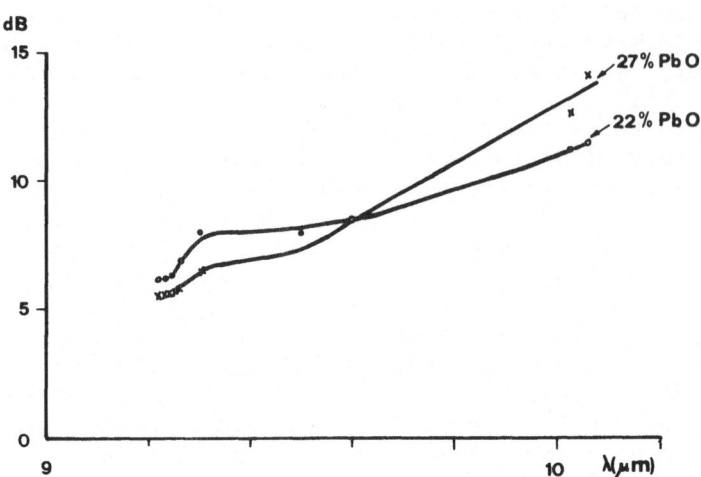

Fig. 4.   Measured attenuation per meter versus wavelength.

lowest emission line. Conversely one can look for materials (for instance GeO glasses[24]) having the best operative wavelength at λ = 10.6 µm. Further such hollow core waveguides present the typical advantages of flexibility and reduced dimensions as the ordinary optical fibers so that they appear very promising especially for endoscopic and cavitational CO_2 laser surgery (at least for moderate power) for which the guidance along a fiber is essential.

REFERENCES

1. K. Yoshida, K. Ono, J. Hiramoto, K. Sunago, S. Takenaka, K. Atsumi, M. Nakajima, A. Ihara, The 4th Congress of International Society for Laser Surgery, Tokyo, Nov. 1981- Proc. of the Conference, Atsemi, N. Nimsakul, (ed.) p. 19-8 (1981).

2. C. K. N. Patel, T. J. Bridges, A. R. Strnad, O. R. Wood, D. B. Karlin, CLEO 1982 Conference Proc. Th 13, p. 78 (1982).

3. T. Miyashita, T. Manabe: IEEE Journal of Quantum Electronics, QE-18, 1432 (1982).

4. D. A. Pinnow, A. L. Gentile, A. G. Standlee, A. Timper: Applied Physics Letters, 33:28 (1978).

5. D. Chen, J. H. Garfunkel, R. A. Skogman, H. Vore, Fifth European Conference on Optical Fiber Communications (Amsterdam, Sept. 1979) Conf.Proc.p. 19-7-1 (1979).

6. T. J. Bridges, J. S. Hasiak, A. Strand: Optics Letters, 5:85 (1980).

7. Y. Mimura, Y. Okamura, Y. Komazawa, C. Ota: Japan J. Applied Physics, 19, L 269 (1980).

8. Y. Mimura, Y. Okamura, Y. Komazawa, C. Ota: Japan J. Applied Physics 20:L17 (1981).

9. J. Y. Boniort, C. Brehm, P. H. Dupont, D. Guignot, C. Le
    Sergent: 6th European Conference on Optical Communications
    (York, Sept. 1980) Conference Publication No. 190, p.61
    (1980).

10. C. Brehm, M. Cornbois, C. Le Sergent, G. P. Parant: J. Non
    Crystalline Solids Nos. 1,2,47: (Part II) p. 251 (1982).

11. E. Garmire, T. McMahon, M. Bass: Applied Optics 15: 145 (1976).

12. E. Garmire, T. McMahon, M. Bass: IEEE J.Quantum Electronics
    QE-16, 23 (1980).

13. M. E. Marhic, L. I. Kwan, M. Epstein: Applied Physics Letters
    33:874 (1978).

14. M. E. Marhic, E. Garmire: Applied Physics Letters 38, 743
    (1981).

15. M. E. MARHIC: Applied Optics 20, 3436 (1981).

16. M. Miyagi, A. Honyo, Y. Aizawa, S. Kawakami: Applied Physics
    Letters 43, 430 (1982).

17. T. Hidaka, T. Morikawa, J. Shimada, J.Applied Physics, 52 4467
    (1981).

18. V. Russo, G. C. Righini, S. Sottini, I. Reali: Proc. III
    National Congress "Elettronica Quantistica e Plasmi" (Como,
    May 1982), p. 520 (1982).

19. V. Russo, G. C. Righini, S. Sottini, I. Reali, G. Papi: Proc.
    "IV Riunione Nazionale di Elettromagnetismo Applicato"
    (Firenze, October 1982), p. 177 (1982).

20. V. Russo, G. C. Righini, S. Sottini, G. Papi: 84th Assembly of
    Italian Electrical and Electrotechnical Association (AEI)
    (Cagliari, October 1983), Paper B. 46 (1983).

21. E. Panzoni, C. Clauser, S. Sottini: Proc. of the 8th Congress of
    the International Association for Maxillo-Facial Surgery
    (Sorrento, 25-28 May 1982), Paper 205 (1982).

22. A. M. Scheggi, M. Brenci, R. Falciai, B. Locardi, F. Nicoletti:
    III National Congress "Elettronica Quantistica e Plasmi"
    (Como, May 1982), p. 525 (1982).

23. A. M. Scheggi, M. Brenci, R. Falciai, B. Locardi, F. Nicoletti,
    F. Barbon: 84th Assembly of Italian Electrical and Electro-
    technical Association (AEI) (Cagliari, October 1983), Paper
    B. 48 (1983).

24. T. Hidaka, K. Kumada, J. Shimada, T. Morikawa, J.Applied
    Physics, 53:5484 (1982).

OPTICAL FIBER TEMPERATURE SENSORS FOR MEDICAL USE

A. M. Scheggi, M. Brenci and A. G. Mignani

Istituto di Ricerca sulle Onde Elettromganetiche of CNR
Florence, Italy

INTRODUCTION

The subject of the present paper, although not strictly related with laser application to biomedicine, presents an increasing interest in several areas of medical procedures and in particular for diagnostic or therapy instrumentation. In fact the optical fibers, besides finding application in endoscopy and laser therapy and surgery, are now employed to sensing a number of physiological parameters[1,2,3].

As is well known optical fiber sensors utilize externally induced changes in the transmission characteristics of the optical fiber and in particular the sensors proposed for medical application field, make in general use of the "incoherent modulation" technique i.e. of induced amplitude modulation of the transmitted light.

Several types have been proposed for a variety of applications such as temperature and pressure measurements, blood pH, O_2 and velocity monitoring[4,5,6,7,8,9]. The optical fibers owing to their small dimensions and flexibility can be introduced in catheters or hypodermic needles and used for highly localized measurement without producing much disturbance. In addition they are safe for the patient and being constructed of chemically inert dielectric materials, they present electrical insulation and electromagnetic interference immunity. In particular at present, an increasing interest is dedicated to thermometers mainly owing to the emphasis which is being put in the use of microwave or RF hyperthermia for cancer treatment, but also in laser surgical application. In fact it is interesting to detect the thermal effect on biological tissues outside the region of direct laser beam impact[10], or to monitor

317

temperature near incisions performed by electric or laser scalpel for
comparing tissue thermic damages. As a result, a group of new
optical temperature measurement technologies are now being developed.
Although these new techniques differ from each other, all can be
grouped into two classes. In the first one the fibers are used as
light duct to or from a transducer. The main optical effects con-
stituting the basis for the measuring system may involve changes in
reflection, absorption, polarization, emission and color of the
sensing material with temperature. In the second category the fiber
itself (or better a portion of the fiber) suitably modified constitutes the sensor. In both cases the optical data are then sent
via fiber to an optoelectronic instrument package that generates
electrical signals for recording and display purposes. Here below
some typical examples of optical fiber temperature sensors for medical purposes will be reported along with their working principle and
main characteristics.

OPTICAL FIBER THERMOMETERS

Birifrangent Crystal Sensor

A first example involves the use of a birifrangent crystal whose
index of refraction differs for orthogonally polarized waves as a
function of the temperature. Cetas[11,12] has developed a small
sensitive temperature sensor consisting of a 0.1 mm thick crystal of
Lithium Tantalate with a polarizing film cemented to one side and a
dielectric mirror at the other side. The light from a LED source
travels along an optical fiber and through the polarizer and crystal
is then reflected back again through the crystal and polarizer to
another optical fiber that conducts the transmitted radiation to the
detector. The birifrangence of the crystal introduces a temperature
dependent phase difference between the two orthogonally polarized
components of the transmitted light which is observed as a change in
intensity of the light passing through the polarizer the second time.
Sensitivities of better than 0.1°C are achievable in the 12-49°C phy-
siologically useful range, with probes of external diameter of 1 mm.

Semiconductor Temperature Sensor

A second effect which was utilized for temperature sensing is
the temperature dependent absorption of a beam of light through an
optical semiconductor material. When the light propagates through
the semiconductor it will be absorbed if its photon energy $E = h\nu$ is
larger than the gap energy (between valence and conduction band).
For most semiconductors such a gap energy decreases with temperature,
hence the amount of absorbed light at a fixed wavelength (and corres-
pondingly transmitted) will vary with temperature. Christensen[13].
developed a device aimed chiefly to temperature monitoring in RF

hyperthermia treatment, where the use of conventional thermocouples
or thermistors, which require metallic components and wires, may
perturb the incident electromagnetic fields and also cause localized
heating spots. In this device the sensor is a small polished GaAs
block having dimensions about 250 x 250 x 125 μm^3. Light (typically
in the near infrared) from a LED is transmitted down two 85 μm
cladding diameter input glass fibers attached to the GaAs crystal,
passes through the sensor and is returned by two receiving fibers
attached to the crystal to a photodiode. With this device an
accuracy of about ±0.1°C has been reported over the range 33-47°C.

Phosphor Temperature Sensor

Another type of sensor uses as sensing material a phosphor or
mixture of phosphors which is excited by ultraviolet light sent
through the optical fiber[5,6]. The same fiber collects the fluorescent emission from the phosphor. Such a fluorescence varies with
temperature differently at different wavelengths; accordingly the
ratio between the intensities of two emission lines results a function of the temperature. This makes the sensor independent of source
fluctuations and fiber bending attenuation. The developed device
uses a small amount of rare earth phosphor on the top of a silica-
plastic fiber of ∿0.7 mm of diameter (400 μm core diameter) (Figure
1): it is capable of covering a temperature range from - 50 to 200°C
with an accuracy of 0.1°C, sensitivity of = .02°C. This sensor is
the only one available on the market (Luxtron Corp., USA).

Thermochromic Transducer Sensor

This sensor is being developed at IROE - CNR, Florence[14]. It
utilizes as transducer a thermochromic Cobalt Chloride salt solution,
whose spectral behavior shows an absorption band strongly dependent
on temperature and, at the same time, constant absorption at other
wavelengths. Consequently the light sent by an optical fiber to

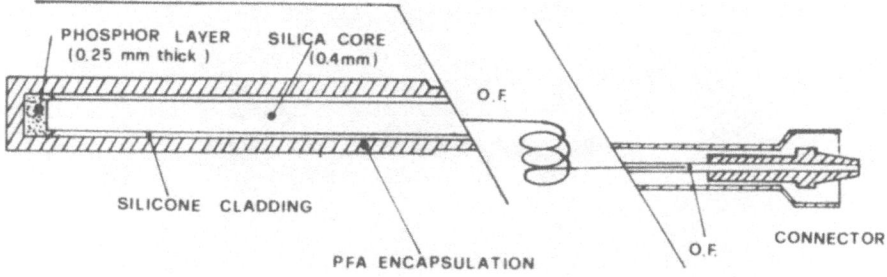

Fig. 1. Layout of the fluoroptic temperature sensor with phosphor
         tip.

a probe containing such thermochromic solution is amplitude modulated
by temperature at the wavelength having a temperature dependent
absorption (λ = 655 nm), while it does not vary at the absorption
wavelength (λ = 800 nm). As the probe has a reflecting bottom the
light is recaptured by a second optical fiber and via a beam splitter
followed by two filters is analyzed at 655 nm (measurement signal S_T)
and at 800 nm (reference signal S_o). The S_T/S_o results independent
either of source fluctuations and of transmission fiber losses due,
for example, to bendings. Figure 2 shows two typical response curves
in a complete heating–cooling cycle in the 30–50°C temperature range.
The sensitivity is less than 0.2°C with a maximum hysteresis of
0.25°C. The probe (Figure 3) containing the thermochromic solution
and the two optical fibers has dimensions: 1.5 mm external diameter
and 1 cm length.

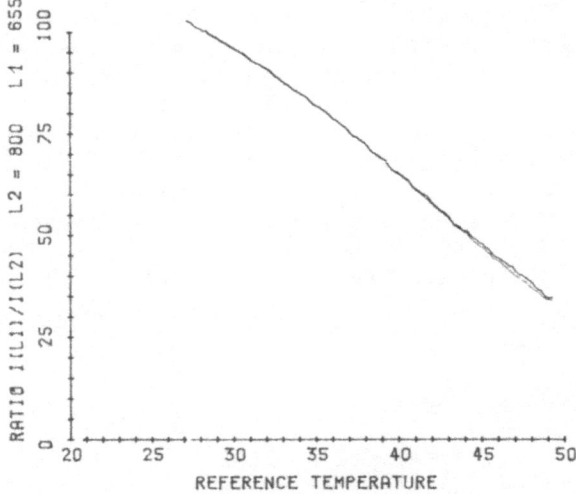

Fig. 2.   Response curves of the thermochromic transducer sensor for a
complete heating cooling cycle.

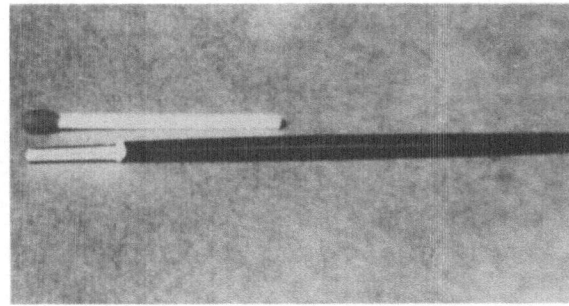

Fig. 3.   Probe of the thermochromic sensor.

All Fiber Sensor

This optical fiber temperature sensor (under study at IROE – CNR, Florence) utilizes a silica-plastic fiber with an unclad terminal portion and reflecting end face immersed in a liquid having a temperature sensitive refractive index (Figure 4)[15,16]. When the light passes through the optical fiber terminal portion is intensity modified by the thermosensitive cladding and the reflecting end face allows the back transmission of the modulated light. More precisely at room temperature the liquid refractive index n_e is higher than that of the core but it decreases increasing temperature. As long as n_e remains higher than the core index, the optical fiber terminal portion guides the light only by partial reflection and the intensity light backward transmitted decreases. But when n_e becomes lower than the core index, also the terminal portion guides the light by total reflection and the light intensity backward transmitted abruptly increases until a saturation is reached when n_e becomes equal to the refractive index of the plastic cladding covering the whole fiber.

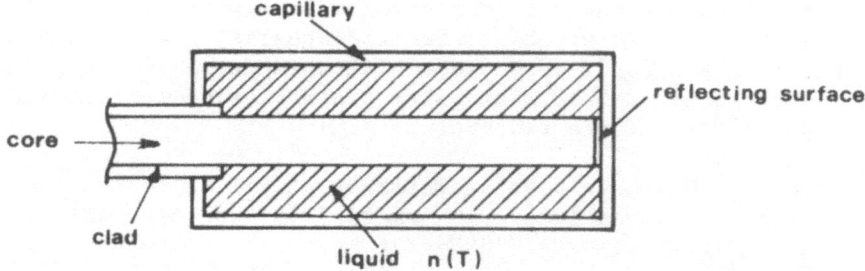

Fig. 4.   Sketch of the optical fiber temperature probe.

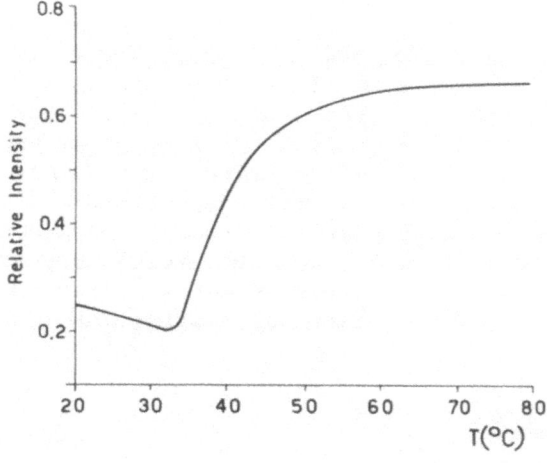

Fig. 5.   Complete response curve of the all fiber thermometer.

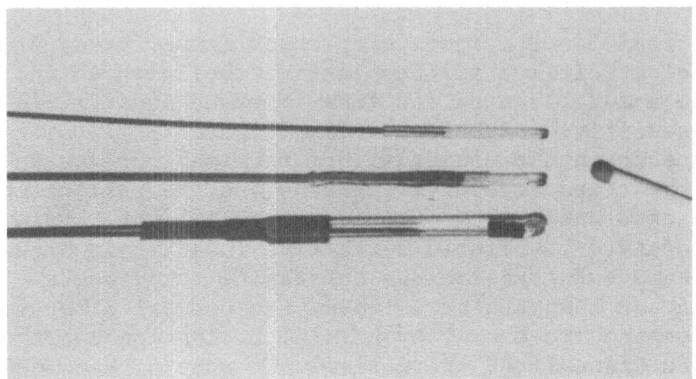

Fig. 6.   Three samples of all fiber temperature probes.

A typical complete response curve of such optical fiber thermometer, utilizing a mineral oil as the thermosensitive cladding is
shown in Figure 5. Because of its better linearity and sensitivity
the intermediate section is employed: with the mineral oil the best
range temperature is 35-50°C, but using other liquids it is possible
to select the desired range temperature.

Different probes have been realized by utilizing 1-2 cm long
glass capillaries (1-2 mm internal diameter 1.5-4 mm external diameter) as liquid containers and 300 or 600 μm core diameter silica-
plastic fibers (Figure 6). The sensitivity reach up to now of about
0.2°C.

CONCLUSIONS

The reported examples show how it is possible to monitor temperature with optical fibers and dielectric probes. The small probe
dimensions and sensitivities already achieved make these sensors very
promising for medical applications. However, many problems remain
still open, which are related to the availability of new optical
components in order to achieve a better compactness and an easier
handling of the whole sensor system. Another problem concerns the
miniaturization of the probe for less invasive measurements. Im-
plementation of these new technologies will stimulate the design of
new sensor types for new applications in diagnostics and therapy.

REFERENCES

1. R. Mack, IFOC 2:21 (1981).
2. P. Dario, D. De Ross, Elettromedicali, 2:76 (1982).

3. T. G. Giallorenzi, J. A. Bucaro, A. Dandridge, G. H. Sigel, Jr. J. H. Cole, S. C. Rashleigh, R. G. Priest, IEEE J. Quantum Elect., QE- 18:626 (1982).

4. T. C. Rozzel, C. C. Johsnon, C. H. Durney, J. L. Lords, R. G. Olsen, J.Microw.Power, 9:241 (1974).

5. K. A. Wickersheim, R. Alves, Ind.Research Develop., Dec. (1979).

6. K. A. Wickersheim: U. S. Patent 4, 075, 493, (Feb. 21 1978).

7. J. I. Peterson, S. R. Goldstein, R. V. Fitzgerald, D. K. Buckhold, Anal.Chem., 52:884 (1980).

8. R. J. Volz, D. A. Christensen, IEEE Trans.Biom.Eng., BME-26:416 (1979).

9. H. Nishihara, J. Koyama, N. Kajiya, M. Hironaga, M. Kano: Appl. Opt., 21:1785 (1982).

10. G. Delfino, E. Casale, Appl.Opt., 20:989 (1981).

11. T. C. Cetas: Proc. 1975, USNC/URSI Symp. HEW Publ. (FDA) 77-8011 Vol II, p. 239.

12. T. C. Cetas, W. G. Connor, Medical Physics Vol. 5(2), p. 79 (1978).

13. D. A. Christensen, J.Biomed.Eng., 1:541 (1977).

14. M. Bacci, M. Brenci, G. Conforti, R. Falciai, A. G. Mignani, A. M. Scheggi, Italian Patent No. 945A/83 (Oct.18, 1983) under extension abroad.

15. M. Brenci, R. Falciai, A. M. Scheggi, Italian Patent No. 84155A/82 (Nov. 1982) under extension abroad.

16. A. M. Scheggi, M. Brenci, G. Conforti, R. Falciai, G. P. Preti, Proc. First Int. Conference on Optical Fiber Sensors, IEE Conference Publication No. 221, p. 13 (1983).

INDEX